Advances in
COMPUTERS
VOLUME 74

Advances in
COMPUTERS

Software Development

EDITED BY

MARVIN V. ZELKOWITZ

Department of Computer Science
University of Maryland
College Park, Maryland

VOLUME 74

Amsterdam • Boston • Heidelberg • London • New York • Oxford
Paris • San Diego • San Francisco • Singapore • Sydney • Tokyo
Academic Press is an imprint of Elsevier

ELSEVIER

ACADEMIC
PRESS

Academic Press is an imprint of Elsevier
84 Theobald's Road, London WC1X 8RR, UK
Radarweg 29, PO Box 211, 1000 AE Amsterdam, The Netherlands
30 Corporate Drive, Suite 400, Burlington, MA 01803, USA
525 B Street, Suite 1900, San Diego, CA 92101-4495, USA

First edition 2008

Copyright © 2008 Elsevier Inc. All rights reserved

No part of this publication may be reproduced, stored in a retrieval system
or transmitted in any form or by any means electronic, mechanical, photocopying,
recording or otherwise without the prior written permission of the publisher.

Permissions may be sought directly from Elsevier's Science & Technology Rights
Department in Oxford, UK: phone (+44) (0) 1865 843830; fax (+44) (0) 1865 853333;
email: permissions@elsevier.com. Alternatively you can submit your request online by
visiting the Elsevier web site at http://elsevier.com/locate/permissions, and selecting
Obtaining permission to use Elsevier material.

Notice
No responsibility is assumed by the publisher for any injury and/or damage to persons or
property as a matter of products liability, negligence or otherwise, or from any use or
operation of any methods, products, instructions or ideas contained in the material herein.

ISBN: 978-0-12-374426-5

ISSN: 0065-2458

For information on all Academic Press publications
visit our website at elsevierdirect.com

Printed and bound in USA

08 09 10 11 12   10 9 8 7 6 5 4 3 2 1

Working together to grow
libraries in developing countries

www.elsevier.com | www.bookaid.org | www.sabre.org

ELSEVIER   BOOK AID International   Sabre Foundation

Contents

CONTRIBUTORS ix
PREFACE xvii

Data Hiding Tactics for Windows and Unix File Systems

Hal Berghel, David Hoelzer and Michael Sthultz

1. The Philosophy of Digital Data Hiding 3
2. Digital Storage and File Systems 4
3. Forensic Implications 10
4. Perspectives 14
5. Conclusion 16
   References 16

Multimedia and Sensor Security

Anna Hać

1. Introduction 20
2. Multimedia Systems and Applications 21
3. Multimedia Security 24
4. Digital Watermarking 26
5. Steganography 28
6. Computer Forensics 29
7. Sensor Networks 30
8. Security Protocols for Wireless Sensor Networks 33
9. Communication Security in Sensor Networks 34
10. Sensor Software Design 35

11.	Trusted Software	35
12.	Hardware Power-Aware Sensor Security	36
13.	Trusted Hardware	36
14.	Sensor Networks and RFID Security	37
15.	Conclusion	37
	References	37

Email Spam Filtering

Enrique Puertas Sanz, José María Gómez Hidalgo and José Carlos Cortizo Pérez

1.	Introduction	47
2.	Technical Measures	51
3.	Content-Based Spam Filtering	56
4.	Spam Filters Evaluation	83
5.	Spam Filters in Practice	92
6.	Attacking Spam Filters	98
7.	Conclusions and Future Trends	109
	References	109

The Use of Simulation Techniques for Hybrid Software Cost Estimation and Risk Analysis

Michael Kläs, Adam Trendowicz, Axel Wickenkamp, Jürgen Münch, Nahomi Kikuchi and Yasushi Ishigai

1.	Introduction	117
2.	Background	119
3.	Related Work	126
4.	Problem Statement	140
5.	Analytical Approaches	143
6.	Stochastic Approaches	145
7.	Experimental Study	149
8.	Summary	166
	Acknowledgments	169
	References	169

An Environment for Conducting Families of Software Engineering Experiments

Lorin Hochstein, Taiga Nakamura, Forrest Shull, Nico Zazworka, Victor R. Basili and Marvin V. Zelkowitz

1. Introduction 176
2. Classroom as Software Engineering Lab 179
3. The Experiment Manager Framework 182
4. Current Status 190
5. Related Work 197
6. Conclusions 198
    Acknowledgments 199
    References 199

Global Software Development: Origins, Practices, and Directions

James J. Cusick, Alpana Prasad and William M. Tepfenhart

1. Introduction 203
2. IT Sourcing Landscape 204
3. Global Software Development 210
4. Current GSD Practice 216
5. A Virtual Roundtable on Outsourcing 246
6. Future Directions in Offshoring 251
7. Conclusions 261
    Acknowledgments 262
    Appendix 1: Interview with K. (Paddy) Padmanabhan
    of Tata Consultancy Services 3/9/07 262
    Appendix 2: List of Acronyms 265
    References 267

AUTHOR INDEX 271
SUBJECT INDEX 279
CONTENTS OF VOLUMES IN THIS SERIES 303

Contributors

Victor R. Basili is a Professor of Computer Science at the University of Maryland, College Park. He holds a Ph.D. in Computer Science from the University of Texas, Austin and is a recipient of two honorary degrees from the University of Sannio, Italy (2004) and the University of Kaiserslautern, Germany (2005). He was Director of the Fraunhofer Center for Experimental Software Engineering - Maryland and a director of the Software Engineering Laboratory (SEL) at NASA/GSFC. He works on measuring, evaluating, and improving the software development process and product. Dr. Basili is a recipient of several awards including the NASA Group Achievement Awards, the ACM SIGSOFT Outstanding Research Award, the IEEE computer Society Harlan Mills Award, and the Fraunhofer Medal. He has authored over 250 journal and refereed conference papers, served as co-Editor-in-Chief of the Springer Journal of Empirical Software Engineering and is an IEEE and ACM Fellow. He can be reached at basili@cs.umd.edu.

Hal Berghel is currently Associate Dean of the Howard R. Hughes College of Engineering, Founding Director of the School of Informatics, and Professor of Computer Science at the University of Las Vegas. He is also the founding Director of the Center for CyberSecurity Research and the Identity Theft and Financial Fraud Research and Operations Center. The author of over 200 publications, his research has been supported by business, industry, and government for over 25 years. He is both an ACM and IEEE Fellow and has been recognized by both organizations for distinguished service to the profession. His consultancy, Berghel.Net, provides technology consulting services to government and industry.

José Carlos Cortizo Pérez (http://www.ainetsolutions.com/jccp) received his M.S. degree in Computer Science and Engineering from Universidad Europea de Madrid. His research areas include data preprocessing in machine learning and data-mining, composite learning methods, AI applications on astrophysics and noisy information retrieval. He is also co-founder of AINetSolutions (http://www.ainetsolutions.com), a data-mining startup.

James Cusick is an Assistant Director of Web Software Engineering in Wolters Kluwer's Corporate Legal Services Division where he provides strategic direction for software development. Previously, James held a variety of roles with AT&T and Bell Labs. James is the author more that 3 dozen papers and talks on Software Reliability, Object Technology, and Software Engineering Technology. He has also held the position of Adjunct Assistant Professor at Columbia University's Department of Computer Science where he taught Software Engineering. James is a graduate of both the University of California at Santa Barbara and Columbia University in New York City and a member of IEEE. James is also a certified PMP and can be reached at j.cusick@computer.org.

José María Gómez Hidalgo holds a Ph.D. in Mathematics, and has been a lecturer and researcher at the Universidad Complutense de Madrid (UCM) and the Universidad Europea de Madrid (UEM), for 10 years, where he has been the Head of the Department of Computer Science. Currently he is Research and Development Director at the security firm Optenet. His main research interests include Natural Language Processing (NLP) and Machine Learning (ML), with applications to Information Access in newspapers and biomedicine, and Adversarial Information Retrieval with applications to spam filtering and pornography detection on the Web. He has taken part in around 10 research projects, heading some of them. José María has co-authored a number of research papers in the topics above, which can be accessed at his home page. He is Program Committee member for CEAS 2007, the spam Symposium 2007 and other conferences, and he has reviewed papers for JASIST, ECIR, and others. He has also reviewed research project proposals for the European Commission.

Anna Hać received the M.S. and Ph.D. degrees in Computer Science from the Department of Electronics, Warsaw University of Technology, Poland, in 1977 and 1982, respectively. She is a Professor in the Department of Electrical Engineering, University of Hawaii at Manoa, Honolulu. She has been a Visiting Scientist at the Imperial College, University of London, England, a Postdoctoral Fellow at the University of California at Berkeley, an Assistant Professor of Electrical Engineering and Computer Science at The Johns Hopkins University, a Member of Technical Staff at AT&T Bell Laboratories, and an ONR/ASEE Senior Summer Faculty Fellow at the Naval Research Laboratory and SPAWAR. Her research contributions include system and workload modeling, performance analysis, reliability, modeling process synchronization mechanisms for distributed systems, distributed file systems, distributed algorithms, congestion control in high-speed networks, reliable software architecture for switching systems, multimedia systems, wireless networks, and network protocols. Her research interests include multimedia, wireless data and sensor networks, and nanotechnology for information processing. She is a member of the Editorial Board of the IEEE Transactions on Multimedia, and is on the Editorial Advisory Board of Wiley's International Journal of Network Management.

Lorin Hochstein received a Ph.D. in computer science from the University of Maryland, an M.S. in electrical engineering from Boston University, and a B.Eng. in computer engineering from McGill University. He is currently an Assistant Professor in the Department of Computer Science and Engineering at the University of Nebraska at Lincoln. He is a member of the Laboratory for Empirically-based Software Quality Research and Development (ESQuaReD). His research interests include combining quantitative and qualitative methods in software engineering research, software measurement, software architecture, and software engineering for high-performance computing. He is a member of the IEEE Computer Society and ACM.

David Hoelzer is well known in the information security industry as an expert in the fields of intrusion detection, incident handling, information security auditing, and forensics. David is the Director of Research for Enclave Forensics (www.enclaveforensics.com). He also serves as the CISO for Cyber-Defense (www.cyber-defense.org). David has been named a SANS Fellow, a Research Fellow with the Internet Forensics Center and an adjunct research associate of the UNLV Center for Cybersecurity Research. In these roles his responsibilities have included acting as an expert witness for the Federal Trade Commission, teaching at major national and international SANS conferences and handling security incident response and forensic investigations for several corporations and financial institutions. David has provided advanced training to security professionals from organizations including NSA, USDA Forest Service, most of the Fortune 500, DHHS, various DoD sites and many universities.

Yasushi Ishigai is Research Director of Research Center for Information Technology at Mitsubishi Research Institute, INC, Tokyo, Japan. And he is also a part time researcher at Software Engineering Center of Information technology Promotion Agency, Japan. Yasushi Ishigai received a degree in mechanical engineering (B.Sc. and M.Sc.) from the University of Tokyo, Japan, in 1988. His research interest and industrial activities include quantitative management, especially software cost modeling.

Nahomi Kikuchi received a B.S. degree in mathematics from Niigata University, Japan, a Master's degree in computer science (MSCS) from Stanford University, and a degree of Doctor of Engineering from Osaka University, Japan. She is a manager for software quality group at Oki Electric Industry Co., Ltd. Japan. Her earlier work at OKI includes the design and verification methods and support systems for telecommunication software systems using ITU-T's SDL Language and Petri net. She has experience of leading a number of software projects. Her recent research and industrial activities include software quality assurance, quality management, and process improvement techniques with emphasis on quantitative methods and measurement and application of software engineering techniques, tools and methods. She has been a part-time researcher at Software Engineering Center, Information-technology Promotion Agency Japan since October 2004. She is a member of IEEE, the IEEE Computer Society, and IPSJ(Japan).

Michael Kläs received his degree in computer science (Diploma) from the University of Kaiserslautern, Germany, in 2005. He is currently a researcher at the Fraunhofer Institute for Experimental Software Engineering (IESE), Kaiserslautern, Germany and member of DASMA, the German member organization of the International Software Benchmarking Standards Group (ISBSG). As a member of the Processes and Measurement department, he works on national and international research and industrial projects. His research and industrial activities include goal-oriented measurement, balancing strategies for quality assurance, and software cost modeling.

Jürgen Münch is Division Manager for Software and Systems Quality Management at the Fraunhofer Institute for Experimental Software Engineering (IESE) in Kaiserslautern, Germany. Before that, Dr. Münch was Department Head for Processes and Measurement at Fraunhofer IESE and an executive board member of the temporary research institute SFB 501, which focused on software product lines. Dr. Münch received his Ph.D. degree (Dr. rer. nat.) in Computer Science from the University of Kaiserslautern, Germany, at the chair of Prof. Dr. Dieter Rombach. Dr. Münch's research interests in software engineering include: (1) modeling and measurement of software processes and resulting products, (2) software quality assurance and control, (3) technology evaluation through experimental means and simulation, (4) software product lines, (5) technology transfer methods. Dr. Münch has significant project management experience and has headed various large research and industrial software engineering projects, including the definition of international quality and process standards. His main industrial consulting activities are in the areas of process management, goal-oriented measurement, quality management, and quantitative modeling. He has been teaching and training in both university and industry environments. Dr. Münch has co-authored more than 80 international publications, and has been co-organizer, program co-chair, or member of the program committee of numerous high-standard software engineering conferences and workshops. Jürgen Münch is a member of ACM, IEEE, the IEEE Computer Society, and the German Computer Society (GI).

Taiga Nakamura received the B.E. and M.E. degrees in aerospace engineering from University of Tokyo in 1997, 1999, respectively. He received the Ph.D. degree in computer science from University of Maryland, College Park in 2007. He is presently a Research Staff Member at IBM Tokyo Research Laboratory, where he works for the Services Software Engineering group. His research interests include software patterns, metrics, software quality engineering and empirical methods in software engineering. Dr. Nakamura is a member of the Association for Computing Machinery, the Institute of Electrical and Electronics Engineers, and the Information Processing Society of Japan.

Alpana Prasad is an Assistant Director at Wolters Kluwer Corporate Legal Services, where she's responsible for technical oversight and management of multiple offshore initiatives. She also works on other corporate initiatives and programs to improve platform and product development standards and best practices. Her research interests include analysis of new and emerging technologies, primarily in the Microsoft domain. She received her master's in business administration from the Indian Institute of Management, Lucknow. She's a member of the IEEE. Contact her at Wolters Kluwer Corporate Legal Services, 111 Eighth Ave., New York, NY 10011; alpana.prasad@wolterskluwer.com.

Enrique Puertas Sanz (www.enriquepuertas.com) is Professor of Computing Science at Universidad Europea de Madrid, Spain. His research interests are broad and include topics like Artificial Intelligence, Adversarial Information Retrieval, Content Filtering, Usability and User Interface Designs. He has co-authored many research papers in those topics and has participated in several research projects in the AI field.

Forrest Shull is a senior scientist at the Fraunhofer Center for Experimental Software Engineering in Maryland (FC-MD), where he serves as Director for the Measurement and Knowledge Management Division. He is project manager for projects with clients that have included Fujitsu, Motorola, NASA, and the U.S. Department of Defense. He has also been lead researcher on grants from the National Science Foundation, Air Force Research Labs, and NASA's Office of Safety and Mission Assurance. Dr. Shull works on projects that help to transfer research results into practice, and has served as a principal member of the NSF-funded national Center for Empirically Based Software Engineering. He is Associate Editor in Chief of IEEE Software. He received a Ph.D. degree in Computer Science from the University of Maryland and can be reached at fshull@fc-md.umd.edu.

Michael Sthultz received his undergraduate education at Claremont Men's College (B.A.) and the University of California, Berkeley (B.S.E.E.). His M.S. in Computer Science was completed at UNLV. He also holds industry certifications in A+, CCNP, CCAI, CEH, CLS, CLP, CNA, MCSE, and MCT. He has extensive industry experience in the areas of systems engineering, programming, network administration, management, and consulting. Michael is currently on the faculty of the College of Southern Nevada, teaching Cisco Networking Academy courses as well as Digital Forensics and Informatics.

William Tepfenhart is author of several books on object orientation. He is currently an Associate Professor in the Software Engineering Department at Monmouth University investigating the potential of software solutions to enhance the effectiveness of collaboration in engineering endeavors. Prior to his entry to the academic world, he was employed as a developer and technologist at AT&T Laboratories where he worked on applications associated with the long distance network, establishment of engineering practices at a corporate level, and working with advanced object-oriented technologies. He had previously worked as a Senior Scientist at Knowledge Systems Concepts investigating the use of artificial intelligence systems for the USAF. Prior to that, he was an Associate Research Scientist at LTV where he worked on applications for manufacturing devices composed of advanced materials.

Adam Trendowicz received a degree in computer science (B.Sc.) and in software engineering (M.Sc.) from the Poznan University of Technology, Poland, in 2000. He is currently a researcher at the Fraunhofer Institute for Experimental Software Engineering (IESE), Kaiserslautern, Germany in the Processes and Measurement department. Before that, he worked as a software engineering consultant at Q-Labs GmbH, Germany. His research and industrial activities include software cost modeling, measurement, and data analysis.

Axel Wickenkamp is working as a scientist at the Fraunhofer Institute for Experimental Software Engineering in the department Processes and Measurement. His main research areas include static code analysis and software process models. Before joining the Fraunhofer Institute, he worked as a freelance consultant and software developer in industrial software development projects. Axel Wickenkamp received his degree in computer science (Diploma) from the University of Kaiserslautern.

Nico Zazworka is currently a Ph.D. student in computer science at the University of Maryland. He received his Diploma Degree from the University of Applied Sciences in Mannheim, Germany in 2006. His research interests are software engineering and information visualization.

Marvin Zelkowitz is a Research Professor of Computer Science at the University of Maryland, College Park. He was one of the founders and principals in the Software Engineering Laboratory at NASA Goddard Space Flight Center from 1976 through 2001. From 1998 through 2007 he was associated with the Fraunhofer Center Maryland, where he was co-Director from 1998 through 2002. He has been involved in environment and tool development, measuring the software process to understand software development technologies, and understanding technology transfer. He has authored over 160 conference, book chapters and journal papers. He has a B.S. in Mathematics from Rensselaer Polytechnic Institute, and an M.S. and Ph.D. in Computer Science from Cornell University. He is a Fellow of the IEEE, recipient of the IEEE Computer Society Golden Core Award and the 2000 ACM SIGSOFT Distinguished Service Award, as well as several ACM and IEEE Certificates of Appreciation. He is also series editor of this series, the **Advances in Computers**. He can be reached at marv@zelkowitz.com.

Preface

Welcome to volume 74 of the **Advances in Computers**, subtitled 'Recent Advances in Software Development.' This series, which began in 1960, is the oldest continuously published series of books that has chronicled the ever-changing landscape of information technology. Each year three volumes are published, each presenting five to seven chapters describing the latest technology in the use of computers today. In this current volume, we present six chapters that give an update on some of the major issues affecting the development of software today.

The six chapters in this volume can be divided into two general categories. The first three chapters deal with the increasing importance of security in the software we write and provide insights into how to increase that security. The three latter chapters look at software development as a whole and provide guidelines for how best to make certain decisions on a project-level basis.

Chapter 1, 'Data Hiding Tactics for Windows and Unix File Systems' by Hal Berghel, David Hoelzer, and Michael Sthultz, describes a form of security vulnerability generally unknown to the casual computer user. Most users today realize that viruses often propagate by embedding their code into the executable files (e.g., .exe files on Windows systems) of the infected system. By removing such programs, most viruses can be eliminated. However, there are storage locations in the computer which are not readily visible. A virus can be hidden in those areas (e.g., parts of the boot sector of a disk are currently unused, so can be used to hide malicious code), which result in them being harder to detect and eradicate. This chapter describes several of these areas where malicious code can hide to help it propagate its infection.

In Chapter 2, 'Multimedia and Sensor Security' by Anna Hać, Dr. Hać looks at the increasingly important area of sensor security. As computers become ubiquitous in our environment, an increasingly larger number of applications collect data from the world around us and process that data, often without human intervention (e.g., the use of RFID tags to track production, as described in the chapter by Roussos in volume 73 of the Advances in Computers). How do we ensure security and integrity of our collected data in an increasingly autonomous environment? In this chapter, various mechanisms to detect vulnerabilities and ensure the integrity of the data are described.

Anyone who has received email, and I am sure if you are reading this volume you either like email or are forced to use it, understands the problems with spam. I receive ~20,000 email messages monthly. Fortunately, spam detectors on my computer systems weed out ~17,000 of those messages giving me a tolerable level of perhaps 100 messages a day, of which maybe 30 are undetected spam. But how do spam detectors work and why are they so successful? Enrique Puertas Sanz, José María Gómez Hidalgo, and José Carlos Cortizo Pérez in Chapter 3, 'Email Spam Filtering,' provide an overview of spam and what can be done about it.

In Chapter 4, 'The Use of Simulation Techniques for Hybrid Software Cost Estimation and Risk Analysis' by Michael Kläs, Adam Trendowicz, Axel Wickenkamp, Jürgen Münch, Nahomi Kikuchi, and Yasushi Ishigai, the authors look at the problems in software cost estimation. Knowing how much a project is going to cost and how long it will take are crucial factors for any software development company. This editor, in 1982, was part of a study where some major developers claimed that understanding this was even more important than lowering the cost. They needed to know how many people to put on a project and not have extra workers doing very little. In this chapter, the authors describe their CoBRA® method for producing such estimates.

Chapter 5, 'An Environment for Conducting Families of Software Engineering Experiments' by Lorin Hochstein, Taiga Nakamura, Forrest Shull, Nico Zazworka, Victor R. Basili, and Marvin V. Zelkowitz, looks at a crucial problem in doing research for improving the software development process. Scientific research in software engineering advances by conducting experiments to measure how effective various techniques are in improving the development process. Data needs to be collected, and in this case usually means defect, effort, and code generation activities from project participants. But programmers and analysts view such data collection as an intrusion on their activities, so the conundrum is to collect this data without the participants' need to 'actively participate.' In this chapter, the authors describe an environment, where they have built an environment that automates, as much as possible, the data collection activities.

In the final chapter, 'Global Software Development: Origins, Practices, and Directions' by James J. Cusick, Alpana Prasad, and William M. Tepfenhart, the authors discuss the current trends in software development becoming international. What do companies have to do in order to provide for a global software development workforce and how do they make the process work successfully? With an emphasis on outsourcing to India, this chapter also discusses outsourcing issues worldwide.

I hope that you find these chapters interesting and useful. I am always looking for new topics to write about, so if you have any ideas in what should be covered in a future volume, please let me know. If you would like to write such a chapter, please contact me. I can be reached at mvz@cs.umd.edu.

Marvin Zelkowitz
University of Maryland
College Park, Maryland

Data Hiding Tactics for Windows and Unix File Systems

HAL BERGHEL

Identity Theft and Financial Fraud Research and Operations Center, University of Las Vegas, Las Vegas, Nevada

DAVID HOELZER

Enclave Forensics, Las Vegas, Nevada

MICHAEL STHULTZ

Identity Theft and Financial Fraud Research and Operations Center, University of Las Vegas, Las Vegas, Nevada

Abstract

The phenomenon of hiding digital data is as old as the computer systems they reside on. Various incarnations of data hiding have found their way into modern computing experience from the storage of data on out-of-standard tracks on floppy disks that were beyond the reach of the operating system, to storage information in non-data fields of network packets. The common theme is that digital data hiding involves the storage of information in places where data is not expected.

Hidden data may be thought of as a special case intentionally 'dark data' versus unintentionally dark data that is concealed, undiscovered, misplaced, absent, accidentally erased, and so on. In some cases, dark and light data coexist. Watermarking provides an example where dark data (an imperceptible watermark) resides within light data. Encryption is an interesting contrast, because it produces light data with a dark message. The variations on this theme are endless.

The focus of this chapter will be intentionally dark or hidden data as it resides on modern file systems – specifically some popular Windows and Unix file systems. We will extrapolate from several examples implications on the practice of modern digital forensics and the tools that are used in support thereof.

One forensically interesting dimension of physical data hiding is those techniques that take advantage of the physical characteristics of formatted storage media to hide data. An early attempt to do this was illustrated by Camouflage (camouflage.unfiction.com) that hid data in the area between the logical end-of-file and the end of the associated cluster in which the file was placed (called file slack or slack space). Although primitive, hiding data in file slack has the dual advantage that the host or carrier file is unaffected while the hidden data is transparent to the host operating system and file managers. The disadvantage is that the hidden message is easily recovered with a basic disk editor.

The ability to hide data on computer storage media is a byproduct of the system and peripheral architectures. If all storage were bit-addressable at the operating system level, there would be no place to hide data, hence no physical concealment. But for efficiency considerations, system addressability has to be at more abstract levels (typically words in primary, and blocks in secondary). Such abstractions create digital warrens where data may go unnoticed or, in some cases, be inaccessible. We investigate some of these warrens below.

1. The Philosophy of Digital Data Hiding 3
   1.1. The Concept of Data Hiding 3
   1.2. Physical Aspect of Data Hiding 3
2. Digital Storage and File Systems 4
   2.1. Disk Structures 4
   2.2. Virtual File Systems 6
   2.3. Partition Organization 7
   2.4. ExtX 8
   2.5. NTFS 8
3. Forensic Implications 10
   3.1. Fat16 10
   3.2. NTFS 12
4. Perspectives 14
5. Conclusion 16
   References 16

1. The Philosophy of Digital Data Hiding

1.1 The Concept of Data Hiding

Digital data hiding is actually a cluster concept that spans many contexts. In modern times, nonphysical data hiding is usually associated with digital forms such as cryptography, steganography, and watermarking. Although related in the sense that they all are means to achieve secure or proprietary communications, there are differences among their three activities at a number of levels – some of which are quite subtle. To illustrate, the cryptographer's interest is primarily with obscuring the content of a message, but not the communication of the message. The steganographer, on the other hand, is concerned with hiding the very communication of the message, while the digital watermarker attempts to add sufficient metadata to a message to establish ownership, provenance, source, and so on. Cryptography and steganography share the feature that the object of interest is embedded, hidden, or obscured, whereas the object of interest in watermarking is the host or carrier which is being protected by the object that is embedded, hidden, or obscured. Further, watermarking and steganography may be used with or without cryptography; and imperceptible watermarking shares functionality with steganography, whereas perceptible watermarking does not. Overviews exist for cryptography [13], steganography [9, 14], and watermarking [3].

1.2 Physical Aspect of Data Hiding

However, there is also a physical aspect of digital data hiding. In this case, digital storage locations are used to hide or conceal data. Obviously, these storage locations must be somewhat obscure or detection would be trivial. Examples include covert channeling (e.g., within TCP/IP packet headers or Loki's use of the ICMP options field) or obscure or infrequently used addressable space in memory media. One of the earliest examples of this arose when microcomputers were first introduced. There was a difference between the number of tracks that a floppy drive controller could access (usually 81 or 82) and the number of tracks that were recognized by the operating system (DOS recognized only 80). The upper two tracks were used to hide data by vendors and hackers alike. For awhile they were even used for product licensing.

A modern analogue might involve taking advantage of the physical characteristics of a storage medium to hide data. The steganographic tool Camouflage was such a program (camouflage.unfiction.com). Camouflage embeds messages in the file slack (see below). This simple approach to data hiding has the advantage that the characteristics of the host or carrier message remain unaffected but are transparent to the host operating system and file managers. The disadvantage is that the hidden message is easily recovered with a hex-editor. Although this approach has not met with great success (and camouflage is no longer supported), it is a great segue into the art of data hiding by taking advantage of the physical characteristics of computer systems.

The ability to hide data in computers is a byproduct of the system and peripheral architectures. If all storage were bit-addressable at the operating system level, there would be no place to hide data. For efficiency considerations, addressability has to be at more abstract levels (typically words in primary, and blocks in secondary). Such abstractions create digital warrens where data may go unnoticed or, in some cases, be inaccessible.

2. Digital Storage and File Systems

2.1 Disk Structures

Digital storage encompasses a wide range of media including diskettes, hard drives, zip disks, USB flash drives, compact flash cards, CD-ROMs, DVDs, and so on. Since the structures of most of these media can be related to hard drives, a discussion of hard drive architecture will serve to illustrate the various data hiding mechanisms.

Hard drive technology existed long before the advent of personal computers (although the early personal computers started out with only cassette tape or diskette storage!). As PCs evolved, the emphasis on backward compatibility and the need for user convenience (i.e., making data structures totally transparent to the user) greatly influenced the development of secondary and tertiary storage technology – both physically and logically. An unintended consequence has been the creation of many places where data can be intentionally hidden or unintentionally left behind.

A functioning hard drive actually consists of a geometric structure and a set of nested data structures: hard drive, partition, file system, file, record, and field. Hidden data can be found at each of these levels. We will assume that the reader is familiar with basic hard disk geometry (cylinders, tracks, blocks, clusters, and sectors). These structures create areas on secondary storage devices where hidden data or data residue could exist. Figure 1 provides a graphical illustration of these digital warrens. In each example, the shaded area represents spaces within the structures where hidden data could reside.

The following describes the various hiding mechanisms (as illustrated in Fig. 1), starting at the level of the hard drive itself and then working down through the set of nested data structures.

DATA HIDING TACTICS FOR WINDOWS AND UNIX FILE SYSTEMS

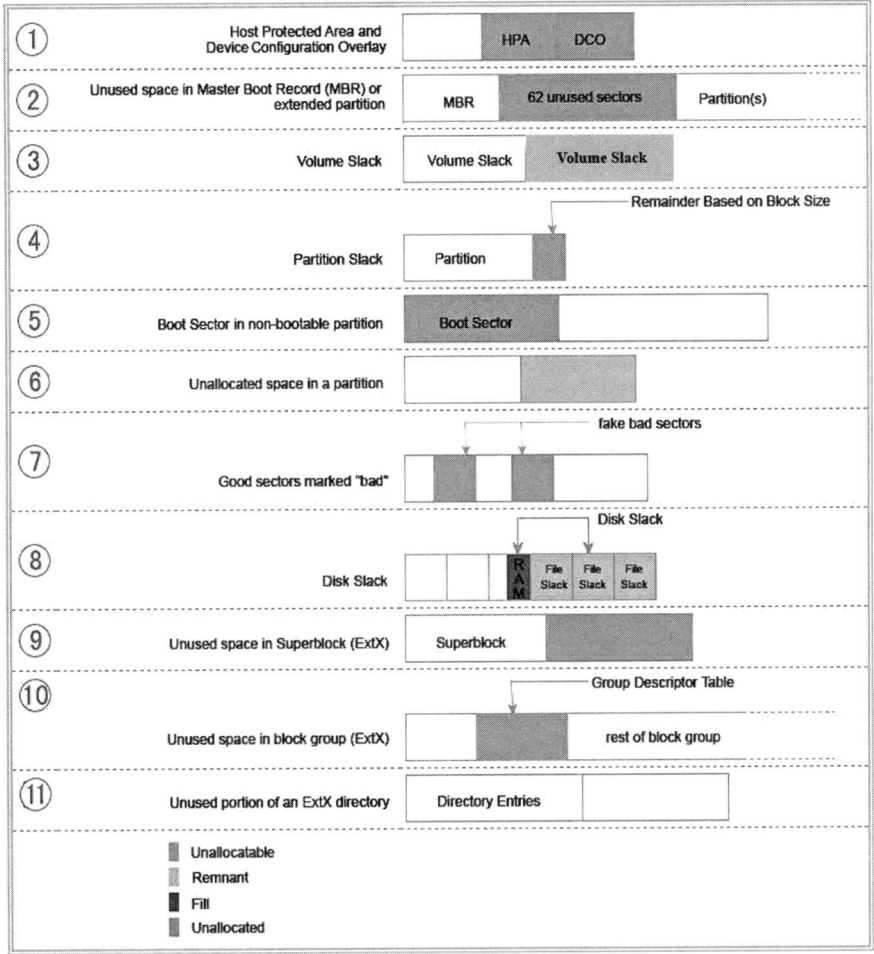

Fig. 1. Digital disk warrens.

Some hard drives can have a reserved area of the disk called the *Host Protected Area (HPA)* (see Fig. 1, item 1). *Device Configuration Overlay* allows modification of the apparent features provided by a hard drive, for example, the number of available clusters. This was designed to be an area where computer vendors could store data that is protected from normal user activities. It is not affected by operating system utilities (format, delete, etc.) and cannot be accessed without the use of a special program that reconfigures the controller to access all physical blocks. It is not difficult, however, to write a program to access these areas, write data to them, and subsequently return the area to an HPA. This is an example of a hiding method that takes advantage of what is more or less a 'physical' feature of the drive architecture.

2.2 Virtual File Systems

At the next layer, common operating systems typically require that a hard drive be partitioned into virtual file systems before it can be used. This is true even if the entire hard drive is to be mapped onto a single partition. A partition is a set of consecutive blocks on a hard disk that appear to the operating system as a separate logical volume (drive for Windows vs directory or mount point for Unix). Note that there are several different types of partition formats. This discussion is confined to the partition format that has evolved from the original PC (often referred to as DOS partitions) and will not apply to Apple, xBSD, Sun Solaris, GPT partitioning, or multiple disk volumes. A recommended source for additional detail appears as a reference [6]. Even so, the various data hiding techniques at the data and file system levels are partition format independent.

Every hard drive using a DOS partition has space reserved at the beginning of the drive for a *Master Boot Record* (*MBR*) (see Fig. 1, item 2). This will often contain the boot code necessary to begin the initial program load of an operating system and will always contain a partition table (provided we are dealing with partitioned media; e.g., floppy disks are not typically partitioned media while fixed disks are) defining the size and location of up to four partitions. Since the MBR requires only a single sector and partitions must start on a cylinder boundary, this results in 62 sectors of *empty MBR* space where data can be hidden.

As disk sizes grew beyond the limitations of existing operating systems, there arose a need for more than four partitions on a single hard disk. *Extended partitions* (as opposed to primary partitions) were then designed that could contain multiple logical partitions. Each of these extended partitions contains a structure similar to the MBR, leaving another 62 sectors within each extended partition with the potential of harboring more hidden data. Extended partitions may contain at most one file system and one extended partition, so this design permits us to nest the extended partitions to satisfy our volume requirements. Of course, each iteration creates yet another convenient hiding place for data.

If the partitions on a hard drive do not use up all of the available space, the remaining area cannot be accessed by the operating system by conventional means (e.g., through Windows Explorer). This wasted space is called *volume slack* (see Fig. 1, item 3). It is possible to create two or more partitions, put some data into them, and then delete one of the partitions. Since deleting the partition does not actually delete the data, that data is now hidden.

2.3 Partition Organization

Once partitions have been defined, we are ready to move up to the next layer and create an organizational structure for each partition. Before an operating system can store and access data within a partition, a file system must be defined. Modern operating systems support one or more native file systems. A file system allocates data in blocks (or clusters) where a block consists of one or more consecutive sectors. This allocation scheme allows a smaller number of references to handle larger amounts of data, but limits us to accessing data within the file system as block-sized chunks rather than sector-sized chunks. Overall, this tends to make storage and access far more efficient than referencing each individual sector. However, if the total number of sectors in a partition is not a multiple of the block size, there will be some sectors at the end of the partition that cannot be accessed by the operating system using any typical means. This is referred to as *partition slack* (see Fig. 1, item 4) and is another place where data can be hidden.

Every partition contains a boot sector, even if that partition is not bootable. The *boot sectors in non-bootable partitions* (see Fig. 1, item 5) are available to hide data.

Any space in a partition not currently allocated (i.e., *unallocated space*) to a particular file (see Fig. 1, item 6) cannot be accessed by the operating system. Until that space has been allocated to a file, it could contain hidden data.

It is possible to manipulate the file system metadata that identifies bad blocks (e.g., the File Allocation Table in a FAT file system or $BadClus in NTFS) so that usable blocks are marked as bad and therefore will no longer be accessed by the operating system. Such *metadata manipulation* (see Fig. 1, item 7) will produce blocks that can store hidden data.

Disk slack (see Fig. 1, item 8) is a byproduct of a strategy to accelerate file management. Modern operating systems write data in complete 'blocks' where a block could be a sector (the minimal addressable unit of a disk) or a cluster (same concept as block in Microsoft's terms). If a file is not an exact multiple of the sector size, the operating system must pad the last sector and, in some cases (with older operating systems), this padding is data from memory (hence the historical term '*RAM slack*'). Modern operating systems tend to pad this area with nulls. If the total amount of data written does not fill an entire block, the remainder of the block from the sector boundary of the last sector within the block actually used by the file to the actual end of the block will remain unused and will likely contain data from a previously deleted file (*file slack*). It may also be effectively used to hide ephemeral data.

All of the above applies to nearly every file system in common use today, including FAT/FAT32, NTFS, and Linux Ext-based file systems. There are some potential data hiding places in Linux file systems that require a more detailed description.

2.4 ExtX

Ext2 and Ext3 (ExtX) file systems are divided into sections called block groups. Block groups are used to store file names, metadata, and file content. Information about block group size and configuration is stored in a superblock at the beginning of the file system, copies of which are scattered throughout the partition. The block following the superblock (if present) or the first block in every group (if not present) contains a group descriptor table with group descriptors describing the layout of each block group.

An ExtX superblock has 1,024 bytes allocated to it and the last 788 bytes are unused. There also might be some reserved area behind the superblock, depending upon the block size. We call this *superblock slack* (see Fig. 1, item 9).

There is a reserved area behind the ExtX group descriptor since the group descriptor is only 32 bytes long and the block bitmap that follows it must start on a block boundary. This means there is a minimum of 992 bytes (1,006 if you count the padding at the end of the group descriptor) where data could be hidden and more if the block size is larger than 1,024 bytes. We refer to this as *ExtX group descriptor slack* (see Fig. 1, item 10).

ExtX directories are like any other file and are allocated in blocks. The space between the last directory entry and the end of the block is unused and can be used to hide data. This is *directory slack* (see Fig. 1, item 11).

Figure 2 is a graphical illustration of the relative volatility of the various data hiding areas discussed above (cf. also [10]). The degree of persistence of the hidden data is dependent upon the characteristics of the particular area where it is hiding and the type of disk activity that has occurred since the data was written there. Figure 2 illustrates the relative persistence through normal disk activity and possible re-partitioning and/or re-formatting. The numbers in the figure correspond to the numbered items in Fig. 1.

2.5 NTFS

NTFS file systems also offer some unique opportunities for data hiding. The NTFS file systems used today contain innovations that provide efficient file access (for instance, B-Tree organization of files within directories) and readily accessible metadata files to manage disk organization (Microsoft's version of resource forks called Alternate Data Streams), and some other small file storage oddities as well.

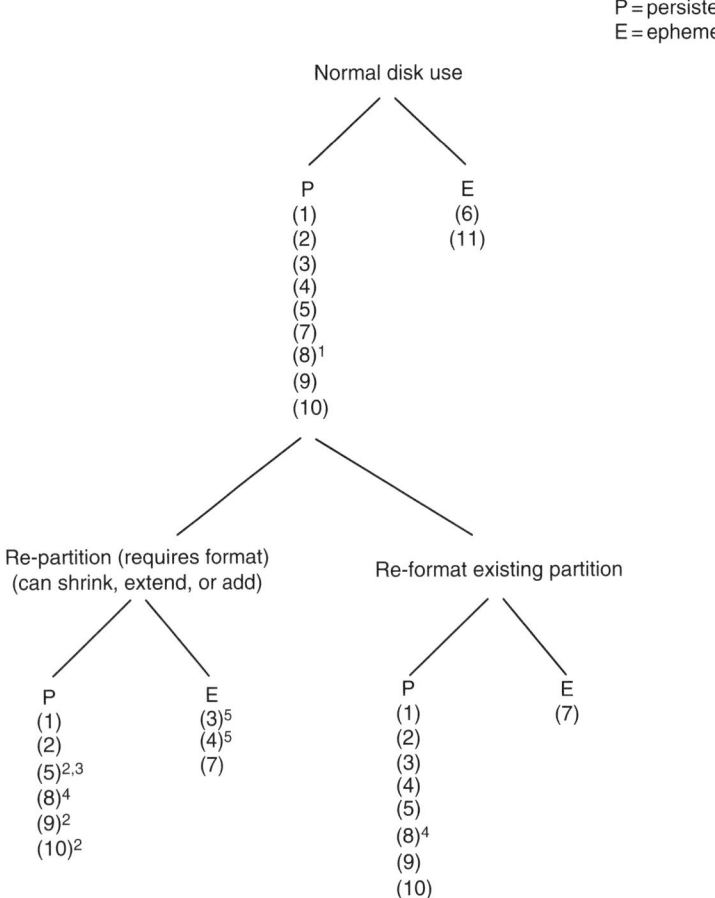

FIG. 2. Relative volatility of data hiding areas.

When seeking to hide data, there are various strategies that might be employed. As mentioned in a more general sense, metadata manipulation may be used to conceal covert data in bad clusters ($BadClus). In fact, this same concept can be extended easily on an NTFS file system by working directly with the $Bitmap file.

The *$Bitmap file* contains a complete map marking the allocation status of every addressable cluster in the partition. Should a consistency check be run, it would become obvious should someone modify this table to hide data, but otherwise this provides a wonderful avenue for hiding data in a way that allows the data to persist for the life of the file system. Depending upon the purpose, these are far better approaches than using file slack which persists only for the life of the file.

NTFS provides some other nooks and crannies. For instance, *Alternate Data Streams* are actually additional $FILE entries associated with a parent file record within the Master File Table. We could not find Microsoft documentation regarding the number of alternate data streams that may be associated with a file or folder, but empirical testing conducted by Jason Fossen (www.fossen.net) suggests that the maximum is 4 106 regardless of the size of the ADSs themselves. These streams are not identified by typical file management tools (e.g., Windows Explorer), and so are hidden at that level. However, several utilities are available that report and manipulate alternate data streams [1]. Alternate data streams persist for the life of the attached file or folder as long as that file or folder remains in an NTFS file structure.

Small files also offer some interesting possibilities, especially when considered in conjunction with alternate data streams. NTFS has a rather unusual capability in that if a file is so small that the entire content of the file can fit within the boundaries of the Master File Table entry for the file, NTFS will store the file there. This is, of course, a convenient place to hide data. But it becomes even more interesting when a file is deleted. When a file is deleted, the clusters that were in use are released to the file system for reallocation. This includes the MFT entry itself. What if one were to create several thousand files, thus consuming several thousand MFT entries; once these files were created, a final file could be created that fits entirely within the MFT. All of these files are now deleted. In essence, we have created a piece of hidden disk space that resides within an allocated file (*$MFT and $MFTMirror*) that will persist until enough files are created to reuse that particular MFT entry. For further detail, see [2].

3. Forensic Implications

3.1 Fat16

The implications of this type of intentional data hiding can be serious in the context of forensic analysis [4], [11], and [12]. Typically, forensic analysis of systems reveals that bad actors do not often take any extraordinary means to hide data beyond, perhaps, either wiping the media or encrypting the data. In fact, in most cases, the data is not even encrypted or wiped. What impact would a deep knowledge of the on-disk structures coupled with a desire to hide data have on an analysis?

To consider this, we have created several sample disk images and have embedded two pieces of data using the methods discussed in this chapter. After embedding the data, we used a tool that is commonly used by law enforcement for forensic analysis to determine how easily an analyst could uncover this hidden data.

The data that was hidden was a text string, 'Hidden Message,' and a GIF image containing an image of the word 'Hidden.' Please understand, of course, that if the investigator already knows something of what he is looking for, for example, the word 'Hidden,' the difficulty of recovering this data is trivial. What we are seeking to measure is whether or not the tool alerts the analyst to the presence of data in unusual locations and how likely it is that the analyst would discover the materials.

The first test was run using FAT16. As a form of control, an image file was copied to the disk using the typical file system tools. Data was hidden using two simple methods: First, the FAT16, when formatted, automatically preferred a cluster size of 4 096 bytes. The result of this was that the boot sector of 512 bytes is immediately followed by three empty sectors. Our 'Hidden Message' text was inserted here. In order to conceal the presence of the GIF image, the first FAT was modified so that clusters 24 through 29 were marked as bad (see Fig. 3). Finally, the GIF image was placed onto the disk beginning in cluster 24.

With our data hidden, we used AccessData's Forensic Toolkit (better known as FTK) to acquire an image of the drive and perform data carving and full text indexing (see Fig. 4).

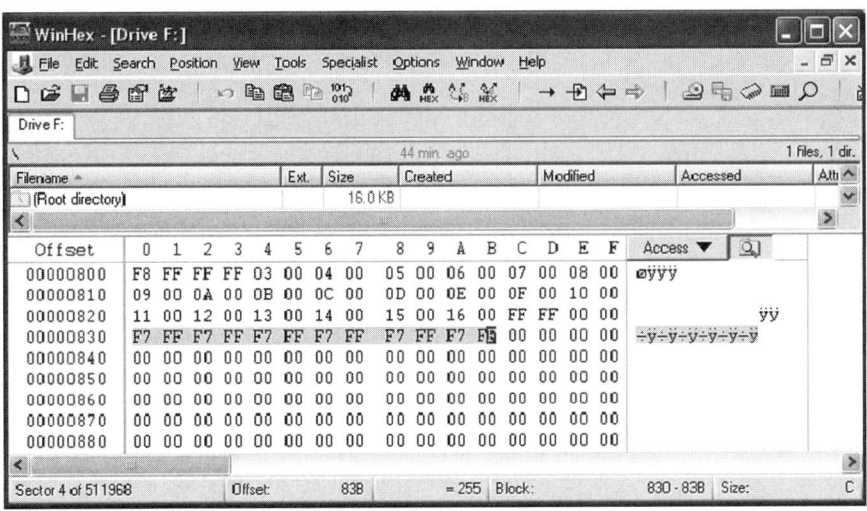

FIG. 3. File allocation table with clusters 24–29 marked 'bad.'

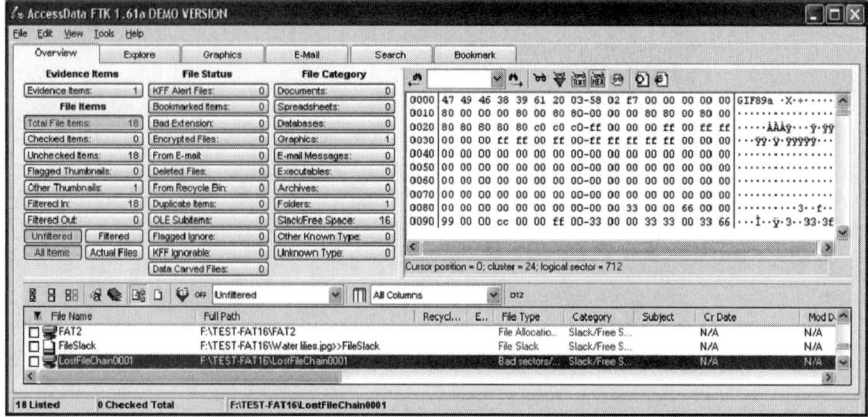

FIG. 4. Forensics toolkit results showing hidden GIF file.

Of particular interest to us were the results from the file carver. File carvers are forensic tools (or data recovery tools) that analyze the data on a drive (typically sector by sector or block by block) without any regard for the actual logical organization of the disk. Each sector or block is then checked for known file signatures and, if one is found, the data is extracted or marked appropriately. One of the reasons that a GIF image was selected is that it has a very standard and easily recognizable 'magic number' or fingerprint: Almost every one of them begins with the marker 'GIF89a.' While FTK did find the file and mark it as a 'Lost File Chain' marked as bad sectors, it did not successfully identify this as an image.

Even though the 'hidden message' is sitting in what should be a completely barren segment of the drive, FTK makes no special mention of this text. Unless the analyst is in the habit of examining all reserved sectors for additional data or already knows to search for a keyword of 'Hidden,' it is very unlikely that this data would be found. From our experience, this is typical behavior for a forensic toolkit. In short, it is unusual for any forensic analysis tool currently on the market to draw the analyst's attention to a piece of data that is residing in an atypical location.

3.2 NTFS

The next set of tests was performed using an NTFS file system with a cluster size of 4096 bytes. Data was hidden using several of the techniques discussed. First, in the reserved area between the primary partition table and the first partition, 'Hidden

Message' was again embedded. Since there was additional space, we also embedded our image file, but this time the file was not embedded beginning at the beginning of a sector. Many file carvers speed up their operation by examining only the first few bytes of a sector or only the first sector in a cluster.

As an additional test, the $Bitmap file was modified to mark the final clusters of the volume as 'In Use' and the GIF image was copied into these now reserved clusters. The 'Hidden Message' was also embedded using the Alternate Data Stream technique discussed.

To accomplish this, a host file was created followed by 1,000 associated streams. The data that we wished to conceal was then copied into what amounts to stream 1,001 and the host file was then deleted. Finally, additional files were copied onto the drive to obscure the original host file MFT entries, but with the cushion of 1,000 alternate data streams, the original hidden data would still be untouched.

With all of these tasks completed, the drive was imaged and analyzed using FTK. FTK was able to identify and carve the hidden data in the slack space between the partition table and the first partition, even though it was not on a sector boundary. Unfortunately, while it was successful with the image file, it still failed to 'notice' the hidden text. What if the file that was hidden did not match a known fingerprint or were encoded in some way? FTK would fail to notify the analyst. It was also successful in identifying the image file tucked away at the tail end of the disk in space that we 'reserved.' While this is good, there was no notification that space was marked 'In Use' that was not actually allocated to any disk structure or file. Obviously, the same problem exists as does for the gap between the partition table and the partition.

What about the alternate data streams? FTK, surprisingly, made no mention of the fact that there were still ~800 remnants of alternate data streams floating around in the MFT and, of course, did not notify us to the presence of the 'Hidden Message' unless we already knew what to search for.

Overall, this means that for us to be successful today with forensic investigations, we need really sharp analysts who are willing to go above and beyond to find data when we are dealing with clever malcontents. Our tests were performed on a small drive (256 MB). Obviously, hunting down data on a 500 GB drive is well-nigh impossible without good tools and good initial information. In the long run, as drives continue to increase in size and those who desire to conceal activities become better at their trade, the forensic community will be at a greater and greater disadvantage unless we become proactive in this 'Arms Race' and move away from strict signature-based analysis of media and start building in some sort of anomaly detection capability.

4. Perspectives

It should be pointed out that this discussion includes methods of hiding data only within the media and file system structure and does not address other data hiding methods such as:

- Altered BIOS parameters

    The system BIOS traditionally stores the physical configuration of the drives that are connected. Typically, the BIOS is set to automatically detect the drive parameters; however, it can be configured manually. By doing so, it is possible to create a sort of pseudo HPA area beyond what is apparently the end of the disk according to the BIOS.

- Registry entries using new keys or unused keys

    There are a variety of technologies available that allow one to create a virtual file system within a file on the system. With some minor modification, it is trivial to create such a file system that is stored, not within a file, but within a set of registry keys. Large disks would be prohibitively slow to access and have a serious impact on system performance due to memory utilization.

- Swap files

    While swap files are in fact files that are known to exist on a system, if the swap mechanism is disabled or otherwise controlled, the swap space can be used for the storage of arbitrary data. While it is true that swap space is frequently analyzed by forensic investigators, it is typically viewed as a memory dump. If a portion of the data is instead formatted as a virtual encrypted disk, it is highly likely that this data will go unnoticed, dismissed as random junk.

- Renamed files (e.g., as .dll)

    Decidedly nontechnical, this remains an extremely effective way to hide almost any kind of data. Malicious code authors have been using this same technique to great effect for many years, concealing the malicious code as what appears to be a system DLL. This will likely not fool a forensic investigator, but a casual user or administrator will often overlook files masked in this way.

- Binding of one executable file to another

    Another very popular technique with malicious code authors is to redirect execution from one application to another. In some ways, this is exactly how appending viruses of the early nineties functioned in that the malicious code was appended to the existing executable and then the executable header was modified or the initial instruction modified to redirect execution to the malicious code. Following execution of the malicious code, control was handed back to the original code. These days redirection is accomplished in a variety of ways, for instance through DLL injection, which could then allow a user to subvert the typical input/output system for file and data access.

- Steganography

    This well-known technique for hiding data in plain sight is fast gaining in popularity. One of the most interesting applications of this is the StegFS driver for Linux (http://www.mcdonald.org.uk/StegFS/) which allows the user to create layers of hidden file systems on top of an existing Linux file system.

- Hiding data within documents (e.g., as metadata or using a white font)
- Hiding data within html files
- Merging Microsoft Office documents

    This is another steganographic technique. While some recent versions of software have made this more difficult (for instance, newer versions of Microsoft Word will actually display the white on white text as a grayed out text), the hiding of data in metadata or through other techniques is still possible. For instance, one tool embeds data by manipulating the white space at the end of existing lines within a document.

- Encrypted files
- Compressed files

    These two are perhaps the most direct way to store data on a system in a somewhat unreadable format, though some may contend with how well hidden the data actually is. From the point of view of someone who is looking for 'unusual' data, a compressed or encrypted file would be of interest. From the point of view of someone who is trying to run a file carver or a string match against a set or large disks, it is quite likely that these methods will be sufficient to hide the data from an investigator. For the compression, simply using a common compression tool would likely not be enough to day. Many tools are smart enough to take these files apart or alert the operator that a password is required. Using a older technology like LHARC could prove quite sufficient, though. This format (and others) could still be recognized, but the investigator's tools are far less likely to understand how to decompress the data.

5. Conclusion

Knowing how data can be hidden within the media and file system structure means knowing how that data can be found. Note that we are talking about only the data hiding mechanisms that have been discussed here; methods such as encryption and steganography present their own sets of analysis problems.

Many of those who attempt to temporarily hide data or use a computer with nefarious intent are aware that there are ways in which this hidden data can be found. They will therefore use one of the available 'disk wiper' utilities in an attempt to eliminate the evidence. What they are not aware of is that many of these utilities are ineffective in eliminating all hidden data. See [2] for more information.

A forensic analysis tool can be as simple as a hex-editor. While effective, this approach is very labor intensive given the size of today's hard drives. There are many commercial and open source digital forensic tools available that will automate this process. However, one must exercise caution when using these tools, since not all of them are equally effective in finding all types of hidden data. It is often advisable to use more than one of these tools to perform a digital forensic investigation. For further information, see [5], [7], and [8].

Finding hidden data is further complicated by the fact that there are a number of antiforensic tools being developed. One source is the research being done by the Metasploit Project (www.metasploit.com/projects/antiforensics). An example is their Slacker tool that automatically encrypts and hides a set of data within the slack space of multiple files that are known to be unlikely to be changed. Data thus hidden cannot be detected by currently available digital forensic analysis software.

Data hiding becomes an increasingly important topic as the sophistication of information technologists on both sides of the law increases.

REFERENCES

[1] Berghel H., and Brajkovska N., 2004. Wading through alternate data streams. *Communications of the ACM*, **47**(4): 21–27.
[2] Berghel H., and Hoelzer D., What does a disk wiper wipe when a disk wiper does wipe disks. *Communications of the ACM*, **49**(8): 17–21.
[3] Berghel H., and O'Gorman L., 1996. Protecting ownership rights through digital watermarks. *IEEE Computer*, **29**(7): 101–103.
[4] Caloyannides M. A., 2001. Computer Forensics and Privacy. Artech House, Norwood, MA.
[5] Carrier B., and Winter B., 2003. Defining digital forensic examination and analysis tools using abstraction layers. *International Journal of Digital Evidence*, **1**(4): 1–12.
[6] Carrier B., 2005. File System Forensic Analysis. Addison-Wesley, Upper Saddle River, NJ.

[7] Carvey H., 2005. Windows Forensics and Incident Recovery. Addison-Wesley, Upper Saddle River, NJ.
[8] Casey E., 2002. Handbook of Computer Crime Investigation. Academic Press, San Diego, CA.
[9] Cole E., 2003. Hiding in Plain Sight: Steganography and the Art of Covert Communication. Wiley Publishing, Indianapolis.
[10] Farmer D., and Venema W., 2005. Forensic Discovery. Addison-Wesley, Upper Saddle River, NJ.
[11] Kruse I. I., Warren G., and Heiser J. G., 2002. Computer Forensics, Incident Response Essentials. Addison-Wesley, Upper Saddle River, NJ.
[12] Nelson B., et al., 2006. Guide to Computer Forensics and Investigations. 2nd edition Course Technology, Florence, KY.
[13] Singh S., 2000. The Code Book. Anchor Books, New York.
[14] Wang H., and Wang S., Cyber warfare: steganography vs. steganalysis. *Communications of the ACM*, **47**(10): 76–82.

Multimedia and Sensor Security

ANNA HAĆ

Department of Electrical Engineering
University of Hawaii at Manoa, Honolulu
Hawaii 96822

Abstract

Multimedia and sensor security emerged as an important area of research largely due to growing accessibility of tools and applications through different media and Web services. There are many aspects of multimedia and sensor security ranging from trust in multimedia and sensor networks; security in ubiquitous databases, middleware, and multi-domain systems; privacy in ubiquitous environment and sensor networks; and radio frequency identification security. Particularly interesting are multimedia information security, forensics, image watermarking, and steganography.

There are different reasons for using security in multimedia and sensors that depend on the method of access, the type of accessed information, the information placement and the availability of copying, sensitivity of protected information, and potential ways for corruption of data and their transmission.

Protection of image and data includes watermarking, steganography, and forensics. Multi-domain systems, middleware, and databases are used for both local and distributed protections. Sensor networks often use radio frequency identification tags for identification. Both trusted hardware and software are needed to maintain multimedia and sensor security.

Different applications require various levels of security often embedded in the multimedia image. In addition, a certain security level is needed for transmission of the application. Sensors use different security for data gathering depending on the type of data collected. Secure transmission is needed for sensor data.

We present a comprehensive approach to multimedia and sensor security from application, system, network, and information processing perspective. Trusted software and hardware are introduced as support base for multimedia and sensor security.

1. Introduction 20
2. Multimedia Systems and Applications 21
3. Multimedia Security 24
4. Digital Watermarking 26
5. Steganography 28
6. Computer Forensics 29
7. Sensor Networks 30
8. Security Protocols for Wireless Sensor Networks 33
9. Communication Security in Sensor Networks 34
10. Sensor Software Design 35
11. Trusted Software 35
12. Hardware Power-Aware Sensor Security 36
13. Trusted Hardware 36
14. Sensor Networks and RFID Security 37
15. Conclusion 37
    References 37

1. Introduction

Multimedia comprises of voice, data, image, and video, protection of which becomes increasingly important in the information age. There are many security levels depending on the application and the method in which the application is executed. Multimedia applications include both software and hardware, which are often the integral parts of the application in addition to the system software and hardware used by the application. Both software and hardware security are crucial to multimedia application performance and execution as well as to the multimedia information delivered. We will describe trusted software and hardware needed to protect multimedia application and its execution in the system.

Internet and Web-based applications have become an integral part of every day life and security in accessing and executing these multimedia applications is necessary for trustworthiness of the information sent and received. The methods used to preserve multimedia information security include forensics, image watermarking, and steganography.

Sensors have also become a part of every day life and of the growing new developments and applications. Sensor applications include data being gathered and transferred as well as the multimedia information. Trusted software and hardware are needed, security and trust in multimedia and sensors is important, multimedia and ubiquitous security is emerging, and the security in ubiquitous databases for information access has been developed. In sensor networks, RFID (radio frequency identification) security and multimedia information security are needed for secure transmission of information.

We describe multimedia systems and applications, sensor networks, security protocols for wireless sensor networks, communication security in sensor networks, sensor software design, trusted software, hardware power-aware sensor security, trusted hardware, sensor networks and RFID security, forensics, image watermarking, and steganography.

2. Multimedia Systems and Applications

Multimedia systems and applications are used in the representation, storage, retrieval, and dissemination of machine-processable information expressed in multimedia, such as voice, image, text, graphics, and video. High capacity storage devices, powerful and economical computers, and high speed integrated services digital networks, providing a variety of multimedia communication services made multimedia applications technically and economically feasible. Multimedia conference systems can help people to interact with each other from their homes or offices while they work as teams by exchanging information in several media, such as voice, text, graphics, and video. Participants can communicate simultaneously by voice to discuss the information they are sharing. The multimedia conference system can be used in a wide variety of cooperative work environments, such as distributed software development, joint authoring, and group decision support [29].

In multimedia conferencing, each participant has a computer that includes high-resolution screen for computer output, keyboard and pointing device, microphone and speaker, and camera and video monitor. Parts of each participant's screen can be dedicated to display shared space where everyone sees the same information. Voice communication equipment can be used by the conference participants for discussion and negotiation. Video communication can add illusion of physical presence by simulation of face-to-face meeting. Video teleconferencing is valuable when nonverbal communication, in the form of gestures and facial expressions, is an important part of the discussion and negotiation that takes place. Video communication can be a useful enhancement to the multimedia conferencing, but is less critical than voice communication and the ability to share information and software.

Electronic blackboards allow joint manipulation of shared image, but are subject to the same limitations as the physical blackboard. The participant's ability to access and manipulate information dynamically by making use of powerful computer-based tools is the distinguishing characteristics of the multimedia conferencing. The goal of the multimedia conference systems is to emulate important characteristics of face-to-face meetings. These systems allow group of users to conduct meeting in real time and the conference participants can jointly view, edit, and discuss relevant multimedia documents.

There are many applications of multimedia conferencing systems, ranging from software development to medical image analysis, and to making assessments and decisions in real time [33, 54]. Security of multimedia conferencing systems is very important since the processing occurs in real time, and in case of some applications, some crucial, even life-saving decisions are made.

Multiple compression algorithms allow for having quality performance in real-time video through radio channels. In these algorithms, the computational complexity is traded against image quality. Multimedia compression includes data and text compressions by using semantic analysis, audio compression by using inaudible audio masks, image compression by using graphic formats and blocks, and video compression by using motion pictures and streaming and mobile applications.

The characteristics of modern embedded systems are the capability to communicate over the networks and to adapt to different operating environments [28]. Embedded systems can be found in consumer devices supporting multimedia applications, for example, personal digital assistants, network computers, and mobile communication devices. The low cost, consumer-oriented, and fast time to market objectives characterize embedded system design. Hardware and software codesigns are used to support growing design complexity [19, 20].

Emerging embedded systems run multiple applications such as Web-browsers, audio and video communication applications, and require network connectivity [10].

Networked embedded systems are divided into

- Multifunction systems that execute multiple applications concurrently, and
- Multimode systems that offer the users a number of alternative modes of operation.

In the multifunction systems, the embedded systems can execute multiple applications concurrently [36]. These applications include capturing video data, processing audio streams, and browsing the Web. Embedded systems must often adapt to the changing operating conditions. For example, multimedia applications adapt to the changing network rate by modifying video frame rate in response to network congestion. Different network rate and compression techniques are often determined by the network load and quality of service feedback from the user applications.

Embedded multimode systems experience a number of alternative modes of operation. A mobile phone is designed to change the way it operates to accommodate different communication protocols supporting various functions and features. Flexible, multimode devices are used in applications such as electronic banking and electronic commerce. Depending on the type of connection and the security level required, devices use different encryption algorithms when transmitting data.

Embedded devices need to communicate over the networks and to adapt to different operating environments [19, 20, 35]. Multimedia systems concurrently execute multiple applications, such as processing audio streams, capturing video data and Web browsing. These multimedia systems need to be adaptive to changing operating conditions. For instance, in multimedia applications, the video frame rate has to be adjusted depending on the network congestion. Audio streams different compression techniques are the function of the network load.

In both home and industrial applications there are sensor and actuator devices which can be remotely controlled and maintained via Internet. These applications need Web-based security to maintain the accessed information and the data changes and updates that may be introduced.

Multimedia is supported by MPEG (Moving Picture Experts Group) standards: MPEG-1, MPEG-2, MPEG-4, MPEG-7, and MPEG-21. MPEG-1 includes video and audio compression standards. MPEG-2 includes transport, video, and audio standards for broadcast quality television. MPEG-4 supports DRM (digital rights management). MPEG-7 is used to describe multimedia content, and MPEG-21 supports multimedia framework. Broadband and streaming video delivery as well as video storage require MPEG-1, MPEG-2, and MPEG-4. Content-based retrieval, adaptation, and multimedia filtering use MPEG-4 and MPEG-7. MPEG-7 allows for semantic-based retrieval and filtering, and MPEG-21 allows for media mining on electronic content by using digital items. Multimedia framework supported by MPEG-21 allows the participants to exchange digital items, which include media resources, learning objects, rights representations, reports, and identifiers.

Network security is critical to multimedia communication and access. There are several security challenges to Internet transmission. Denial of service (DoS) attacks may occur when the session protocol proxy server or voice-gateway devices receive many unauthentic packets. A hijacking attack occurs when a participant registers with an attacker's address instead of the user's legitimate address. Eavesdropping is an unauthorized interception of voice packets or decoding of signaling messages. Packet spoofing occurs when an attacker impersonates a legitimate participant transmitting multimedia information. Replay occurs when a message is retransmitted causing the network device to reprocess the same message again. Message integrity ensures that the message received is the same as the message that was sent.

3. Multimedia Security

Multimedia security includes protection of information during transmission as well as content security of the image and data, protection of intellectual property, and support for trustworthiness. Multimedia content is protected through data authentication and confidentiality that ensures privacy of multimedia transmission.

Protection of a multimedia image is important once it is posted on a Web site. The multimedia image can be easily downloaded by unauthorized users in a very short period of time after it has been posted. Having watermarks embedded in multimedia images does not prevent unauthorized downloading of a file. Putting copyright notices into image source code does not protect the multimedia image from being downloaded by unauthorized users. Encryption of the image ensures that it cannot be saved to a disk or linked directly from the other Web sites. Protection of multimedia images by using encryption is available for both JPEG (joint photographic experts group) and non-animated GIF (graphics interchange format) files.

We present multimedia security frameworks including image and communication security. DRM describes, protects, and monitors all forms of access and use to protect multimedia from illegal distribution and theft. A DRM system links user rights to multimedia access for viewing, duplication, and sharing. Security services preserve data integrity and confidentiality, support non-repudiation by protecting against denial by the users during communication, and maintain access control and authentication. Multimedia confidentiality is preserved by using encryption supporting real-time access and communication as well as image compression [4, 51, 62].

Encryption algorithms use keys to maintain security of the system. The strength of a cryptographic system depends on the encryption algorithm used, the system design, and the length of the key. The length of the key reflects the difficulty for an attacker to find the right key to decrypt a message. The longer key requires more combinations to be checked in a brute force attack to match the key. The key length is given in bits and the number of possible keys that can be used grows exponentially with the increase of key length.

DES (data encryption standard) is based on Feistel Cipher Structure, which is a block cipher with the plaintext of 64-bit blocks and 56-bits key [83]. The key length is increased to 112/168 bits in triple DES and to 128/192/256 bits in AES (advanced encryption standard) with 128-bit data. Both DES and AES require significant computational resources and use block-based structures that introduce delay in real-time communications. In RSA (Rivest, Shamir, and Adleman), the key length is variable, and commonly used key length is 512 bits [70]. The block size in RSA is variable. In MD5 (message digest standard), the message is processed in 512-bit blocks, and the message digest is a 128-bit quantity [69].

Lossless compression is used in applications where both the original file and decompressed information need to be identical. The original file can be derived from the compressed information after lossless compression algorithms have been applied. In many multimedia applications, lossless compression, which requires large compressed files, is expensive and unnecessary. Examples of files using lossless compression are GIF, BMP (bitmap), PNG (portable network graphics), and TIFF (tagged image file format).

Lossy compression is used to compress multimedia image, audio, and video, in such a way that the retrieved decompressed multimedia information is a fraction of lossless uncompressed multimedia while retaining acceptable quality. Because lossy compression eliminates permanently redundant information from the multimedia image, thus, repeatedly compressing and decompressing a multimedia file progressively decreases decompressed file quality. JPEG file uses lossy compression. Examples of lossy compression methods include DCT (discrete cosine transform), VQ (vector quantization), fractal compression, and DWT (discrete wavelet transform).

Multimedia authentication requires sensitivity, robustness for lossy compression, security, portability, localization to detect changed area, and recovery to approximate the original in its particular area that was changed. Robust digital signature verifies information integrity endorsed by the signed owner. Digital signature can use content-related feature codes and distinguish content-preserving operations from malicious attacks. SARI (self authentication and recovery images) detect the manipulated area in multimedia work, and recover approximate values in the changed area. Content-related feature codes use syntax in the form of objects' composition and semantics to specify multimedia objects. Syntax describes shape, texture, and color of multimedia objects. Semantic authentication describes types of objects and their relationships. Semantic authentication also includes events and the scene, where the objects are located.

A multimedia security framework using randomized arithmetic coding [24, 25] allows for protection, encryption, and conditional access. The randomized arithmetic coding is employed in JPEG 2000 security applications.

Image authentication, robustness, and security are explored in [80] by using an image hash function based on Fourier transform. A framework for generating a hash consists of feature extraction that generates a key-dependent feature vector from the multimedia image, quantization of the generated feature vector, and finally compression of the quantized vector.

A multimedia encryption scheme that does not rely on the security strength of an underlying cryptographic cipher is introduced in [91]. A pseudo-random key hoping sequence is generated that prevents a chosen-plaintext attack. The method ensures high semantic security by enhancing MHT (Multiple Huffman Tables) encryption scheme. An evaluation of MHT schemes by using chosen-plaintext attacks is described in [102].

An encryption scheme for MPEG-4 FGS (fine granularity scalability) video coding stream is proposed in [95]. Compression efficiency is preserved by processing encrypted FGS stream on the cipher text.

A peer-to-peer (P2P) interactive multimedia communication is described in [8]. Streaming multimedia traffic in wireless communication under SRTP (secure real-time transport protocol) and the key management protocol MIKEY (multimedia Internet KEYing) are explained. Authentication methods use message authentication code or a signature.

An agent-based multimedia security system is proposed in [46]. A hierarchy of hosts supports multilevel security groups. Security service agents are created and dispatched to security groups. Security problems related to multimedia multicast are described in [59, 60, 93]. Additional bandwidth for multicast transmission with multimedia security protection increases demand on distributed system resources. A multicast packet authentication scheme is presented in [71]. Mobile multicast security is discussed in [22]. Multimedia security in FTTH (fiber to the home) connectivity in passive optical networks is explained in [73]. Proposed security enhancements include wavelength hopping and codes sequencing in downstream traffic in P2MP (point to multi-point) network.

Multilevel 2D (2-dimensional) bar codes for high-capacity storage applications are described in [84]. A print-and-scan channel is adapted to multilevel 2D bar codes communication. Fully format compliant scrambling methods for compressed video are proposed in [37]. A decoder can decode the secured multimedia stream in fully format compliant scrambling method. Security problems for video sensors are discussed in [42]. Multimedia security for sensor networks includes scaling down security features to accommodate smaller sensors' size and processing capabilities.

4. Digital Watermarking

Watermark is an additional information used to protect images. A watermark can be a copyright notice on the picture, a date-stamp on photographs taken, additional names and comments on the pictures, and a company logo on a document.

Digital watermarking uses hidden messages to copyright digital images, video, and audio by employing sequence of bits carrying information characterizing the owner of a multimedia item. A watermark can contain any useful information that identifies copyright protection. In addition, there are watermarks that indicate other types of ownerships.

Watermarking is used to embed visible and invisible information in the form of code into multimedia to protect the image and to identify its ownership. Watermark must be very difficult or impossible to remove. Visible watermark must be non-obstructive. The invisible watermark should not visually affect the image content, and must resist image manipulations.

Watermarking on multimedia content embeds watermark information into the multimedia image by using coding and modulation. Coding is done by scrambling with the use of cryptographic keys and error correction coding. Modulation employs TDMA (time division multiple access), FDMA (frequency division multiple access), CDMA (code division multiple access), or spread spectrum. Spread spectrum uses DFT (discrete Fourier transform), DCT, and wavelet transforms.

JND (just noticeable distortion) is the maximum amount of invisible changes in a specific pixel (or frequency coefficients) of an image. JND employs luminance masking with the threshold specified by light adaptation of human cortex, where the brighter background increases the luminance masking threshold. JND also applies contrast masking, in which the visibility of one image component is reduced by the presence of another image component. The contrast masking is the strongest when both image components are of the same spatial frequency, orientation, and location.

Multimedia security can be compromised through unauthorized embedding, where the attacker is able to compose and embed the original watermark image, which is either given away by the owner or extracted by the attacker. Standard cryptographic techniques and content-related watermarks help to prevent unauthorized embedding. Multimedia security can also be compromised by unauthorized detection of watermark image that was intended to remain sealed from the attacker. This unauthorized detection can be prevented by using encryption and decryption techniques. Unauthorized removal is another multimedia security attack where the watermark is eliminated or masked. Spread spectrum techniques can be used to prevent unauthorized removal. Finally, in the system-level multimedia security attack, the weakness in using the watermark is exploited. By compromising the system control over the watermark image, the attacker can gain access to the watermarked multimedia work. To prevent system-level attack, a device scrambling watermark multimedia work makes the watermark image undetectable to the attacker. A descrambling device inverts the watermark multimedia work. There are also attacks distorting watermark image and affecting synchronization between multimedia audio and video. Ambiguity attacks pretend that the watermark image is embedded where no such embedding occurred. The owner of the multimedia work can use non-invertible embedding techniques to prevent ambiguity attack.

Digital watermarking can be fragile, which does not resist modifications of the multimedia, or it can be robust, which is secure against common modifications [43]. Fragile watermarks can be used for multimedia authentication whereas robust watermark can be used in multimedia leakage. Semi-fragile watermarks can be used for both media authentication and recovery. An image authentication scheme that uses relative similarity between neighboring image blocks is introduced in [16]. A fragile watermarking technique to increase security by embedding watermark into the image frequency domain by using genetic algorithms is described in [74]. A method to vary quantization factor in DCT domain to weight robustness of the watermark versus the watermark's image quality is presented in [47]. A public and private watermark extraction methods are described in [1] by using asymmetric and symmetric techniques, respectively. A watermark image, parts of which are embedded into different video scenes is described in [14]. In addition, the error correcting code is embedded in the audio channel.

A multimedia gateway security protocol is introduced in [53]. The protocol provides authentication, a watermark detection at a server or gateway, and the system response to security breach. Digital watermarking parameters and security measures are described in [9]. The embedding algorithms and procedures are introduced in [89]. Discussions to improve the performance of watermarking algorithms are conducted in [43]. A visible watermarking technique by using a contrast-sensitive function is introduced in [34]. An algorithm to embed watermark coefficients into the image blocks is presented in [67]. This nonuniform embedding allows for robustness against cropping. A method to embed a watermark using neural networks is presented in [97].

5. Steganography

Steganography is used to write hidden messages inside the image in such a way that nobody except for the intended recipient knows about the message. The hidden message appears as something different from the image in which it is embedded, for example, it can be a picture or text. A steganographic message can be encrypted in many different ways, and only the recipient who knows the method used can decrypt the message. In computer communication networks, the steganographic coding occurs at the transport layer.

Steganography allows for hiding files or encrypted messages in several file types, for example, JPEG, PNG, BMP, HTML (hypertext markup language), and WAV (waveform). Steganalysis is used to detect modified files by comparing their contents. Many files accessible through the Internet-use GIF image. A method to detect secret messages in multimedia GIF format image by using statistical steganalysis is presented in [40].

In files using lossless compression, for example, GIF and BMP files, a simple data hiding within the image can employ LSB (least significant bit) insertion. Changing the LSB in the binary image according to a predefined pattern allows for inserting hidden data.

Steganography methods can use other patterns to include or to extract a hidden message. File transformation can be done by using DCT coefficients in JPEG images to hide the information in lossy compression.

Steganography is employed to detect hidden communication channels used by adversaries over the Internet. Intrusion detection systems for VoIP (voice over Internet protocol) applications are described in [18, 31, 41].

6. Computer Forensics

Computer forensics is a technological inspection of a computer, its files, and the devices for its present and past content that may be used as evidence of a crime or other computer use that is the subject of an inspection. This computer inspection must be both technical and legal to meet the standards of evidence that are admissible in a court of law. The examination of a computer includes its hardware, for instance, memory, disk, and other I/O (input/output) devices, as well as the operating systems actions, for example, system calls, including system traps, and finally the software that was executed on the computer. In addition, the user programs and electronic messages (E-mails) need to be reviewed.

The information recovered to show the crime depends on whether the computer was a target of a crime or was a computer involved to commit a crime. Forensics investigation includes computer break-ins by an unauthorized user, disallowed data duplication, theft, espionage, inappropriate Internet use or abuse, etc. The information recovered often resides on the disk drives and other storage devices that retain traces of unauthorized access.

Multimedia forensics uses authentication and identification to recognize the ownership and possible media manipulation. Multimedia forensics can be applied to video and audio, can be used for facial recognition, speaker identification, handwriting comparison, and image authentication. DNA profiling is used in forensic analysis and investigations. Image forensics analyzes the condition of an image to detect forgery or to identify the source of the image. Image forgery may occur through photomontage, retouching, or removing some objects. The image source can be identified as the camera, computer graphics, printer, scanner, etc. The image scene can be identified as 2D or 3D (3-dimensional). The image forensics searches image authenticity by defining characteristics of the imaging device and the scene of the image. The device characteristics include specific effects of a particular camera, for instance, optical low-pass, color filter array interpolation, and lens distortion. The scene characteristics include lighting and shadows on the image, and the reflectance of the objects present in the image.

Digital fingerprinting embeds a unique identifier into each copy of the multimedia work. In a collusion attack, a group of users generates a new version of multimedia work by combining these users' copies of the same content, which originally had different fingerprints. These fingerprints are attenuated or removed by a collusion attack.

A theoretical framework for the linear collusion attack and analysis is presented in [78]. A multiple-frame linear collusion resulting in an approximation of the original watermark is created by scaling and adding watermarked video frames. Collusion-resistant watermarks are created by changing watermarking key every number of frames that ensures certain frames to have pair-wise correlations.

The purpose of fingerprinting multimedia systems is to prevent collusion attacks and other single-copy attacks. Secure video streaming of fingerprinted applications over the networks to reduce bandwidth for large number of participants is presented in [100]. A fingerprinting image makes each multimedia application slightly different and prevents traditional multicast technology, for instance, one-to-many or many-to-many transfer, to be implemented directly. Spread spectrum embedding-based fingerprinting is a data hiding method through coefficient embedding to protect the multimedia image against collusion attacks.

An analysis of behavior pattern of attackers during multiuser collusion is discussed in [98]. Timing of processing a fingerprinted copy by a single colluder impacts collusion detection. The quality of the fingerprinted copy, particularly a higher resolution, increases likelihood of collusion attack [99].

7. Sensor Networks

Wireless sensor networks use small devices equipped with sensors, microprocessor, and wireless communication interfaces [63, 64, 92]. Wireless sensor nodes are deployed in areas and environments where they may be hard to access, yet these nodes need to provide information about measurements of temperature, humidity, biological agents, seismic activity, and sometimes take pictures and perform many other activities. Macrosensor nodes usually provide accurate information about measured activity. The accuracy of information provided by individual microsensor nodes is lower, yet a network of hundreds or thousands of nodes deployed in an area enables to achieve fault tolerant high quality measurements.

Wireless sensor nodes are designed by using MEMS (microelectromechanical systems) technology and its associated interfaces, signal processing, and RF (radio frequency) circuitry. Communication occurs within a wireless microsensor network, which aggregates the data to provide information about the observed environment.

Low energy dissipation is particularly important for wireless microsensor nodes, which are deployed in hundreds or thousands, and are often hard to read in inhospitable terrain. A power-aware system employs a network design whose energy consumption adapts to constraints and changes in the environment [66, 75, 76]. These power-aware design methods offer scalable energy savings in wireless microsensor environment [3]. There is a trade off between battery lifetime and quality performance of data collection and transmission [2].

Activity in the observed environment may lead to significant measurement diversity in the sensor node microprocessor. Node functionality may also vary, for instance, a sensor networking protocol may request the node to act as a data gatherer, aggregator, relay, or any combination of these. This way, the microprocessor can adjust the energy consumption depending on the activity in the measured environment.

Several devices have been built to perform sensor node functions. A software and hardware framework includes a microprocessor, low-power radio, battery, and sensors [11, 65, 87]. Data aggregation and network protocols are processed by using a micro-operating system. Data aggregation is used as a data reduction tool [44]. Aggregates summarize current sensor values in sensor network. Computing aggregates in sensor network preserves network performance and saves energy by reducing the amount of data routed through the network [17].

The computation of aggregates can be optimized by using SQL (Structured Query Language). The data are extracted from the sensor network by using declarative queries. Examples of database aggregates (COUNT, MIN, MAX, SUM, and AVERAGE) can be implemented in a sensor network. Aggregation in SQL-based database systems is defined by an aggregate function and a grouping predicate. The aggregate function specifies how to compute an aggregate [50].

Aggregation can be implemented in a centralized network by using a server-based approach where all sensor readings are sent to the host PC (personal computer), which computes the aggregates. A distributed system approach allows in-network computing of aggregates, thus decreasing the number of forwarded readings that are routed through the network to the host PC.

The small battery-powered sensor devices have limited computational and communication resources [30]. This makes it impractical or even impossible to use secure algorithms designed for powerful workstations. A sensor node memory is not capable of holding the variables required in asymmetric cryptographic algorithms, and perform operations by using these variables.

The sensor nodes communicate by using RF, thus trust assumptions and minimal use of energy are important for network security. The sensor network communication patterns include sensor readings, which involve node to base station communication, specific requests from the base station to the node, and routing or queries from the base station to all sensor nodes. Because of sensor nodes limitations, the sensor nodes are not trusted. The base stations on the other hand, belong to the TCPA trusted computing base. The senor nodes trust the base station and are given a master key which is shared with the base station. The possible threats to network communication security are an insertion of malicious code, an interception of the messages, and injecting false messages.

Wireless technologies enabling communications with sensor nodes include Bluetooth and LR-WPAN (Low-Rate Wireless Personal Area Network) [27].

Bluetooth enables seamless voice and data communication via short-range radio links. Bluetooth provides a nominal data rate of 1 Mbps for a piconet, which consists of one master and up to seven slaves. The master defines and synchronizes the frequency hop pattern in its piconet. Bluetooth operates in the 2.4 GHz ISM (Industrial, Scientific, and Medical) band.

Low-Rate Wireless Personal Area Network (LR-WPAN) is defined by the IEEE 802.15.4 standard. This network has ultralow complexity, cost, and power for low data-rate sensor nodes [13]. The IEEE 802.15.4 offers two physical layer options: the 2.4 GHz physical layer and the 868/915 MHz physical layer. The 2.4 GHz physical layer specifies operation in the 2.4 GHz ISM band. The 868/915 MHz physical layer specifies operation in the 868 MHz band in Europe and in 915 MHz band in the United States.

The main features of the IEEE 802.15.4 standard are network flexibility, low cost, and low-power consumption. This standard is suitable for many applications in the home requiring low-data-rate communications in an ad hoc self-organizing network.

The major resource constraint in sensor networks is power, due to the limited battery life of sensor devices [30, 32]. Data-centric methodologies can be used to solve this problem efficiently. Data-centric storage (DCS) is used as a data-dissemination paradigm for sensor networks. In DCS, data is stored, according to event type, at corresponding sensornet nodes. All data of a certain event type is stored at the same node. A significant energy saving benefit of DCS is that queries for data of a certain type can be sent directly to the node storing data of that type. Resilient Data-Centric Storage (R-DCS) is a method to achieve scalability and resilience by replicating data at strategic locations in the sensor network [21]. This scheme leads to significant energy savings in networks and performs well with increasing node density and query rate.

Sensor network management protocol has to support control of individual nodes, network configuration updates, location information data exchange, network clustering, and data aggregation rules. Sensor network gateway has to provide tools and functions for presentation of network topology, services, and characteristics to the users and to connect the network to other networks and users.

Sensor networks are vulnerable to security attacks due to the wireless nature of the transmission medium. In wireless sensor networks, the nodes are often placed in a hostile or dangerous environment where they are not physically protected.

The standard approach to maintain data secrecy is to encrypt the data with a secret key that is known only to the intended recipients. The secure channels can be set up between sensor nodes and base stations based on their communications patters.

8. Security Protocols for Wireless Sensor Networks

Sensor networks become more commonly used in many fields and applications, causing the security issues to become more important. Security in sensor networks needs to be optimized for resource-constrained environments and wireless communication. SPINS (Security Protocol for Sensor Networks) [61] has two secure building blocks: SNEP (Secure Network Encryption Protocol) and μTESLA (the micro version of the Timed, Efficient, Streaming, Loss-tolerant Authentication Protocol). SNEP provides the following security primitives: data confidentiality, two-party data authentication, and data freshness. Efficient broadcast authentication is an important mechanism for sensor networks. μTESLA is a protocol which provides authenticated broadcast for resource-constrained environments making it suitable for sensor nodes. An authenticated routing protocol uses SPINS building blocks. Communications of sensor nodes uses RF that consumes node energy. There is a trade off between computation and communication in energy consumption.

Devices usually employ a small battery as the energy source. The other sources of energy are also limited. Communication over radio is the highest energy-consuming function performed by the devices. This imposes limits on security used by these devices. The lifetime of security keys is limited by the power supply of the microprocessor performing security functions. On the other hand, the base stations have large energy supplies often wired and connected to the communications network.

Wireless sensor network connects to a base station, to which information is routed from the sensor nodes. The base station has significant battery power, large memory to store cryptographic keys, and communicates with other wired and wireless networks. The communication between wireless sensor nodes and base station includes sensor readings from sensor node to the base station, requests from base station to sensor node, and routing beacons and queries from base station to sensor nodes.

Several encryption algorithms are evaluated in [49]. SEAL (software-optimized encryption algorithm) is stream cipher, making it much faster than block cipher RC5, but having longer initialization phase. SEAL is a very safe algorithm using 160-bit key encryption [49].

RC4 is a fast stream cipher algorithm with key size of up to 2,048 bits. This algorithm can be used with a new key selected for each message. RC5 is a fast symmetric block cipher algorithm [68, 69].

Tiny encryption algorithm (TEA) [55] is a block cipher operating on 64-bit blocks and using 128-bit key. TEA is fast and simple, but its security weakness is having equivalent keys, which reduces the effective key size to 126 bits. There are extensions of TEA [56], which correct some of the weaknesses of original TEA. XTEA (eXtended TEA) has a more complex key schedule, and XXTEA (corrected block TEA) operates on variable length blocks that are 32 bits. A fast software encryption is described in [96].

The comparison of encryption algorithms in [49] suggests TEA as the most suitable for sensor networks with limited memory and high speed requirements.

9. Communication Security in Sensor Networks

A possible threat to network communication security is an insertion of malicious code, which the network could spread to all nodes, potentially destroying the whole network, or even worse, taking over the network on behalf of an adversary. The sensor nodes location needs to be hidden from potential attackers. In addition, the application specific content of messages needs to be protected. An adversary can inject false messages that provide incorrect information.

The security overhead in wireless sensor network should be related to sensitivity of the encrypted information. Three levels of information security rank the most sensitive the mobile code, then the sensor location information in messages, and finally the application specific information [77]. The strength of the encryption for each of security rank corresponds to the sensitivity of the encrypted information. The encryption applied at the top rank is stronger than the encryption applied at next rank, and the weakest encryption is at the lowest security rank. RC6 (symmetric block cipher) meets the requirements of AES, and uses an adjustable parameter, which is the number of rounds that affects its strength. The overhead for the RC6 encryption algorithm increases with the strength of the encryption.

A framework for a sensor network to protect data and communicate with the operating system is introduced in [82]. The proposed framework detects attacks on the sensor data and software. A model that captures system and user behavior is developed and introduced as a control behavioral model (CBM). An application of group key management scheme to sensor networks by using a clustering approach is described in [94].

10. Sensor Software Design

Embedded systems require hardware and software codesign tools and frameworks. A complete design environment for embedded systems should include dynamically reconfigurable hardware components [5]. Designing configurable hardware and software systems requires specification, initial profiling, and the final implementation of the system's software components. Designing an embedded system's digital hardware requires a hardware description language and synthesis tools, which are used for an abstract circuit design. In order to reduce development time and effort, embedded systems' designers employ abstract specification, and apply the reuse of hardware and software components.

In embedded system design, a major share of functionality is implemented in software. This allows for faster implementation, more flexibility, easier upgradability, and customization with additional features [19, 20]. Networked embedded systems are equipped with communication capabilities and can be controlled over networks.

Networked embedded system design includes Internet mobility, network programming, security, and synchronization [79, 81]. Programmable architecture is an efficient way of implementing the multiple-application support.

Several system-level design languages and codesign frameworks have been proposed. Designing run-time reconfigurable hardware and software systems requires a complete design environment for embedded systems including dynamically reconfigurable hardware components.

11. Trusted Software

Software is an important part of multimedia systems and applications. Reliable and secure software design is often based on design diversity.

Software security is crucial to maintain run-time environment for multimedia applications. Intrusion detection is a method to verify that the running software is not prone to a particular attack. Intrusion prevention makes an adversary to solve a computationally difficult (preferably intractable) task to create a binary program that can be executed. SPEF (Secure Program Execution Framework) [39] is an intrusion prevention system that installs software binary programs by encoding constraints using secure one-way hash function, where the secret key is hidden in the processor hardware and can only be accessed by the software installer.

12. Hardware Power-Aware Sensor Security

Energy consumption is an important challenge to networks of distributed microsensors [85, 86]. Microsensors are used for data gathering, they are small, and the life of microsensor nodes is an important factor. Nodes' lifetimes [6] depend on the algorithms and protocols that provide the option of trading quality for energy savings. Dynamic voltage scaling on the sensor node's processor enables energy savings from these scalable algorithms [12, 52, 58]. Energy consumption depends on the available resources and communication between sensor nodes [23]. The architecture for a power-aware microsensor node employs collaboration between software and hardware [15].

The node's processor is capable of scaling energy consumption gracefully with computational workload. This scalability allows for energy-agile algorithms of scalable computational complexity. Energy and quality are important characteristics of DSP (Digital Signal Processing) algorithms.

Power awareness becomes increasingly important in VLSI (Very Large Scale Integration) systems to scale power consumption in response to changing operating conditions [7, 88]. These changes are caused by the changes in inputs, the output quality, or the environmental conditions. Low-power system design allows the system's energy consumption to scale with changing conditions and quality requirements [26].

13. Trusted Hardware

Hardware is part of multimedia systems and applications. Reliable hardware design is based on duplication.

TCPA (trusted computing platforms) use hardware processor architecture [48] that provides copy protection and tamper-resistance functions. In this hardware, the processor is the trusted part of the architecture. Other architecture parts are not trusted, for example, main memory. The trustworthiness of the management system for the applications that do not trust that system creates a specialized set of functions to maintain security between untrusted manager and trusted hardware processor.

14. Sensor Networks and RFID Security

RFID is a compact wireless technology using devices called RFID tags or transponders [72]. Those RFID tags are attached to a product, an animal, or a person in a form of an integrated circuit communicating via radio frequency signal with the receiving antenna. A chipless RFID allows for discrete identification of tags without an integrated circuit, a decrease in cost of printed tags. The RFID tags are used to store and retrieve data about the products, animals, or people; these tags are attached to or printed onto.

Applications of different RFID tags in telemedicine and in banking are discussed in [90]. An architecture of RFID network is presented in [101]. The RFID middleware is employed to provide resilience for diversified RFIDs and increase the architecture scalability and security. An analysis of middleware quality is introduced in [57]. The concept of mobile RFID is introduced in [45] along with related problems of privacy threats. A multi-domain RFID system security is described in [38].

15. Conclusion

We have discussed several different aspects and requirements for multimedia and sensor security. These aspects include communication security to transfer the information over the communication medium, as well as access security to identify the objects that were not subject to tampering.

The method to protect the information depends on the application, and on the type of device, on which the application is executed. Multimedia applications rely on security in real time, compression for transmission, and on digital watermarking and steganography for access. Sensor security explores limited device size, power, memory, and processing capability to maintain secure access and communication.

REFERENCES

[1] Ahmed F., and My S., December 2005. A hybrid-watermarking scheme for asymmetric and symmetric watermark extraction. In *Proceedings of the IEEE 9th International Multitopic Conference*, pp. 1–6.

[2] Amirthajah R., Xanthopoulos T., and Chandrakasan A. P., 1999. Power scalable processing using distributed arithmetic. In *Proceedings of the International Symposium on Low Power Electronics and Design*, pp. 170–175.
[3] Asada G., et al., September 1998. Wireless integrated network sensors: Low power systems on a chip. In *Proceedings of the ESSCIRC*.
[4] Bellovin S., and Merrit M., 1993. Augmented encrypted key exchange: A password-based protocol secure against dictionary attacks and password file compromise. In *First ACM Conference on Computer and Communications Security CCS-1*, pp. 244–250.
[5] Benini L., and Micheli G. D., 1997. Dynamic Power Management: Design Techniques and CAD Tools. Kluwer, Norwell, MA.
[6] Bhardwaj M., Garnett T., and Chandrakasan A., June 2001. Upper bounds on the lifetime of sensor networks. In *Proceedings of the ICC*, vol. 3, pp. 785–790.
[7] Bhardwaj M., Min R., and Chandrakasan A., December 2001. Quantifying and enhancing power-awareness of VLSI systems. *IEEE Transaction on Very Large Scale Integration (VLSI) Systems*, **9**(6): 757–772.
[8] Blom R., Carrara E., Lindholm F., Norrman K., and Naslund M., September 9–11, 2002. Conversational IP multimedia security. In *Proceedings of the IEEE 4th International Workshop on Mobile and Wireless Communications Network*, pp. 147–151.
[9] Bojkovic Z., and Milovanovic D., October 1–3, 2003. Multimedia contents security: Watermarking diversity and secure protocols. In *Proceedings of the 6th International Conference on Telecommunications in Modern Satellite, Cable and Broadcasting Service*, vol. 1, pp. 377–383.
[10] Borriello G., and Want R., May 2000. Embedding the internet: Embedded computation meets the world wide web. *Communications of the ACM*, **43**(5): 59–66.
[11] Bult K., et al., 1996. Low power systems for wireless microsystems. In *Proceedings of the ISLPED*, pp. 17–21.
[12] Burd T., Pering T., Stratakos A., and Brodersen R., 2000. A dynamic voltage scaled microprocessor system. In *Proceedings of the ISSCC*, pp. 294–295.
[13] Callaway E., Gorday P., Hester L., Gutierrez J. A., Naeve M., Heile B., and Bahl V., August 2002. Home networking with IEEE 802.15.4: A developing standard for low-rate wireless personal area networks. *IEEE Communications Magazine*, **40**(8): 70–77.
[14] Chan P. W., Lyu M. R., and Chin R. T., December 2005. A novel scheme for hybrid digital video watermarking: Approach, evaluation and experimentation. *IEEE Transactions on Circuits and Systems for Video Technology*, **15**(12): 1638–1649.
[15] Chandrakasan A. P., et al., May 1999. Design considerations for distributed microsensor systems. In *Proceedings of the Custom Integrated Circuits Conference*, pp. 279–286. San Deigo, CA.
[16] Chotikakamthorn N., and Sangiamkun W., August 19–22, 2001. Digital watermarking technique for image authentication by neighboring block similarity measure. In *Proceedings of the IEEE Region 10th International Conference on Electrical and Electronic Technology*, vol. 2, pp. 743–747.
[17] Czerwinski S. E., Zhao B. Y., Hodes T. D., Joseph A. D., and Katz R. H., August 1999. An architecture for a secure service discovery service. In *Fifth Annual ACM/IEEE International Conference on Mobile Computing and Networking*, pp. 24–35. Seattle, WA.
[18] Dittmann J., and Hesse D., September 29–October 1, 2004. Network based intrusion detection to detect steganographic communication channels: On the example of audio data. In *Proceedings of the IEEE 6th Workshop on Multimedia Signal Processing*, pp. 343–346.
[19] Fleischmann J., Buchenrieder K., and Kress R., 1998. A hardware/software prototyping environment for dynamically reconfigurable embedded systems. In *Proceedings of the International IEEE Workshop on Hardware/Software Codesign (CODES/CASHE'98)*, pp. 105–109.

[20] Fleischmann J., and Buchenrieder K., 1999. Prototyping networked embedded systems. *Computer*, **32**(2): 116–119.
[21] Ghose A., Grossklags J., and Chuang J., 2003. Resilient data-centric storage in wireless Ad-hoc sensor networks. In *Proceedings of the 4th International Conference on Mobile Data Management (MDM 2003)*, Melbourne, Australia, January 21–24, 2003, pp. 45–62. Lecture Notes in Computer Science 2574. Springer.
[22] Gong L., and Shacham N., 1995. Multicast security and its extension to a mobile environment. *Wireless Networks*, **1**(3): 281–295.
[23] Goodman J., Dancy A., and Chandrakasan A., November 1998. An energy/security scalable encryption processor using an embedded variable voltage DC/DC converter. *Journal of Solid State Circuits*, **33**(11): 1799–1809.
[24] Grangetto M., Grosso A., and Magli E., September 29–October 1, 2004. Selective encryption of JPEG 2000 images by means of randomized arithmetic coding. In *Proceedings of the IEEE 6th Workshop on Multimedia Signal Processing*, pp. 347–350.
[25] Grangetto M., Magli E., and Olmo G., October 2006. Multimedia selective encryption by means of randomized arithmetic coding. *IEEE Transactions on Multimedia*, **8**(5): 905–917.
[26] Gutnik V., and Chandrakasan A. P., December 1997. Embedded power supply for low-power DSP. *IEEE Transactions on Very Large Scale Integration (VLSI) Systems*, **5**(4): 425–435.
[27] Haartsen J., Naghshineh M., Inouye J., Joeressen O. J., and Allen W., October 1998. Bluetooth: Vision, goals, and architecture. *Mobile Computing and Communications Review*, **2**(4): 38–45.
[28] Hać A., June 13–16, 2005. Embedded systems and sensors in wireless networks. In *Proceedings of the International IEEE Conference on Wireless Networks, Communications, and Mobile Computing WirelessCom*, pp. 330–335. Maui, Hawaii.
[29] Hać A., and Lu D., 1997. Architecture, design, and implementation of a multimedia conference system. *International Journal of Network Management*, **7**(2): 64–83.
[30] Heinzelman W., Chandrakasan A., and Balakrishnan H., January 2000. Energy-efficient communication protocol for wireless microsensor networks. In *Proceedings of the 33rd Hawaii International Conference on System Sciences (HICSS)*, pp. 3005–3014.
[31] Hesse D., Dittmann J., and Lang A., 2004. Network based intrusion detection to detect steganographic communication channels – on the example of images. In *Proceedings of the 30th Euromicro Conference*, pp. 453–456.
[32] Hill J., Szewczyk R., Woo A., Hollar S., Culler D., and Pister K., November 2000. System architecture directions for networked sensors. In *Proceedings of the 9th ACM International Conference on Architectural Support for Programming Languages and Operating Systems*, pp. 93–104. Cambridge, Massachusetts.
[33] Hsu C., May 2002. WaveNet processing brassboards for live video via radio. In *Proceedings of the IEEE International Joint Conference on Neural Networks*, vol. 3, pp. 2210–2213.
[34] Huang B.-B., and Tang S.-X., April–June 2006. A contrast-sensitive visible watermarking scheme. *IEEE Multimedia*, **13**(2): 60–66.
[35] Kalavade A., and Moghe P., 1998. A tool for performance estimation of networked embedded endsystems. In *Proceedings of the IEEE Design Automation Conference (DAC)*.
[36] Kalavade A., and Subrahmanyam P. A., 1997. Hardware/software partitioning for multi-function systems. In *Proceedings of the IEEE International Conference on Computer Aided Design*, pp. 516–521.
[37] Kiaei M. S., Ghaemmaghami S., and Khazaei S., February 19–25, 2006. Efficient fully format compliant selective scrambling methods for compressed video streams. In *Proceedings of the International Conference on Internet and Web Applications and Services/Advanced International Conference on Telecommunications*, pp. 42–49.

[38] Kim D. S., Shin T.-H., and Park J. S., April 10–13, 2007. A Security framework in RFID multi-domain system. In *Proceedings of the Second IEEE International Conference on Availability, Reliability and Security*, pp. 1227–1234.

[39] Kirovski D., Drinic M., and Potkonjak M., 2002. Enabling trusted software integrity. In *Proceedings of the 10th ACM International Conference on Architectural Support for Programming Languages and Operating Systems*, pp. 108–120. ACM Press, San Jose, CA.

[40] Kong X., Wang Z., and You X., December 6–9, 2005. Steganalysis of palette images: Attack optimal parity assignment algorithm. In *Proceedings of the Fifth IEEE International Conference Information, Communications and Signal Processing*, pp. 860–864.

[41] Kratzer C., Dittmann J., Vogel T., and Hillert R., May 21–24, 2006. Design and evaluation of steganography for voice-over-IP. In *Proceedings of the IEEE International Symposium on Circuits and Systems*, p. 4.

[42] Kundur D., Zourntos T., and Mathai N. J., November 2004. Lightweight security principles for distributed multimedia based sensor networks. In *Proceedings of the Thirty-Eighth Asilomar Conference on Signals, Systems and Computers*, vol. 1, pp. 368–372.

[43] Kundur D., October–December 2001. Watermarking with diversity: Insights and implications. *IEEE Multimedia*, **8**(4): 46–52.

[44] Larson P.-A., 2002. Data reduction by partial preaggregation. In *Proceedings of the International Conference on Data Engineering*, pp. 706–715.

[45] Lee H., and Kim J., April 20–22, 2006. Privacy threats and issues in mobile RFID. In *Proceedings of the IEEE First International Conference on Availability, Reliability and Security*, p. 5.

[46] Li H., and Dhawan A., June 2004. Agent based multiple level dynamic multimedia security system. In *Proceedings of the Fifth Annual IEEE SMC Information Assurance Workshop*, pp. 291–297.

[47] Li X., and Xue X., May 2–5, 2004. A novel blind watermarking based on lattice vector quantization. In *Proceedings of the Canadian Conference on Electrical and Computer Engineering*, vol. 3, pp. 1823–1826.

[48] Lie D., Thekkath C. A., and Horowitz M., October 2003. Implementing an untrusted operating system on trusted hardware. In *Proceedings of the 19th ACM Symposium on Operating Systems Principles*, pp. 178–192. ACM Press.

[49] Luo X., Zheng K., Pan Y., and Wu Z., October 2004. Encryption algorithms comparisons for wireless networked sensors. In *Proceedings of the International Conference on Systems, Man and Cybernetics*, vol. 2, pp. 1142–1146.

[50] Madden S., Szewczyk R., Franklin M. J., and Culler D., June 2002. Supporting aggregate queries over Ad-Hoc wireless sensor networks. In *Proceedings of the Fourth IEEE Workshop on Mobile Computing and Systems Applications*, pp. 49–58.

[51] Menezes A. J., van Oorschot P., and Vanstone S., 1997. Handbook of Applied Cryptography. CRC Press.

[52] Min R., Furrer T., and Chandrakasan A., April 2000. Dynamic voltage scaling techniques for distributed microsensor networks. In *Proceedings of the IEEE Computer Society Annual Workshop on VLSI (WVLSI 2000)*, pp. 43–46.

[53] Narang S., Grover P. S., and Koushik S., July 30–August 2, 2000. Multimedia security gateway protocol to achieve anonymity in delivering multimedia data using watermarking. In *Proceedings of the IEEE International Conference on Multimedia and Expo*, vol. 1, pp. 529–532.

[54] Nawab S. H., et al., January 1997. Approximate signal processing. *Journal of VLSI Signal Processing Systems for Signal, Image, and Video Technology*, **15**(1/2): 177–200.

[55] Needham R., and Wheeler D., 1994. TEA, a tiny encryption algorithm. In *Fast Software Encryption: Second International Workshop*, vol. 1008, pp. 14–16. Springer LNCS.

[56] Needham R., and Wheeler D., 1996. TEA Extension. Computer Laboratory Cambridge University.
[57] Oh G., Kim D., Kim S., and Rhew S., November 2006. A quality evaluation technique of RFID middleware in ubiquitous computing. In *Proceedings of the IEEE International Conference on Hybrid Information Technology*, vol. 2, pp. 730–735.
[58] Pering T., Burd T., and Broderson R., 1998. The simulation and evaluation of dynamic voltage scaling algorithms. In *Proceedings of the International Symposium on Low Power Electronics and Design*, pp. 76–81.
[59] Perrig A., Canetti R., Song D., and Tygar J. D., February 2001. Efficient and secure source authentication for multicast. In *Proceedings of the Network and Distributed System Security Symposium, NDSS '01*.
[60] Perrig A., Canetti R., Tygar J. D., and Song D., May 2000. Efficient authentication and signing of multicast streams over lossy channels. In *Proceedings of the IEEE Symposium on Security and Privacy*.
[61] Perrig A., Szewczyk R., Wen V., Culler D., and Tygar J. D., June 2001. SPINS: Security protocols for sensor networks. In *Proceedings of the ACM/IEEE International Conference on Mobile Computing and Networking (MobiCom 2001)*, pp. 189–199.
[62] Pfleeger C. P., 1997. Security in Computing. Prentice Hall.
[63] Pottie G., 1998. Wireless sensor networks. In *Proceedings of the Information Theory Workshop*, pp. 139–140.
[64] Pottie G. J., and Kaiser W. J., May 2000. Embedding the internet: Wireless integrated network sensors. *Communications of the ACM*, **43**(5): 51–58.
[65] Rabaey J. M., Ammer M. J., da Silva J. L., Patel D., and Roundy S., July 2000. PicoRadio supports ad hoc ultra-low power wireless networking. *IEEE Computer*, 42–48.
[66] Raghunathan V., Schurgers C., Park S., and Srivastava M. B., March 2002. Energy-aware wireless microsensor networks. *IEEE Signal Processing Magazine*, **19**(2): 40–50.
[67] Retsas I., Pieper R., and Cristi R., March 18–19, 2002. Watermark recovery with a DCT based scheme employing nonuniform embedding. In *Proceedings of the 2002 IEEE Thirty-Fourth Southeastern Symposium on System Theory*, pp. 157–161.
[68] Rivest R. L., 1995. The RC5 encryption algorithm. In *Proceedings of the First Workshop on Fast Software Encryption*, pp. 86–96.
[69] Rivest R. L., April 1992. The MD5 message-digest algorithm. Internet Request for Comments, RFC 1321.
[70] Rivest R. L., Shamir A., and Adleman L. M., 1978. A method for obtaining digital signatures and public-keycrypto systems. *Communications of the ACM*, **21**(2): 120–126.
[71] Rohatgi P., November 1999. A compact and fast hybrid signature scheme for multicast packet authentication. In *Sixth ACM Conference on Computer and Communications Security*.
[72] Roussos G., 2008. Computing with RFID: Drivers, technology and implications. In M. Zelkowitz, editor, Advances in Computers Elsevier.
[73] Shawbaki W., February 19–25, 2006. Multimedia security in passive optical networks via wavelength hopping and codes cycling technique. In *Proceedings of the International Conference on Internet and Web Applications and Services/Advanced International Conference on Telecommunications*, pp. 51–56.
[74] Shih F. Y., and Wu Y.-T., June 27–30, 2004. A novel fragile watermarking technique. In *Proceedings of the International Conference on Multimedia and Expo*, vol. 2, pp. 875–878.
[75] Sinha A., and Chandrakasan A., January 2001. Operating system and algorithmic techniques for energy scalable wireless sensor networks. In *Proceedings of the Second International Conference on Mobile Data Management (MDM 2001)*, Hong-Kong, Lecture Notes in Computer Science, vol. 1987, Springer Verlag.

[76] Sinha A., and Chandrakasan A., Mar/April 2001. Dynamic power management in wireless sensor networks. *IEEE Design and Test of Computers*, **18**(2): 62–74.
[77] Slijepcevic S., Potkonjak M., Tsiatsis V., Zimbeck S., and Srivastava M. B., 2002. On communication security in wireless ad-hoc sensor networks. In *Proceedings of the Eleventh IEEE International Workshops on Enabling Technologies: Infrastructure for Collaborative Enterprises (WET ICE 2002)*, pp. 139–144.
[78] Su K., Kundur D., and Hatzinakos D., February 2005. Statistical invisibility for collusion-resistant digital video watermarking. *IEEE Transactions on Multimedia*, **7**(1): 43–51.
[79] Sukhatme G. S., and Mataric M. J., May 2000. Embedding the internet: Embedding robots into the internet. *Communications of the ACM*, **43**(5): 67–73.
[80] Swaminathan A., Mao Y., and Wu M., June 2006. Robust and secure image hashing. *IEEE Transactions on Information Forensics and Security*, **1**(2): 215–230.
[81] Tennenhouse D., May 2000. Embedding the internet: Proactive computing. *Communnications of the ACM*, **43**(5): 43–50.
[82] Uppuluri P., and Basu S., LASE: Layered approach for sensor security and efficiency. In *Proceedings of the International IEEE Conference on Parallel Processing Workshops*, pp. 346–352.
[83] U. S. National Institute of Standards and Technology (NIST) January 1999. Data encryption standard (DES), draft federal information processing standards publication 46–3 (FIPS PUB 46–3).
[84] Villan R., Voloshynovskiy S., Koval O., and Pun T., December 2006. Multilevel 2-D Bar Codes: toward high-capacity storage modules for multimedia security and management. *IEEE Transactions on Information Forensics and Security*, **1**(4): 405–420.
[85] Wang A., Cho S.-H., Sodini C., and Chandrakasan A., August 2001. Energy efficient modulation and MAC for asymmetric RF microsensor systems. In *Proceedings of the ISLPED*, pp. 106–111.
[86] Wang A., Heinzelman W., and Chandrakasan A., October 1999. Energy-scalable protocols for battery-operated microsensor networks. In *Proceedings of the SiPS*, pp. 483–492.
[87] Wei G., and Horowitz M., A low power switching supply for self-clocked systems. In *Proceedings of the ISLPED*, pp. 313–317.
[88] Weiser M., Welch B., Demers A., and Shenker S., 1998. Scheduling for reduced CPU energy. In A. Chandrakasan and R. Brodersen, editors, Low Power CMOS Design, pp. 177–187.
[89] Wu Y.-T., Shih F. Y., and Wu Y.-T., May 2006. A robust high-capacity watermarking algorithm. In *Proceedings of the IEEE International Conference on Electro/information Technology*, pp. 442–447.
[90] Xiao Y., Shen X., Sun B., and Cai L., April 2006. Security and privacy in RFID and applications in telemedicine. *IEEE Communications Magazine*, **44**(4): 64–72.
[91] Xie D., and Kuo C.-C. J., May 23–26, 2004. Enhanced multiple huffman table (MHT) encryption scheme using key hopping. In *Proceedings of the International Symposium on Circuits and Systems*, vol. 5, pp. V-568–V-571.
[92] Yao K., et al., October 1998. Blind beam forming on a randomly distributed sensor array system. *IEEE Journal on Selected Topics in Communications*, **16**(8): 1555–1567.
[93] Yongliang L., Gao W., and Liu S., August 29–September 1, 2004. Multimedia security in the distributed environment. In *Proceedings of the Joint Conference of the 10th Asia-Pacific Conference on Communications and the 5th International Symposium on Multi-Dimensional Mobile Communications*, vol. 2, pp. 639–642.
[94] Younis M. F., Ghumman K., and Eltoweissy M., August 2006. Location-aware combinatorial key management scheme for clustered sensor networks. *IEEE Transactions on Parallel and Distributed Systems*, **17**(8): 865–882.

[95] Yuan C., Zhu B. B., Wang Y., Li S., and Zhong Y., May 25–28, 2003. Efficient and fully scalable encryption for MPEG-4 FGS. In *Proceedings of the International Symposium on Circuits and Systems*, vol. 2, pp. II-620–II-623.

[96] Yuval G., 1997. Reinventing the travois: Encryption/MAC in 30 ROM bytes. In *Proceedings of the Fourth Workshop on Fast Software Encryption*.

[97] Zhang J., Wang N.-C., and Feng Xiong, November 4–5, 2002. A novel watermarking for images using neural networks. In *Proceedings of the International Conference on Machine Learning and Cybernetics*, vol. 3, pp. 1405–1408.

[98] Zhao H. V., and Liu K. J. R., December 2006. Traitor-within-traitor behavior forensics: strategy and risk minimization. *IEEE Transactions on Information Forensics and Security*, **1**(4): 440–456.

[99] Zhao H. V., and Liu K. J. R., September 2006. Behavior forensics for scalable multiuser collusion: Fairness versus effectiveness. *IEEE Transactions on Information Forensics and Security*, **1**(3): 311–329.

[100] Zhao H. V., and Liu K. J. R., January 2006. Fingerprint multicast in secure video streaming. *IEEE Transactions on Image Processing*, **15**(1): 12–29.

[101] Zhao Y. Z., and Gan O. P., August 2006. Distributed design of RFID network for large-scale RFID deployment. In *Proceedings of the IEEE International Conference on Industrial Informatics*, pp. 44–49.

[102] Zhou J., Liang Z., Chen Y., and Au O. C., March 2007. Security analysis of multimedia encryption schemes based on multiple huffman table. *IEEE Signal Processing Letters*, **14**(3): 201–204.

Email Spam Filtering

ENRIQUE PUERTAS SANZ

Universidad Europea de Madrid
Villaviciosa de Odón, 28670 Madrid, Spain

JOSÉ MARÍA GÓMEZ HIDALGO

Optenet
Las Rozas
28230 Madrid, Spain

JOSÉ CARLOS CORTIZO PÉREZ

AINet Solutions
Fuenlabrada 28943, Madrid, Spain

Abstract

In recent years, email spam has become an increasingly important problem, with a big economic impact in society. In this work, we present the problem of spam, how it affects us, and how we can fight against it. We discuss legal, economic, and technical measures used to stop these unsolicited emails. Among all the technical measures, those based on content analysis have been particularly effective in filtering spam, so we focus on them, explaining how they work in detail. In summary, we explain the structure and the process of different Machine Learning methods used for this task, and how we can make them to be cost sensitive through several methods like threshold optimization, instance weighting, or MetaCost. We also discuss how to evaluate spam filters using basic metrics, TREC metrics, and the receiver operating characteristic convex hull method, that best suits classification problems in which target conditions are not known, as it is the case. We also describe how actual filters are used in practice. We also present different methods used by spammers to attack spam filters and what we can expect to find in the coming years in the battle of spam filters against spammers.

1. Introduction 47
   1.1. What is Spam? 47
   1.2. The Problem of Email Spam 47
   1.3. Spam Families 48
   1.4. Legal Measures Against Spam 50
2. Technical Measures 51
   2.1. Primitive Language Analysis or Heuristic Content Filtering 51
   2.2. White and Black Listings 51
   2.3. Graylisting 52
   2.4. Digital Signatures and Reputation Control 53
   2.5. Postage 54
   2.6. Disposable Addresses 54
   2.7. Collaborative Filtering 55
   2.8. Honeypotting and Email Traps 55
   2.9. Content-Based Filters 56
3. Content-Based Spam Filtering 56
   3.1. Heuristic Filtering 57
   3.2. Learning-Based Filtering 63
   3.3. Filtering by Compression 80
   3.4. Comparison and Summary 83
4. Spam Filters Evaluation 83
   4.1. Test Collections 84
   4.2. Running Test Procedure 87
   4.3. Evaluation Metrics 88
5. Spam Filters in Practice 92
   5.1. Server Side Versus Client Side Filtering 93
   5.2. Quarantines 95
   5.3. Proxying and Tagging 96
   5.4. Best and Future Practical Spam Filtering 98
6. Attacking Spam Filters 98
   6.1. Introduction 98
   6.2. Indirect Attacks 99
   6.3. Direct Attacks 101
7. Conclusions and Future Trends 109
   References 109

1. Introduction

1.1 What is Spam?

In literature, we can find several terms for naming unsolicited emails. Junk emails, bulk emails, or unsolicited commercial emails (UCE) are a few of them, but the most common word used for reference is 'spam.' It is not clear where do the word spam comes from, but many authors state that the term was taken from a Monty Python's sketch, where a couple go into a restaurant, and the wife tries to get something other than spam. In the background are a bunch of Vikings that sing the praises of spam: 'spam, spam, spam, spam . . . lovely spam, wonderful spam.' Pretty soon the only thing you can hear in the skit is the word 'spam.' That same idea would happen to the Internet if large-scale inappropriate postings were allowed. You could not pick the real postings out from the spam.

But, what is the difference between spam and legitimate emails? We can consider an email as spam if it has the following features:

- *Unsolicited*: The receiver is not interested in receiving the information.
- *Unknown sender*: The receiver does not know and has no link with the sender.
- *Massive*: The email has been sent to a large number of addresses.

In the next subsections, we describe the most prominent issues regarding spam, including its effects, types, and main measures against it.

1.2 The Problem of Email Spam

The problem of email spam can be quantified in economical terms. Many hours are wasted everyday by workers. It is not just the time they waste reading spam but also the time they spend deleting those messages.

Let us think in a corporate network of about 500 hosts, and each one receiving about 10 spam messages every day. If because of these emails 10 min are wasted we can easily estimate the large number of hours wasted just because of spam. Whether an employee receives dozens or just a few each day, reading and deleting these messages takes time, lowering the work productivity. As an example, the United Nations Conference on Trade and Development estimates the global economic impact of spam could reach $20 b in lost time and productivity. The California legislature found that spam costs United States organizations alone more than $10 b in 2004, including lost productivity and the additional equipment, software, and manpower needed to combat the problem. A repost made by Nucleus Research[1] in 2004 claims that spam will cost US employers $2K per employee in lost productivity. Nucleus found that unsolicited email reduced employee productivity by a staggering 1.4%. Spam-filtering solutions have been doing little to control this situation, reducing spam levels by only 26% on average, according to some reports.

There are also problems related to the technical problems caused by spam. Quite often spam can be dangerous, containing virus, trojans, or other kind of damaging software, opening security breaches in computers and networks. In fact, it has been demonstrated that virus writers hire spammers to disseminate their so-called malware. Spam has been the main means to perform 'phishing' attacks, in which a bank or another organization is supplanted in order to get valid credentials from the user, and steal his banking data leading to fraud.

Also, network and email administrators have to employ substantial time and effort in deploying systems to fight spam. As a final remark, spam is not only dangerous or a waste of time, but also it can be quite disturbing. Receiving unsolicited messages is a privacy violation, and often forces the user to see strongly unwanted material, including pornography. There is no way to quantify this damage in terms of money, but no doubt it is far from negligible.

1.3 Spam Families

In this subsection, we describe some popular spam families or genres, focusing on those we have found most popular or damaging.

1.3.1 Internet Hoaxes and Chain Letters

There are a whole host of annoying hoaxes that circulate by email and encourage you to pass them on to other people. Most hoaxes have a similar pattern. These are some common examples to illustrate the language used:

- Warnings about the latest nonexistent virus dressed up with impressive but nonsensical technical language such as 'nth-complexity binary loops.'
- Emails asking to send emails to a 7-year-old boy dying of cancer, promises that one well-known IT company's CEO or president will donate money to charity for each email forwarded.
- Messages concerning the Helius Project, about a nonexistent alien being communicating with people on Earth, launched in 2003 and still online. Many people who interacted with Helius argue that Helius is real.

---

[1] See http://www.nucleusresearch.com for more information.

In general, messages that says 'forward this to everyone you know!' are usually hoaxes or chain letters. The purpose of these letters is from joking to generating important amounts of network traffic that involves economic losses in ISPs.

1.3.2 Pyramid Schemes

This is a common attempt to get money from people. Pyramid schemes are worded along the lines of 'send ten dollars to this address and add yourself to the bottom of the list. In six weeks you'll be a millionaire!' They do not work (except from the one on the top of the pyramid, of course). They are usually illegal and you will not make any money from them.

1.3.3 Advance Fee Fraud

Advance fee fraud, also known as Nigerian fraud or 419 fraud, is a particularly dangerous spam. It takes the form of an email claiming to be from a businessman or government official, normally in a West African state, who supposedly has millions of dollars obtained from the corrupt regime and would like your help in getting it out of the country. In return for depositing the money in your bank account, you are promised a large chunk of it.

The basic trick is that after you reply and start talking to the fraudsters, they eventually ask you for a large sum of money up front in order to get an even larger sum later. You pay, they disappear, and you lose.

1.3.4 Commercial Spam

This is the most common family of spam messages. They are commercial advertisements trying to sell a product (that usually cannot be bought in a regular store). According to a report made by Sophos about security threats in 2006,[2] health- and medical-related spam (which primarily covers medication which claims to assist in sexual performance, weight loss, or human growth hormones) remained the most dominant type of spam and rose during the year 2006. In the report we find the top categories in commercial spam:

- Medication/pills – Viagra, Cialis, and other sexual medication.
- Phishing scams – Messages supplanting Internet and banking corporations like Ebay, Paypal, or the Bank of America, in order to get valid credentials and steal users' money.
- Non-English language – an increasing number of commercial spam is translated or specifically prepared for non-English communities.

---

[2] See the full report at http://www.sophos.com.

- Software – or how you can get very cheap, as it is OEM (Original Equipment Manufacturer), that is, prepared to be served within a brand new PC at the store.
- Mortgage – a popular category, including not only mortgages but also specially debt grouping.
- Pornography – one of the most successful businesses in the net.
- Stock scams – interestingly, it has been observed that promoting stock corporations via email has had some impact in their results.

The economics behind the spam problem are clear: if users did not buy products marketed through spam, it would not be such a good business. If you are able to send 10 million messages for a 10 dollars product, and you get just one sell among every 10 thousand messages, you will be getting 10 thousand dollars from your spam campaign. Some spammers have been reported earning around five hundred thousand dollars a month, for years.

1.4 Legal Measures Against Spam

Fighting spam requires uniform international laws, as the Internet is a global network and only uniform global legislation can combat spam.

A number of nations have implemented legal measures against spam. The United States of America has both a federal law against spam and a separate law for each state. Something similar can be found in Europe: the European Union has its anti-spam law but most European countries have its own spam law too. There are specially effective or very string antispam laws like those in Australia, Japan, and South Korea. There are also bilateral treaties on spam and Internet fraud, as those between the United States and Mexico or Spain. On the other side, there are also countries without specific regulation about spam so it is an activity that is not considered illegal.

With this scenario, it is very difficult to apply legal measures against spammers. Besides that, anonymity is one of the biggest advantages of spammers. Spammers frequently use false names, addresses, phone numbers, and other contact information to set up 'disposable' accounts at various Internet service providers. In some cases, they have used falsified or stolen credit card numbers to pay for these accounts. This allows them to quickly move from one account to the next as each one is discovered and shut down by the host ISPs. While some spammers have been caught (a noticeable case is that of Jeremy Jaynes), there are many spammers that have avoided their capture for years. A trustable spammers' hall of fame is maintained by The Spamhaus Project, and it is known as the Register Of Known Spam Operations (ROKSO).[3]

---

[3] The ROKSO can be accessed at http://www.spamhaus.org/rokso/index.lasso.

2. Technical Measures

Among all the different techniques used for fighting spam, technical measures have become the most effective. There are several approaches used to filter spam. In the next section, we will comment some of the most popular approaches.

2.1 Primitive Language Analysis or Heuristic Content Filtering

The very first spam filters used primitive language analysis techniques to detect junk email. The idea was to match specific texts or words to email body or sender address. In the mid 1990s when spam was not the problem that it is today, users could filter unsolicited emails by scanning them, searching for phrases or words that were indicative of spam like 'Viagra' or 'Buy now.' Those days spam messages were not as sophisticated as they are today and this very simplistic approach could filter ~80% of the spam.

The first versions of the most important email clients included this technique that it worked quite well for a time, before spammers started to use their tricks to avoid filters. The way they obfuscated messages made this technique ineffective.

Another weakness of this approach was the high false-positive rate: any message containing 'forbidden words' was sent to trash. Most of those words were good for filtering spam, but sometimes they could appear in legitimate emails. This approach is not used nowadays because of the low accuracy and the high error rates it has.

This primitive analysis technique is in fact a form of content analysis, as it makes use of every email content to decide if it is spam or legitimate. We have called this technique heuristic filtering, and it is extensively discussed below.

2.2 White and Black Listings

White and black lists are extremely popular approaches to filter spam email [41]. White lists state which senders' messages are never considered spam, and black lists include those senders that should always be considered spammers.

A white list contains addresses that are supposed to be safe. These addresses can be individual emails, domain names, or IP addresses, and it would filer an individual sender or a group of them. This technique can be used in the server side and/or in the client side, and is usually found as a complement to other more effective approaches.

In server-side white lists, an administrator has to validate the addresses before they go to the trusted list. This can be feasible in a small company or a server with a small number of email accounts, but it can turn into a pain if pretended to be used in large corporate servers with every user having his own white list. This is because the task of validating each email that is not in the list is a time-consuming job. An extreme use of this technique could be to reject all emails coming from senders that are not in the white list. This could sound very unreasonable, but it is not. It can be used in restricted domains like schools, where you prefer to filter emails from unknown senders but want to keep the children away from potentially harmful content, because spam messages could contain porn or another kind of adult content. In fact, this aggressive antispam technique has been used by some free email service providers as Hotmail, in which a rule can be stated preventing any message coming from any other service to get into their users mailboxes.

White listings can also be used in the client side. In fact, one of the first techniques used to filters spam consisted of using user's address book as a white list, tagging as potential spam all those emails that had in the FROM: field an address that was not in the address book. This technique can be effective for those persons who use email just to communicate with a limited group of contacts like family and friends.

The main problem of white listings is the assumption that trusted contacts do not send junk email and, as we are going to see, this assumption could be erroneous. Many spammers use computers that have been compromised using trojans and viruses for sending spam, sending them to all the contacts of the address book, so we could get a spam message from a known sender if his computer has been infected with a virus. Since these contacts are in the white list, all messages coming from them are flagged as safe.

Black listings, most often known as DNS Blacklists (DNSBL), are used to filter out emails which are sent by known spam addresses or compromised servers. The very first and most popular black list has been the trademarked Realtime Blackhole List (RBL), operated by the Mail Abuse Prevention System. System administrators, using spam detection tools, report IP addresses of machines sending spam and they are stored in a common central list that can be shared by other email filters. Most antispam softwares have some form of access to networked resources of this kind.

Aggressive black listings may block whole domains or ISPs having many false positives. A way to deal with this problem is to have several distributed black listings and contrast sender's information against some of them before blocking an email. Current DNS black lists are dynamic, that is, not only grow with new information, but also expire entries, maintaining fresh reflection of current situation in the address space.

2.3 Graylisting

As a complement to white and black listings, one could use graylistings [94]. The core behind this approach is the assumption that junk email is sent using spam bots, this is specific software made to send thousands of emails in a short time. This software differs from traditional email servers and does not respect email RFC standards. In particular, emails that fail to reach its target are not sent again, as a real system would do. This is the right feature used by graylistings. When the system receives an email from an unknown sender that is not in a white listing, it creates a tupla sender–receiver. The first time that tupla occurs in the system, the email is rejected so it is bounced back to the sender. A real server will send that email again so the second time the system finds the tupla, the email is flagged as safe and delivered to the recipient.

Graylistings have some limitations and problems. The obvious ones are the delay we could have in getting some legitimate emails when using this approach because we have to wait until the email is sent twice, and the waste of bandwidth produced in the send–reject–resend process.

Other limitations are that this approach will not work when spam is sent from open relays as they are real email servers and the easy way for spammer to work-around graylistings, just adding a new functionality to the software, allowing it to send bounced emails again.

2.4 Digital Signatures and Reputation Control

With the emergence of Public Key Cryptography, and specifically, its application to email coding and signing, most prominently represented by Pretty Good Privacy (PGP) [103] and GNU Privacy Guard (GPG) [64], there exists the possibility of filtering out unsigned messages, and in case they are signed, those sent by untrusted users. PGP allows keeping a nontrivial web/chain of trust between email users, the way that trust is spread over a net of contacts. This way, a user can trust the signer of a message if he or she is trusted by another contact of the email receiver.

The main disadvantage of this approach is that PGP/GPG users are rare, so it is quite risky to consider legitimate email coming only from trusted contacts. However, it is possible to extend this idea to email servers.

As we saw in previous section, many email filters use white listings to store safe senders, usually local addresses and addresses of friends. So if spammers figure out who our friends are, they could forge the FROM: header of the message with that information, avoiding filters because senders that are in white listings are never filtered out. Sender Policy Framework (SPF), DomainKeys, and Sender ID try to prevent forgery by registering IPs of machines used to send email from for every valid email sender in the server [89]. So if someone is sending an email from a particular domain but it does not match the IP address of the sender, you can know the email has been forged. The messages are signed by the public key of the server, which makes its SPF, DomainKeys, or Sender ID record public. As more and more email service providers (specially the free ones, like Yahoo!, Hotmail, or Gmail) are making their IP records public, the approach will be increasingly effective.

Signatures are a basic implementation of a more sophisticated technique, which is reputation control for email senders. When the system receives an email from an unknown sender, the message is scanned and classified as legitimate or spam. If the email is classified as legitimate, the reputation of the sender is increased, and decreased if classified as spam. The more emails are sent from that address, the more positive or negative the sender is ranked. Once reputation crosses a certain threshold, it can be moved to a white or black list. The approach can be extended to the whole IP space in the net, as current antispam products by IronPort feature, named SenderBase.[4]

2.5 Postage

One of the main reasons of spam success is the low costs of sending spam. Senders do not have to pay for sending email and costs of bandwidth are very low even if sending millions of emails.[5] Postage is a technique based upon the principle of senders of unsolicited messages demonstrating their goodwill by paying some kind of postage: either a small amount of money paid electronically, a sacrifice of a few seconds of human time at answering a simple question, or some time of computation in the sender machine.

As the email services are based on the Simple Mail Transfer Protocol, economic postage requires a specific architecture over the net, or a dramatic change in the email protocol. Abadi et al. [1] describes a ticket-based client–server architecture to provide postage for avoiding spamming (yet other applications are suitable).

An alternative is to require the sender to answer some kind of question, to prove he is actually a human being. This is the kind of Turing Test [93] that has been implemented in many web-based services, requiring the user to type the hidden word in a picture. These tests are named CAPTCHAs (Completely Automated Public Turing test to tell Computers and Humans Apart) [2]. This approach can be especially effectively used to avoid *outgoing* spam, that is, preventing the spammers to abuse of free email service providers as Hotmail or Gmail.

A third approach is requiring the sender machine to solve some kind of computationally expensive problem [32], producing a delay and thus, disallowing spammers to send millions of messages per day. This approach is, by far, the less annoying of the postage techniques proposed, and thus, the most popular one.

2.6 Disposable Addresses

Disposable addresses [86] are a technique used to prevent a user to receive spam. It is not a filtering system itself but a way to avoid spammers to find out our address. To harvest email addresses, spammers crawl the web searching for addresses in web

---

[4] The level of trust of an IP in SenderBase can be checked at: http://www.senderbase.org/.

[5] Although the case against the famous spammer Jeremy Jaynes has revealed he has been spending more than one hundred thousand dollars per month in high speed connections.

pages, forums, or Usenet groups. If we do not publish our address on the Internet, we can be more or less protected against spam, but the problem is when we want to register in a web page or an Internet service and we have to fill in our email address in a form. Most sites state that they will not use that information for sending spam but we cannot be sure and many times the address goes to a list that is sold to third party companies and used for sending commercial emails. Moreover, these sites can be accessed by hackers with the purpose of collecting valid (and valuable) addresses.

To circumvent this problem, we can use disposable email addresses. Instead of letting the user prepare his own disposable addresses, he can be provided with an automatic system to manage them, like the channels' infrastructure by ATT [50]. The addresses are temporary accounts that the user can use to register in web services. All messages sent to disposable address are redirected to our permanent safe account during a configurable period of time. Once the temporary address is no longer needed, it is deleted so even if that account receives spam, this is not redirected.

2.7 Collaborative Filtering

Collaborative filtering [48] is a distributed approach to filter spam. Instead of having each user to have his own filter, a whole community works together. Using this technique, each user shares his judgments of what is spam and what is not with the other users. Collaborative filtering networks take advantage of the problem of some users that receive spam to build better filters for those that have not yet received those spam messages. When a group of users in the same domain have tagged an email coming from a common sender as spam, the system can use the information in those emails to learn to classify those particular emails so the rest of users in the domain will not receive them.

The weakness of this approach is that what is spam for somebody could be a legitimate content for another. These collaborative spam filters cannot be more accurate as a personal filter in the client side but it is an excellent option for filtering in the server side. Another disadvantage of this approach is that spammers introduce small variations in the messages, disallowing the identification of a new upcoming spam email as a close variation of one receiver earlier by another user [58].

2.8 Honeypotting and Email Traps

Spammers are known to abuse vulnerable systems like open mail relays and public open proxies. In order to discover spam activities, some administrators have created honeypot programs that simulate being such vulnerable systems. The existence of such fake systems makes more risky for spammers to use open relays and open proxies for sending spam. Honeypotting is a common technique used for system administrators to detect hacking activities on their servers. They create fake vulnerable servers in order to burst hackers while protecting the real servers. Since the term honeypotting is more appropriate for security environments, the terms 'email traps' or 'spam traps' can be used instead of referring to these techniques when applied to prevent spam.

Spam traps can used to collect instances of spam messages on keeping a fresh collection of spammer techniques (and a better training collection in learning-based classifiers), to build and deploy updated filtering rules in heuristic filters, and to detect new spam attacks in advance, avoiding them reach, for example, a corporate network in particular.

2.9 Content-Based Filters

Content-based filters are based on analyzing the content of emails. These filters can be hand-made rules, also known as heuristic filters, or learned using Machine Learning algorithms. Both approaches are widely used these days in spam filters because they can be very accurate when they are correctly tuned up, and they are going to be deeply analyzed in next section.

3. Content-Based Spam Filtering

Among the technical measures to control spam, content-based filtering is one of the most popular ones. Spam filters that analyze the contents of the messages and take decisions on that basis have spread among the Internet users, ranging from individual users at their home personal computers, to big corporate networks. The success of content-based filters is so big that spammers have performed increasingly complex attacks designed to avoid them and to reach the users' mailbox.

This section covers the most relevant techniques for content-based spam filtering. Heuristic filtering is important for historical reasons, although the most popular modern heuristic filters have some learning component. Learning-based filtering is the main trend in the field; the ability to learn from examples of spam and legitimate messages gives these filters full power to detect spam in a personalized way. Recent TREC [19] competitive evaluations have stressed the importance of a family of learning-based filters, which are those using compression algorithms; they have scored top in terms of effectiveness, and so they deserve a specific section.

3.1 Heuristic Filtering

Since the very first spam messages, users (that were simultaneously their own 'network administrators') have coded rules or heuristics to separate spam from their legitimate messages, and avoid reading the first [24]. A content-based heuristic filter is a set of hand-coded rules that analyze the contents of a message and classify it as spam or legitimate. For instance, a rule may look like:

$$\text{if}(P\text{'Viagra'} \in M) \text{or}(\text{'VIAGRA'} \in M) \text{then } class(M) = \text{spam}$$

This rule means that if any of the words 'Viagra' or 'VIAGRA' (that are in fact distinct characters strings) occur in a message M, then it should be classified as spam. While first Internet users were often privileged user administrators and used this kind of rules in the context of sophisticated script and command languages, most modern mail user clients allow writing this kind of rules through simple forms. For instance, a Thunderbird straightforward spam filter is shown in Fig. 1. In this example, the users has prepared a filter named 'spam' that deletes all messages in which the word '**spam**' occurs in the Subject header.

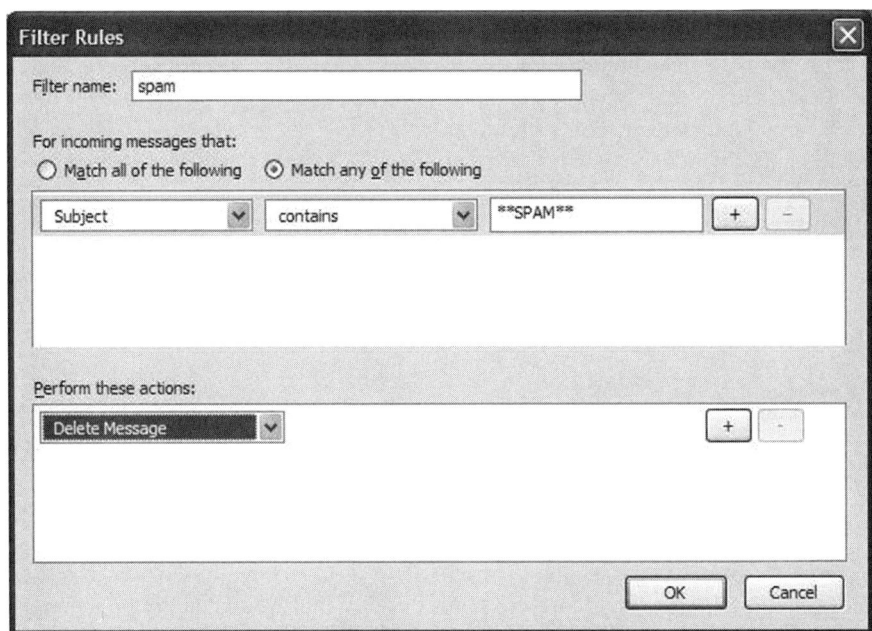

FIG. 1. A simple spam filter coded as a Thunderbird mail client rule. If the word '**spam**' occurs in the Subject of a message, it will be deleted (sent to trash).

However, these filtering utilities are most often used to classify the incoming messages into folders, according to their sender, their topic, or the mail list they belong to. They can also be used in conjunction with a filtering solution out of the mail client, which may tag spam messages (for instance, with the string '**spam**' in the subject, or being more sophisticate, by adding a specific header like, for example, X-mail report, which can include a simple tag or a rather informative output with even a score), that will be later processed by the mail client by applying the suitable filters and performing the desired action. A sensible action is to send the messages tagged as spam to a quarantine folder in order to avoid false positives (legitimate messages classified as spam).

It should be clear that maintaining an effective set of rules can be a rather time-consuming job. Spam messages include offers of pharmacy products, porn advertisements, unwanted loans, stock recommendations, and many other types of messages. Not only their content, but their style is always changing. In fact, it is hard to find a message in which the word 'Viagra' occurs without alterations (except for a legitimate one!). In other words, there are quite many technical experts highly committed to make the filter fail: the spammers. This is why spam filtering is considered a problem of 'adversarial classification' [25].

Neither a modern single network administrator nor even an advanced user will be writing their own handmade rules to filter spam. Instead, a list of useful rules can be maintained by a community of expert users and administrators, as it has been done in the very popular open-source solution SpamAssassin or in the commercial service Brightmail provided by Symantec Corporation. We discuss these relevant examples in the next subsections, finishing this section with the advantages and disadvantages of this approach.

3.1.1 The SpamAssassin Filter

While it is easy to defeat a single network administrator, it is harder to defeat a community. This is the spirit of one of the most spread heuristic filtering solutions: SpamAssassin [87]. This filter has received a number of prices, and as a matter of example, it had more than 3600 downloads in the 19 months the project was hosted at Sourceforge[6] (February 2002–September 2003).

SpamAssassin is one of the oldest still-alive filters in the market, and its main feature (for the purpose of our presentation) is its impressive set of rules or heuristics, contributed by tens of administrators and validated by the project

---

[6] Sourceforge is the leading hosting server for open-source projects, providing versioning and downloading services, statistics, and more. See: http://sourceforge.net.

committee. The current (version 3.2) set of rules (named 'tests' in SpamAssassin) has 746 tests.[7] Some of them are administrative, and a number of them are not truly 'content-based,' as they, for example, check the sender address or IP against public white lists. For instance, the test named 'RCVD_IN_DNSWL_HI' checks if the sender is listed in the DNS Whitelist.[8] Of course, this is a white listing mechanism, and it makes nearly no analysis of the very message content. On the other side, the rule named 'FS_LOW_RATES' tests if the Subject field contains the words 'low rates,' which is very popular in spam messages dealing with loans or mortgages. Many SpamAssassin tests address typing variability by using quite sophisticated regular expressions. We show a list of additional examples in the Fig. 2, as they are presented in the project web page.

A typical SpamAssassin content matching rule has the structure shown in the next example:

```
body DEMONSTRATION_RULE /test/
score DEMONSTRATION_RULE 0.1
describe DEMONSTRATION_RULE This is a simple test rule
```

The rule starts with a line that describes the test to be performed, it goes on with line presenting the score, and it has a final line for the rule description. The sample rule name is 'DEMONSTRATION_RULE,' and it checks the (case sensitive) occurrence of the word 'test' in the body section of an incoming email message. If the condition is

AREA TESTED	LOCALE	DESCRIPTION OF TEST	TEST NAME	DEFAULT SCORES (local, net, with bayes, with bayes+net)
header		Envelope sender listed in bl.open-whois.org.	DNS_FROM_OPENWHOIS	0 2.431 0 1.130
body		Provision for income taxes	DOS_PROVISION4	1.5
body		Report of financial income	DOS_REPORT_FIN_INC	0.5
body		Pump and dump stock spam	DOS_STOCK_CDYV_GENERIC	2.5
uri		Found an asterisk in a URI	DOS_URI_ASTERISK	1
header		Subject =~ /\bhoodia\b/i	DRUGS_HDIA	2.529 2.501 2.483 2.697
body		Add / Gain inches	FB_ADD_INCHES	2.999 2.999 2.620 2.131
body		It's almost sex, but not!	FB_ALMOST_SEX	3.099 3.096 2.841 2.110

FIG. 2. A sample of SpamAssassin test or filtering rules. The area tested may be the header, the body, etc. and each test is provided with one or more scores that can be used to set a suitable threshold and vary the filter sensitivity.

---

[7] The lists of tests used by SpamAssassin are available at: http://spamassassin.apache.org/tests.html.

[8] The DNS Whitelist is available at: http://www.dnswl.org/.

satisfied, that is, the word occurs, then the score 0.1 is added to the message global score. The score of the message may be incremented by other rules, and the message will be tagged as spam if the global score exceeds a manually or automatically set threshold. Of course, the higher the score of a rule, the more it contributes to the decision of tagging a message as spam.

The tests performed in the rules can address all the parts in a message, and request preprocessing or not. For instance, if the rule starts with 'header,' only the headers will be tested:

```
header DEMONSTRATION_SUBJECT Subject =~ /test/
```

In fact, the symbols '=~' preceding the test, along with the word 'Subject,' mean that only the subject header will be tested. This case, the subject field name is case insensitive.

The tests performed allow complex expressions written in the Perl Regular Expressions (Regex) Syntax. A slightly more complex example may be:

```
header DEMONSTRATION_SUBJECT Subject =~ /\btest\b/i
```

In this example, the expression '/\btest\b/i' means that the word 'test' will be searched as a single word (and not as a part of others, like 'testing'), because it starts and finishes with the word-break mark '\b,' and the test will be case insensitive because of the finishing mark '/i.' Of course, regular expressions may be much more complex, but covering them in detail is beyond the scope of this chapter. We suggest [54] for the interested reader.

Manually assigning scores to the rules is not a very good idea, as the rule coder must have a precise and global idea of all the scores in all the rest of the rules. Instead, an automated method is required, which should be able to look at all the scores and a set of testing messages, and compute the scores that minimize the error of the filter. In versions 2.x, the scores of the rules have been assigned using a Genetic Algorithm, while in (current) versions 3.x, the scores are assigned using a neural network trained with error back propagation (a perceptron). Both systems attempt to optimize the effectiveness of the rules that are run in terms of minimizing the number of false positives and false negatives, and they are presented in [87] and [91], respectively.

The scores are optimized on a set of real examples contributed by volunteers. The SpamAssassin group has in fact released a corpus of spam messages, the so-named SpamAssassin Public Corpus.[9] This corpus includes 6047 messages, with ~31% spam ratio. As it has been extensively used for the evaluation of content-based spam filter, we leave a more detailed description of it for Section 4.

---

[9] The SpamAssassin Public Corpus is available at: http://spamassassin.apache.org/publiccorpus/.

3.1.2 The Symantec Brightmail Solution

Some years ago, Brightmail emerged as an antispam solution provider based on an original business model in the antispam market. Instead of providing a final software application with filtering capabilities, it focused more on the service, and took the operation model from antivirus corporations: analysts working 24 h a day on new attacks, and frequent delivering of new protection rules. The model succeeded, and on June 21st, 2004, Symantec Corporation acquired Brightmail Incorporated, with its solution and its customers. Nowadays, Symantec claims that Brightmail Anti-spam protects more than 300 million users, and filters over 15% of the worldwide emails.

The Brightmail Anti-spam solution works at the clients' gateway, scanning incoming messages to the corporation, and deciding if they are spam or not. The decision is taken on the basis of a set of filtering rules provided by the experts working in the Symantec operations center, named BLOC (Brightmail Logistics Operations Center). The operational structure of the solution is shown in Fig. 3. In this figure, circles are used to denote the next processing steps:

1. The Probe Network (TM) is a network of fake email boxes ('spam traps' or 'honeypots') that have been seeded with the only purpose of receiving spam. These email addresses can be collected only by automatic means, as spammers

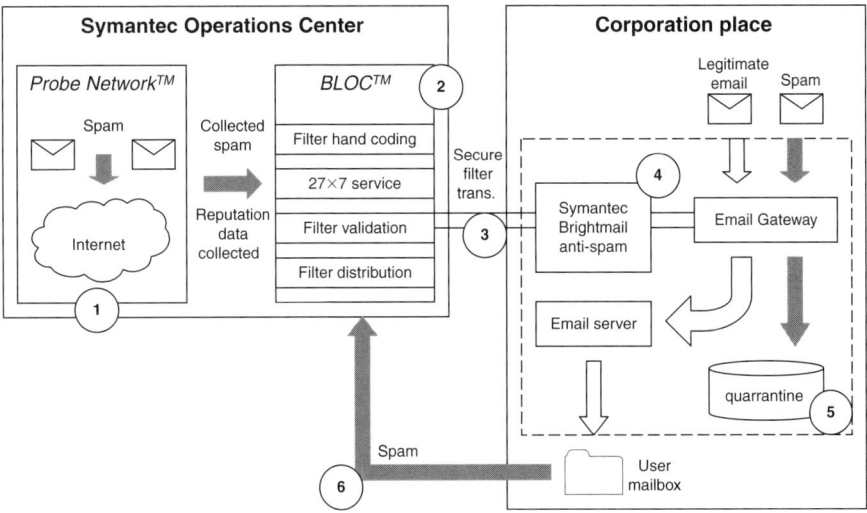

FIG. 3. Operational structure of the Symantec Brightmail Anti-spam solution. The numbers in circles denote processes described.

do, and in consequence, they can receive only spam emails. The spam collected is sent to the BLOC.
2. At the BLOC, groups of experts and, more recently, content-based analysis automatic tools build, validate, and distribute antispam filters to Symantec corporate customers.
3. Every 10 min, the Symantec software downloads the most recent version of the filters from the BLOC, in order to keep them as updated as possible.
4. The software filters the incoming messages using the Symantec and the user-customized filters.
5. The email administrator determines the most suitable administration mode for the filtered email. Most often, the (detected as) spam email is kept into a quarantine where users can check if the filter has mistakenly classified a legitimate message as spam (a false positive).
6. The users can report undetected spam to Symantec for further analysis.

The processes at the BLOC were once manual, but the ever-increasing number of spam attacks has progressively made impossible to approach filter building as a hand-made task. In the recent times, spam experts in the BLOC have actually switched their role to filter adapters and tuners, as the filters are being produced by using the automatic, learning-based tools described in the next section.

3.1.3 Problems in Heuristic Filtering

Heuristic content-based filtering has clear advantages over other kinds of filtering, especially those based on black and white listing. The most remarkable one is that it filters not only on the 'From' header, but also it can make use of the entire message, and inconsequence, to make a more informed decision. Furthermore, it offers a lot of control on the message information that is scanned, as the filter programmer decides which areas to scan, and what to seek.

However, heuristic filtering has two noticeable drawbacks. First, writing rules is not an easy task. Or it has to be left on the hands of an experienced email administrator, or it has to be simplified via the forms in commercial mail clients as described above. The first case usually involves some programming, probably including a bit of regular expression definition, which is hard and error-prone. The second one implies a sacrifice of flexibility to gain simplicity.

The second drawback is that, even being the rules written by a community of advanced users or administrators, the number of spammers is bigger, and moreover, they have a strong economic motivation to design new methods to avoid detection. In this arms race, the spammers will be always having the winning hand if the work of administrators is not supported with automatic (learning-based) tools as those we describe in the next section.

3.2 Learning-Based Filtering

During the past 9 years, a new paradigm of content-based spam filtering has emerged. Bayesian filters, or more in general, learning-based filters, have the ability to learn from the email flow and to improve their performance over time, as they can adapt themselves to the actual spam and legitimate email a particular user receives. Their impressive success is demonstrated by the deep impact they have had on spam email, as the spammers have to costly change their techniques in order to avoid them. Learning-based filters are the current state of the art of email filtering, and the main issue in this chapter.

3.2.1 Spam Filtering as Text Categorization

Spam filtering is an instance of a more general text classification task named Text Categorization [85]. Text Categorization is the assignment of text documents to a set of predefined classes. It is important to note that the classes are preexistent, instead of being generated on the fly (what corresponds to the task of Text Clustering). The main application of text categorization is the assignment of subject classes to text documents. Subject classes can be web directory categories (like in Yahoo![10]), thematic descriptors in libraries (like the Library of Congress Subject Headings[11] or, in particular domains, the Medical Subject Headings[12] by the National Library of Medicine, or the Association for Computing Machinery ACM's Computing Classification System descriptors used in the ACM Digital Library itself), personal email folders, etc. The documents may be Web sites or pages, books, scientific articles, news items, email messages, etc.

Text Categorization can be done manually or automatically. The first method is the one used in libraries, where expert catalogers scan new books and journals in order to get them indexed according to the classification system used. For instance, the National Library of Medicine employs around 95 full-time cataloguers in order to index the scientific and news articles distributed via MEDLINE.[13] Obviously this is a time- and money-consuming task and the number of publications is always increasing.

The increasing availability of tagged data has allowed the application of Machine Learning methods to the task. Instead of hand classifying the documents, or manually building a classification system (as what we have named a heuristic filter

---

[10] Available at: http://www.yahoo.com/.
[11] Available at: http://www.loc.gov/cds/lcsh.html.
[12] Available at: http://www.nlm.nih.gov/mesh/.
[13] James Marcetich, Head of the Cataloguing Section of the National Library of Medicine, in personal communication (July 18, 2001).

above), it is possible to automatically build a classifier by using a Machine Learning algorithm on a collection of hand-classified documents suitably represented. The administrator or the expert does not have to write the filter, but let the algorithm learn the document properties that make them suitable for each class. This way, the traditional expert system 'knowledge acquisition bottleneck' is alleviated, as the expert can keep on doing what he or she does best (that is, in fact, classifying), and the system will be learning from his decisions.

The Machine Learning approach has achieved considerable success in a number of tasks, and in particular, in spam filtering. In words by Sebastiani:

> (Automated Text Categorization) has reached effectiveness levels comparable to those of trained professionals. The effectiveness of manual Text Categorization is not 100% anyway (...) and, more importantly, it is unlikely to be improved substantially by the progress of research. The levels of effectiveness of automated TC are instead growing at a steady pace, and even if they will likely reach a plateau well below the 100% level, this plateau will probably be higher than the effectiveness levels of manual Text Categorization. [85] (p. 47)

Spam filtering can be considered an instance of (Automated) Text Categorization, in which the documents to classify are the user email messages, and the classes are spam and legitimate email. It may be considered easy, as it is a single-class problem, instead of the many classes that are usually considered in a thematic TC task.[14] However, it shows special properties that makes it a very difficult task:

1. Both the spam and the complementary class (legitimate email) are not thematic, that is, they can contain messages dealing with several topics or themes. For instance, as of 1999, a 37% of the spam email was 'get rich quick' letters, a 25% was pornographic advertisements, and an 18% were software offers. The rest of the spam included Web site promos, investment offers, (fake) health products, contests, holidays, and others. Moreover, some of the spam types can overlap with legitimate messages, both commercial and coming from distribution lists. While the percentages have certainly changed (health and investment offers are now most popular), this demonstrates that current TC systems that relay on words and features for classification may have important problems because the spam class is very fragmented.
2. Spam has an always changing and often skewed distribution. For instance, according to the email security corporation MessageLabs [66], spam has gone from 76.1% of the email sent in the first quarter of 2005, to 56.9% in the first quarter of 2006, and back to 73.5% in the last quarter of 2006. On one side,

---

[14] The Library of Congress Subject Headings 2007 edition has over 280000 total headings and references.

Machine Learning classifiers expect the same class distribution they learnt from; any variation of the distribution may affect the classifier performance. On the other, skewed distributions like 90% spam (reached in 2004) may make a learning algorithm to produce a trivial acceptor, that is, a classifier that always classifies a message as spam. This is due to the fact that Machine Leaning algorithms try to minimize the error or maximize the accuracy, and the trivial acceptor is then 90% accurate. And even worse, the spam rate can vary from place to place, from company to company, and from person to person; in that situation, is very difficult to build a fit-them-all classifier.

3. Like many other classification tasks, spam filtering has imbalanced misclassification costs. In other words, the kinds of mistakes the filter makes are significant. No user will be happy with a filter that catches 99% of spam but that deletes a legitimate message once-a-day. This is because false positives (legitimate messages classified as spam) are far more costly than false negatives (spam messages classified as legitimate, and thus, reaching the users' inbox). But again, it is not clear which proportion is right: a user may accept a filter that makes a false positive per 100 false negatives or per 1,000, etc. It depends on the user's taste, the amount of spam he or she receives, the place where he or she lives, the kind of email account (work or personal), etc.

4. Perhaps the most difficult issue with spam classification is that it is an instance of adversarial classification [25]. An adversarial classification task is one in which there exists an adversary that modifies the data arriving at the classifier in order to make it fail. Spam filtering is perhaps the most representative instance of adversarial classification, among many others like computer intrusion detection, fraud detection, counter-terrorism, or a much related one: web spam detection. In this latter task, the system must detect which webmasters manipulate pages and links to inflate their rankings, after reverse engineering the ranking algorithm. The term spam, although coming from the email arena, is so spread that it is being used for many other fraud problems: mobile spam, blog spam, 'spim' (spam over Instant Messaging), 'spit' (spam over Internet Telephony), etc. Regarding spam email filtering, standard classifiers like Naïve Bayes were initially successful [79], but spammers soon learned to fool them by inserting 'nonspam' words into emails, breaking up 'spam' ones like 'Viagra' with spurious punctuation, etc. Once spam filters were modified to detect these tricks, spammers started using new ones [34].

In our opinion, these issues make spam filtering a very unusual instance of Automated Text Categorization. Being said this, we must note that the standard structure of an Automated Text Categorization system is suited to the problem of spam filtering, and so we will discuss this structure in the next subsections.

3.2.2 Structure of Processing

The structure of analysis and learning in modern content-based spam filters that make use of Machine Learning techniques, is presented in Fig. 4. In this figure,[15] we represent processes or functions as rounded boxes, and information items as plain boxes. The analysis, learning, and retraining (with feedback information) of the classifier are time- and memory-consuming processes, intended to be performed offline and periodically. The analysis and filtering of new messages must be a fast process, to be performed online, as soon as the message arrives at the system. We describe these processes in detail below.

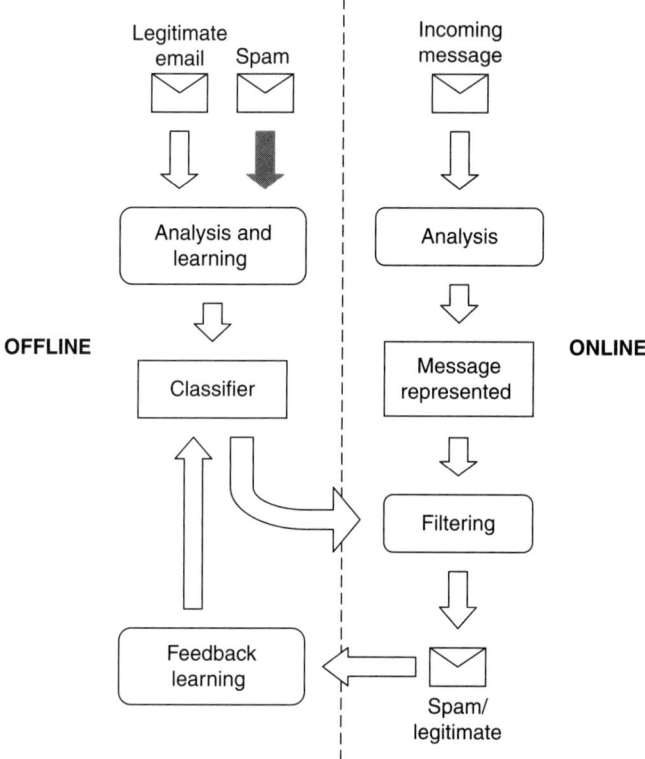

FIG. 4. Processing structure of a content-based spam filter that uses Machine Learning algorithms.

---

[15] That figure is a remake of that by Belkin and Croft for text retrieval in [7].

The first step in a learning-based filter is getting a collection of spam and legitimate messages and training a classifier on it. Of course, the collection of messages (the training collection) must represent the operating conditions of the filter as accurately as possible. Machine Learning classifiers often perform poorly when they have been trained on noisy, inaccurate, and insufficient data. There are some publicly available spam/legitimate email collections, discussed in Section 4 because they are most often used as test collections.

The first process in the filter is learning a classifier on the training messages, named instances or examples. This process involves analyzing the messages in order to get a suitable representation for learning from them. In this process, the messages are often represented as attribute-value vectors, in which attributes are word tokens in the messages, and values are, for example, binary (the word token occurs on the message or not). Next, a Machine Learning algorithm is fed with the represented examples, and it produces a classifier, that is a model of the messages or a function able to classify new messages if they follow the suitable representation. Message representation and classifier learning are the key processes in a learning-based filter, and so they are discussed in detail in the next subsections.

Once the classifier has been trained, it is ready to filter new messages. As they arrive, they are processed in order to represent them according to the format used in the training messages. That typically involves, for instance, ignoring new words in the messages, as classification is made on the basis of known words in the training messages. The classifier receives the represented message and classifies it as spam or legitimate (probably with a probability or a score), and tagged accordingly (or routed to the quarantine, the user mailbox, or whatever).

The main strength of Machine Learning-based spam filters is their ability to learn from user relevance judgments, adapting the filter model to the actual email received by the user. When the filter commits a mistake (or depending on the learning mode, after every message), the correct output is submitted to the filter, which stores it for further re-training. This ability is a very noticeable strength, because if every user receives different email, every filter is different (in terms of stored data and model learned), and it is very complex to prepare attacks able to avoid the filters of all users simultaneously. As spammers benefit relies on the number of messages read, they are forced to prepare very sophisticated attacks able to break different vendor filters with different learned models. As we discuss in Section 6, they sometimes succeed using increasingly complex techniques.

3.2.3 Feature Engineering

Feature engineering is the process of deciding, given a set of training instances, which properties will be considered for learning from them. The properties are the features or attributes, and they can take different values; this way, every training instance (a legitimate or email message) is mapped into a vector in a multidimensional space, in which the dimensions are the features. As many Machine Learning are very slow or just unable to learn in highly dimensional spaces, it is often required to reduce the number of features used in the representation, performing attribute selection (determining a suitable subset of the original attributes) or attribute extraction (mapping the original set of features into a new, reduced set). These tasks are also a part of the feature engineering process.

3.2.3.1 *Tokens and Weights.*

In Text Categorization and spam filtering, the most often used features are the sequences of characters or strings that minimally convey some kind of meaning in a text, that is, the words [39, 85]. More generally, we speak of breaking a text into tokens, a process named tokenization. In fact, this just follows the traditional Information Retrieval Vector Space Model by Salton [81]. This model specifies that, for the purpose of retrieval, texts can be represented as term-weight vectors, in which the terms are (processed) words (our attributes), and weights are numeric values representing the importance of every word in every document.

First, learning-based filters have taken relatively simple decisions in this sense, following what was the state of the art on thematic text categorization. The simplest definition of features is words, being a word any sequence of alphabetic characters, and considering any other symbol as a separator or blank. This is the approach followed in the seminal work in this field, by Sahami et al. [79]. This work has been improved by Androutsopoulos et al. [4–6], in which they make use of a lemmatizer (or stemmer), to map words into their root, and a stoplist (a list of frequent words that should be ignored as they bring more noise than meaning to thematic retrieval: pronouns, prepositions, etc.). The lemmatizer used it Morph, included in the text analysis package GATE,[16] and the stoplist includes the 100 most frequent words in the British National Corpus.[17]

In the previous work, the features are binary, that is, the value is one if the token occurs in the message, and zero otherwise. There are several more possible definitions of the weights or values, traditionally coming in the Information Retrieval field. For instance, using the same kind of tokens or features, the authors of [39] and [40] make use of TF.IDF weights. Its definition is the following one:

$$w_{ij} = tf_{ij} \times \log_2\left(\frac{N}{df_i}\right)$$

---

[16] The GATE package is available at: http://gate.ac.uk/.
[17] The British National Corpus statistics are available at: http://www.natcorp.ox.ac.uk/.

tf_{ij} being the number of times that the i-th token occurs in the j-th message, N the number of messages, and df_i the number of messages in which the i-th token occurs. The TF (Term Frequency) part of the weight represents the importance of the token or term in the current document or messages, while the second part IDF (Inverse Document Frequency) gives an *ad hoc* idea of the importance of the token in the entire document collection. TF weights are also possible.

Even relatively straightforward decisions like lowercasing all words, can strongly affect the performance of a filter. The second generation of learning filters has been much influenced by Graham's work [46, 47], who took advantage of the increasing power and speed of computers to ignore most preprocessing and simplifying decisions taken before. Graham makes use of a more complicated definition of a token:

1. Alphanumeric characters, dashes, apostrophes, exclamation points, and dollar signs are part of tokens, and everything else is a token separator.
2. Tokens that are all digits are ignored, along with HTML comments, not even considering them as token separators.
3. Case is preserved, and there is neither stemming nor stoplisting.
4. Periods and commas are constituents if they occur between two digits. This allows getting IP addresses and prices intact.
5. Price ranges like $20–$25 are mapped to two tokens, $20 and $25.
6. Tokens that occur within the To, From, Subject, and Return-Path lines, or within URLs, get marked accordingly. For example, 'foo' in the Subject line becomes 'Subject*foo.' (The asterisk could be any character you do not allow as a constituent.)

Graham obtained very good results on his personal email by using this token definition and an *ad hoc* version of a Bayesian learner. This definition has inspired other more sophisticated works in the field, but has also led spammers to focus on tokenization as one of the main vulnerabilities of learning-based filters. The current trend is just the opposite: making nearly no analysis of the text, considering any white-space separated string as a token, and letting the system learn from a really big number of messages (tens of thousands instead of thousands). Even more, HTML is not decoded, and tokens may include HTML tags, attributes, and values.

3.2.3.2 *Multi-word Features.*
Some researchers have investigated features spanning over two or more tokens, seeking for 'get rich,' 'free sex,' or 'OEM software' patterns. Using statistical word phrases has not resulted into very good results in Information Retrieval [82], leading to even decreases in effectiveness. However, they have been quite successful in spam filtering. Two important works in this line are the ones by Zdziarski [102] and by Yerazunis [100, 101].

Zdziarski has first used case-sensitive words in his filter Dspam, and latter added what he has called 'chained tokens.' These tokens are sequences of two adjacent words, and follow the additional rules:

- There are no chains between the message header and the message body.
- In the message header, there are no chains between individual headers.
- Words can be combined with nonword tokens.

Chained tokens are not a replacement for individual tokens, but rather a complement to be used in conjunction with them for better analysis. For example, if we are analyzing an email with the phrase 'CALL NOW, IT's FREE!,' there are four tokens created under standard analysis ('CALL,' 'NOW,' 'IT's,' and 'FREE!') and three more chained tokens: 'CALL NOW,' 'NOW IT's,' 'IT's FREE!.' Chained tokens are traditionally named word bigrams in the fields of Language Modeling and Information Retrieval.

In Table I, we can see how chained tokens may lead to better (more accurate) statistics. In this table, we show the probability of spam given the occurrence of a word, which is the conditional probability usually estimated with the following formula[18]:

$$P(\text{spam}|w) \approx \frac{N(\text{spam}, w)}{N(w)}$$

where $N(\text{spam}, w)$ is the number of times that w occurs in spam messages, and $N(w)$ is the number of times the word w occurs. Counting can be done also *per message*: the number of spam messages in which w occurs, and the number of messages in which w occurs.

In the table, we can see that the words 'FONT' and 'face' have probabilities next to 0.5, meaning that they neither support spam nor legitimate email. However, the probability of the bigram is around 0.2, what represents a strong support to the legitimate class. This is due to the fact that spammers and legitimate users use

TABLE I
SOME EXAMPLES OF CHAINED TOKENS ACCORDING TO [102]

Token words	$P(\text{spam}|w_1)$	$P(\text{spam}|w_2)$	$P(\text{spam}|w_1*w_2)$
w_1=FONT, w_2=face	0.457338	0.550659	0.208403
w_1=color, w_2=#000000	0.328253	0.579449	0.968415
w_1=that, w_2=sent	0.423327	0.404286	0.010099

---

[18] Please note that this is the Maximum Likelihood Estimator.

different patterns of HTML code. While the firsts use the codes *ad hoc*, the seconds generate HTML messages with popular email clients like Microsoft Outlook or Mozilla Thunderbird, that always use the same (more legitimate) patterns, like putting the face attribute of the font next to the FONT HTML tag. We can also see how being 'color' and '#000000' quite neutral, the pattern 'color=#000000' (the symbol '=' is a separator) is extremely guilty. Zdziarski experiments demonstrate noticeable decreases in the error rates and especially in false positives, when using chained tokens plus usual tokens.

Yerazunis follows a slightly more sophisticated approach in [100] and [101]. Given that spammers had already begun to fool learning-based filters by disguising spam-like expressions with intermediate symbols, he proposed to enrich the feature space with bigrams obtained by combining tokens in a sliding 5-words window over the training texts. He has called this Sparse Binary Polynomial Hash (SBPH), and it is implemented in his CRM114 Discriminator filter. In a window, all pairs of sorted words with the second word being the final one are built. For instance, given the phrase/window 'You can get porn free,' the following four bigrams are generated: 'You free,' 'can free,' 'get free,' 'porn free.' With this feature and what the author calls the Bayesian Chain Rule (a simple application of the Bayes Theorem), impressive results have been obtained on his personal email, claiming that the 99.9% (of accuracy) plateau has been achieved.

3.2.3.3 Feature Selection and Extraction.

Dimensionality reduction is a required step because it improves efficiency and reduces overfitting. Many algorithms perform very poorly when they work with a large amount of attributes (exceptions are k Nearest Neighbors or Support Vector Machines), so a process to reduce the number of elements used to represent documents is needed. There are mainly two ways to accomplish this task: feature selection and feature extraction.

Feature selection tries to obtain a subset of terms with the same or even greater predictive power than the original set of terms. For selecting the best terms, we have to use a function that selects and ranks terms according to how good they are. This function measures the quality of the attributes.

In the literature, terms are often selected with respect to their information gain (IG) scores [6, 79, 80], and sometimes according to *ad hoc* metrics [42, 71]. Information gain can be described as:

$$\text{IG}(X,C) = \sum_{x=0,1;c=s,l} P(X=x, C=c) \log_2 \frac{P(X=x, C=c)}{P(X=x) \times P(C=c)}$$

s being the spam class and l the legitimate email class in the above equation. Interestingly, IG is one of the best selection metrics [78]. Other quality metrics are

Mutual Information [35, 59], χ^2 [13, 99], Document Frequency [85, 99], or Relevancy Score [95].

Feature extraction is a technique that aims to generate an artificial set of terms different and smaller than the original one. Techniques used for feature extraction in Automated Text Categorization are Term Clustering and Latent Semantic Indexing.

Term Clustering creates groups of terms that are semantically related. In particular, cluster group words then can be synonyms (like thesaurus classes) or just in the same semantic field (like 'pitcher,' 'ball,' 'homerun,' and 'baseball'). Term Clustering, as far as we know, has not been used in the context of spam filtering.

Latent Semantic Indexing [27] tries to alleviate the problem produced by polysemy and synonymy when indexing documents. It compresses index vectors creating a space with a lower dimensionality by combining original vectors using patterns of terms that appear together. This algebraic technique has been applied to spam filtering by Gee and Cook with moderate success [37].

3.2.4 Learning Algorithms

One of the most important parts in a document classification system is the learning algorithm. Given an ideally perfect classification function Φ that assigns each message a T/F value,[19] learning algorithms have the goal to build a function $\bar{\Phi}$ that approximates the function Φ. The approximation function is usually named a classifier, and it often takes the form of model of the data it has been trained in. The most accurate the approximation is, the better the filter will perform.

A wide variety of learning algorithms families have been applied to spam classification, including the probabilistic Naïve Bayes [4, 6, 42, 71, 75, 79, 80], rule learners like Ripper [34, 71, 75], Instance Based k-Nearest Neighbors (kNN) [6, 42], Decision Trees like C4.5 [14, 34], linear Support Vector Machines (SVM) [30], classifiers committees like stacking [80] and Boosting [14], and Cost-Sensitive learning [39]. By far, the most often applied learner is the probability-based classifier Naive Bayes. In the next sections, we will describe the most important learning families.

To illustrate some of the algorithms that we are going to describe in following sections, we are going to use the public corpus SpamBase.[20] The SpamBase collection contains 4,601 messages, being 1,813 (39.4%) spam. This collection has been preprocessed, and the messages are not available in raw form (to avoid privacy problems). Each message is described in terms of 57 attributes, being the first 48

---

[19] Being T equivalent to spam; that is, spam is the "positive" class because it is the class to be detected.
[20] This collection can be accessed at: ftp://ftp.ics.uci.edu/pub/machine-learning-databases/spambase/.

continuous real [0,100] attributes of type word_freq_WORD (percentage of words in the email that match WORD). A word is a sequence of alphanumeric characters, and the words have been selected as the most unbalanced ones in the collection. The last 9 attributes represent the frequency of special characters like '$' and capital letters.

In order to keep examples as simple as possible, we have omitted the nonword features and selected the five attributes with higher Information Gain scores. We have also binarized the attributes, ignoring the percentage values. The word attributes we use are 'remove,' 'your,' 'free,' 'money,' and 'hp.' The first four words correspond to spammy words, as they occur mainly in spam messages (within expressions like 'click here to remove your email from this list,' 'get free porn,' or 'save money'), and the fifth word is a clue of legitimacy, as it is the acronym of the HP corporation in which the email message donors work.

3.2.4.1 Probabilistic Approaches.
Probabilistic filters are historically the first filters and have been frequently used in recent years [46, 79]. This approach is mostly used in spam filters because of its simplicity and the very good results it can achieve.

This kind of classifiers is based on the Bayes Theorem [61], computing the probability for a document d to belong to a category c_k as:

$$P(c_k|d) = \frac{P(c_k) \cdot P(d|c_k)}{P(d)}$$

This probability can be used to make the decision about whether a document should belong to the category. In order to compute this probability, estimations about the documents in the training set are made. When computing probabilities for spam classification, we can obviate the denominator because we have only two classes (spam and legitimate), and one document cannot be classified in more than one of them, so denominator is the same for every k. $P(c_k)$ can be estimated as the number of documents in the training set belonging to the category, divided by the total number of documents.

Estimating $P(d|c_k)$ is a bit more complicated because we need in the training set some documents identical to the one we want to classify. When using Bayesian learners, it is very frequent to find the assumption that terms in a document are independent and the order they appear in the document is irrelevant. When this happens, the learner is called 'Naïve Bayes' learner [61, 65]. This way probability can be computed in the following way:

$$P(d|c_k) = \prod_{i=1}^{T} P(t_i|c_k)$$

T being the number of terms considered in the documents representation and t_i the i-th term (or feature) in the representation.

It is obvious that this assumption about term independency is not found in a real domain, but it helps to compute probabilities and rarely affects accuracy [28]. That is the reason why this is the approach used in most works.

The most popular version of a Naïve Bayes classifier is that by Paul Graham [46, 47]. Apart from using an *ad hoc* formula for computing terms probabilities, it makes use of only the 15 most extreme tokens, defined as those occurring in the message that have a probability far from the half point (0.5). This approach leads to more extreme probabilities, and has been proven more effective than using the whole set of terms occurring or not in a message. So strong is the influence of this work in the literature that learning-based filters have been quite often named Bayesian Filters despite using radically different learning algorithms.

3.2.4.2 Decision Trees.

One of the biggest problems of probabilistic approaches is that results are not easy to understand for human beings, speaking in terms of legibility. In the Machine Learning field, there are some families of learners that are symbolic algorithms, with results that are easier to understand for people. One of those is the family of Decision Tree learners.

A Decision Tree is a finite tree with branches annotated with tests, and leaves being categories. Tests are usually Boolean expressions about term weights in the document. When classifying a document, we move from top to down in the tree, starting in the root and selecting conditions in branches that are evaluated as true. Evaluations are repeated until a leaf is reached, assigning the document to the category that has been used to annotate the leaf.

There are many algorithms used for computing the learning tree. The most important ones are ID3 [35], C4.5 [18, 56, 61], and C5 [62].

One of the simplest ways to induce a Decision Tree from a training set of already classified documents is:

1. Verify if all documents belong to the same category. Otherwise, continue.
2. Select a term t_i from the document representation and, for every feasible r weight values w_{ir} (i.e., 0 or 1), build a branch with a Boolean test $t_i = w_{ir}$, and a node grouping all documents that satisfy the test.
3. For each document group, go to 1 and repeat the process in a recursive way.

The process ends in each node when all grouped documents in it belong to the same category. When this happens, the node is annotated with the category name.

A critical aspect in this approach is how terms are selected. We can usually find functions that measure the quality of the terms according to how good they are separating the set of documents, using Information Gain, or Entropy metrics. This is basically the algorithm used by ID3 system [76]. This algorithm has been greatly improved using better test selection techniques, and tree pruning algorithms. C4.5 can be considered a state of the art in Decision Tree induction. In Fig. 5, we show a portion of the tree learned by C4.5 on our variation of the SpamBase collection. The figure shows how a message that does not contain the words 'remove,' 'money,' and 'free' is classified as legitimate (often called ham), and if does not contain 'remove,' 'money,' and 'hp' but 'free,' it is classified as spam. The triangles represent other parts of the tree, omitted for the sake of readability.

3.2.4.3 *Rule Learners.*
Rules of the type 'if-then' are the base of one of the concept description languages most popular in Machine Learning field. On one hand, they allow one to present the knowledge extracted using learning algorithms in an easy to understand way. On the other hand, they allow the experts to exanimate and validate that knowledge, and combine it with other known facts in the domain.

Rule learners algorithms build this kind of conditional rules, with a logic condition on the left part of it, the premise, and the class name as the consequent, on the

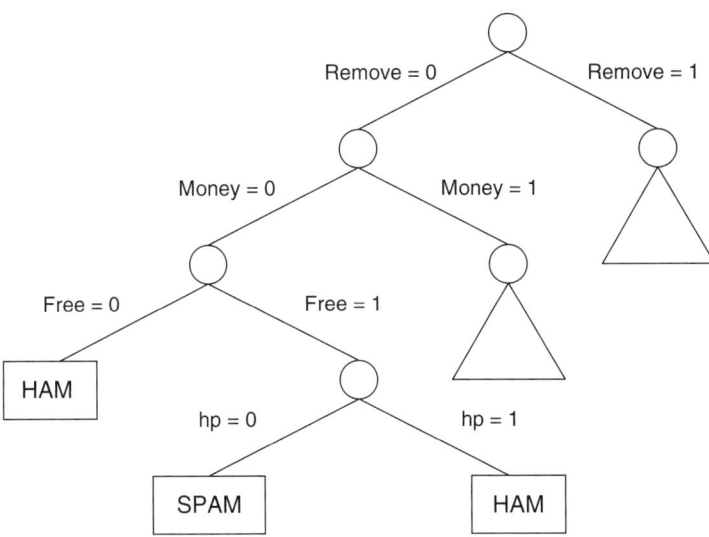

FIG. 5. Partial Decision Tree generated by C4.5 using the SpamBase corpus.

right part. The premise is usually built as a Boolean expression using weights of terms that appear in the document representation. For binary weights, conditions can be simplified to rules that look for if certain combination of terms appears or not in the document.

Actually there are several techniques used to induce rules. One of the most popular ones is the algorithm proposed in [68]. It consists of:

1. Iterate building one rule on each step, having the maximum classify accuracy over any subset of documents.
2. Delete those documents that are correctly classified by the rule generated in the previous step.
3. Repeat until the set of pending documents is empty.

As in other learners, the criteria for selecting the best terms to build rules in step 1 can be quality metrics like Entropy or Information Gain. Probably the most popular and effective rule learner is Ripper, applied to spam filtering in [34, 71, 75]. If the algorithm is applied to our variation of the SpamBase collection, we get the next four rules:

(remove = 1) => SPAM
((free = 1) and (hp = 0)) => SPAM
((hp = 0) and (money = 1)) => SPAM
() => HAM

The rules have been designed to be applied sequentially. For instance, the second rule is fired by a message that has not fired the first rule (and in consequence, does not contain the word 'remove'), and that contains 'free' but not 'hp.' As it can be seen, the fourth rule is a default one that covers all the instances that are not covered by the previous rules.

3.2.4.4 Support Vector Machines.

Support Vector Machines (SVM) have been recently introduced in Automatic Classification [55, 56], but they have become very popular rather quickly because of the very good results obtained with these algorithms especially in spam classification [22, 30, 39].

SVMs are an algebraic method, in which maximum margin hyperplanes are built in order to attempt to separate training instances, using, for example, Platt's sequential minimal optimization algorithm (SMO) with polynomial kernels [72].

Training documents do not need to be linearly separable. Thus, the main method is based on calculation of an arbitrary hyperplane for separation. However, the simplest form of hyperplane is a plane, that is, a linear function of the attributes. Fast to learn, impressively effective in Text Categorization in general, and in spam classification in particular, SVMs represent one of the leading edges in learning-based spam filters.

The linear function that can be obtained when using the SMO algorithm on our version of the SpamBase collection is:

$$f(m) = (-1.9999 * w\_remove) + (-0.0001 * w\_your) + (-1.9992 * w\_free)$$
$$+ (-1.9992 * w\_money) + (2.0006 * w\_hp) + 0.9993$$

This function means that given a message m, in which the weights of the words are represented by 'w_word,' being 0 or 1, the message is classified as legitimate if replacing the weights in the functions leads to a positive number. So, negative factors like -1.9 (for 'remove') are spammy, and positive factors like 2.0 (for 'hp') are legitimate. Note that there is an independent term (0.9993) that makes a message without any of the considered words being classified as legitimate.

3.2.4.5 k-Nearest Neighbors.
Previous learning algorithms were based on building models about the categories used for classification. An alternative approach consists of storing training documents once they have been preprocessed and represented, and when a new instance has to be classified, it is compared to stored documents and assigned to the more appropriate category according to the similarity of the message to those in each category.

This strategy does not build an explicit model of categories, but it generates a classifier know as 'instance based,' 'memory based,' or 'lazy' [68]. The most popular one is kNN [99]. The k parameter represents the number of neighbors used for classification. This algorithm does not have a training step and the way it works is very simple:

1. Get the k more similar documents in the training set.
2. Select the most often category in those k documents.

Obviously, a very important part of this algorithm is the function that computes similarity between documents. The most common formula to obtain the distance between two documents is the 'cosine distance' [82], which is the cosine of the angle between the vectors representing the messages that are being compared. This formula is very effective, as it normalizes the length of the documents or messages.

In Fig. 6, we show a geometric representation of the operation of kNN. The document to be classified is represented by the letter D, and instances in the positive class (spam) are represented as X, while messages in the negative class are represented as O. In the left pane, we show the case of a linearly separable space. In that case, a linear classifier and a kNN classifier would give the same outcome (spam in this case). In the right pane, we can see a more mixed space, where kNN can show its

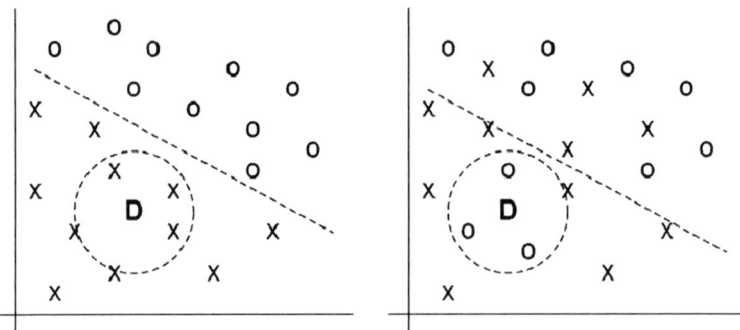

FIG. 6. Geometric representation of k-Nearest Neighbors (kNN) classifier. For making figure simpler, Euclidean distance has been used instead of 'cosine distance.'

full power by selecting the locally most popular class (legitimate), instead of the one a linear classifier would learn (spam).

3.2.4.6 Classifier Committees.
Another approach consists of applying different models to the same data, combining them to get better results. Bagging, boosting, and stacking are some of the techniques used to combine different learners.

The concept of bagging (voting for classification, averaging for regression-type problems with continuous dependent variables of interest) combines the predicted classifications (prediction) from multiple models, or from the same type of model for different learning data. Note that some weighted combination of predictions (weighted vote, weighted average) is also possible, and commonly used. A sophisticated (Machine Learning) algorithm for generating weights for weighted prediction or voting is the boosting procedure. The concept of boosting (applied to spam detection in [14]) is used to generate multiple models or classifiers (for prediction or classification), and to derive weights to combine the predictions from those models into a single prediction or predicted classification.

A simple algorithm for boosting works like this: Start by applying some method to the learning data, where each observation is assigned an equal weight. Compute the predicted classifications, and apply weights to the observations in the learning sample that are inversely proportional to the accuracy of the classification. In other words, assign greater weight to those observations that were difficult to classify (where the misclassification rate was high), and lower weights to those that were easy to classify (where the misclassification rate was low). Then apply the classifier again to the weighted data (or with different misclassification costs), and continue with the next iteration (application of the analysis method for classification to the re-weighted data).

If we apply boosting to the C4.5 learner, with 10 iterations, we obtain 10 decision trees with weights, which are applied to an incoming message. The first tree is that of Fig. 5, with weight 1.88, and the last tree has got only a test on the word 'hp': if it does not occur in the message, it is classified as spam, and as legitimate otherwise. The weight of this last tree is only 0.05.

The concept of stacking (short for Stacked Generalization) [80] is used to combine the predictions from multiple models. It is particularly useful when the types of models included in the project are very different. Experience has shown that combining the predictions from multiple methods often yields more accurate predictions than can be derived from any one method [97]. In stacking, the predictions from different classifiers are used as input into a meta-learner, which attempts to combine the predictions to create a final best predicted classification.

3.2.4.7 Cost-Sensitive Learning.

When talking about spam filtering, we have to take into account that costs of misclassifications are not balanced in real life, as the penalty of a false positive (a legitimate message classified as spam) is much higher than the one of a false negative (a spam message classified as legitimate). This is due to the risk of missing important valid messages (like those from the users' boss!) because messages considered spam can be immediately purged or, in a more conservative scenario, conserved in a quarantine that the user rarely screens. The algorithms commented above assume balanced misclassification costs by default, and it is wise to use techniques to make those algorithms cost-sensitive, in order to build more realistic filters [39].

Thresholding is one of the methods used for making algorithms cost-sensitive. Once a numeric-prediction classifier has been produced using a set of pre-classified instances (the training set), one can compute a numeric threshold that optimizes cost on another set of pre-classified instances (the validation set). When new instances are to be classified, the numeric threshold for each of them determines if the instances are classified as positive (spam) or negative (legitimate). The cost is computed in terms of a cost matrix that typically assigns 0 cost to the hits, a positive cost to false negatives, and a much bigger cost to false positives. This way, instead of optimizing the error or the accuracy, the classifier optimizes the cost.

The weighting method consists of re-weighting training instances according to the total cost assigned to each class. This method is equivalent to stratification by oversampling as described in [29]. The main idea is to replicate instances of the most costly class, to force the Machine Learning algorithm to correctly classify that class instances. Another effective cost-sensitive meta-learner is the MetaCost method [29], based on building an ensemble of classifiers using the bagging method, relabeling training instances according to cost distributions and the ensemble outcomes, and finally training a classier on the modified training collection.

In [39], the experiments with a number of Machine Learning algorithms and the three previous cost-sensitive schemas have shown that the combination of weighting and SVM is the most effective one.

3.3 Filtering by Compression

Compression has recently emerged as a new paradigm for Text Classification in general [90], and for spam filtering in particular [11]. Compression demonstrates high performance in Text Categorization problems in which classification depends on nonword features of a document, such as punctuation, word stems, and features spanning more than one word, like dialect identification and authorship attribution. In the case of spam filtering, they have emerged as the top performers in competitive evaluations like TREC [19].

An important problem of Machine Learning algorithms is the dependence of the results obtained with respect to their parameter settings. In simple words, a big number of parameters can make it hard to find the optimal combination of them, that is, the one that leads to the most general and effective patterns. Keogh et al. discuss the need of a parameter-free algorithm:

> Data mining algorithms should have as few parameters as possible, ideally none. A parameter-free algorithm prevents us from imposing our prejudices and presumptions on the problem at hand, and let the data itself speak to us. [57] (p. 206)

Keogh presents data compression as a Data Mining paradigm that realizes this vision. Data compression can be used as an effective Machine Learning algorithm, especially on text classification tasks. The basic idea of using compression in a text classification task is to assign a text item to the class that best compresses it. This can be straightforwardly achieved by using any state-of-the-art compression algorithm and a command line process. As a result, the classification algorithm (the compression algorithm plus a decision rule) is easy to code, greatly efficient, and it does not need any preprocessing of the input texts. In other words, there is no need to represent it as a feature vector, avoiding one of the most difficult and challenging tasks, that is, text representation. As a side effect, this makes especially hard to reverse engineer the classification process, leading to more effective and stronger spam detection systems.

The rule of classifying a message into the class that best compresses it is a straightforward application of the Minimum Description Length Principle (MDL) [11] that favors the most compact (short) explanation of the data. The recent works by Sculley and Brodley [84] and by Bratko et al. [11] formalize this intuition in a different way; we will follow partially the work by Sculley and Brodley.

Let $C(.)$ be a compression algorithm. A compression algorithm is a function that transforms strings into (shorter) strings.[21] A compression algorithm usually generates a (possibly implicit) model. Let $C(X|Y)$ also be the compression of the string Y using the model generated by compressing the string X. We denote by $|S|$ the length of a string S, typically measured as a number of bits, and by XY the string Y appended to the string X.

The MDL principle states that, given a class A of text instances, a new text X should be assigned to A if it compresses X better than A^C. If we interpret the class A as a sequence of texts (and so it is A^C), the decision rule may be:

$$\text{class}(X) = \arg\min_{c=A,A^C} \{|C(c|X)|\}$$

This formula is one possible decision rule for transforming a compression algorithm into a classifier. The decision rule is based on the 'approximate' distance[22] $|C(A|X)|$. Sculley and Brodley [84] review a number of metrics and measures that are beyond the scope of this presentation.

The length of the text X compressed with the model obtained from a class A can be approached by compressing AX:

$$|C(A|X)| \approx |C(AX)| - |C(A)|$$

This way, any standard compressor can be used to predict the class of a text X, given the classes A and A^C.

Following [11], there are two basic kinds of compressors: two part and adaptive coders. The first class of compressors first trains a model over the data to encode, and then encode the data. These require two passes over the data. These kinds of encoders append the data to the model, and the decoder reads the model and then decodes the data. The most classical example of this kind of compressors is a double pass Huffman coder, which accumulates the statistics for the observed symbols, builds a statistically optimal tree using a greedy algorithm, and builds a file with the tree and the encoded data.

---

[21] This presentation can be made in terms of sequences of bits instead of strings as sequences of characters.

[22] This is not a distance in the proper sense, as it does not satisfy all the formal requirements of a distance.

Adaptive compressors instead start with an empty model (e.g., a uniform distribution over all the symbols), and update it as they are encoding the data. The decoder repeats the process, building its own version of the model as the decoding progresses. Adaptive coders require a single pass over the data, so they are more efficient in terms of time. They have also reached the quality of two-part compressors in terms of compressing ratio, and more interestingly for our purposes, they make the previous approximation an equality. Examples of adaptive compressors include all of those used in our work, which we describe below. For instance, the Dynamic Markov Compression (DMC), the Prediction by Partial Matching (PPM), and the family of Lempel-Ziv (LZ) algorithms are all adaptive methods (see [11] and [84]).

Let us get back to the parameter issue. Most compression algorithms do not require any preprocessing of the input data. This clearly avoids the steps of feature representation and selection usually taken when building a learning-based classifier. Also, most often compressors have a relatively small number of parameters, approaching the vision of a parameter-free or parameter-light Data Mining. However, a detailed analysis performed by Sculley and Brodley in [84] brings light to this point, as they may depend on explicit parameters in compression algorithms, the notion of distance used, and the implicit feature space defined by each algorithm. On the other hand, it is extremely easy to build a compression-based classifier by using the rules above and a standard out-of-the-shelf compressor, and they have proven to be effective on a number of tasks, and top-performer in spam filtering. In words by Cormack (one of the organizers of the reputed TREC spam Track competition) and Bratko:

> "At TREC 2005, arguably the best-performing system was based on adaptive data compression methods", and "one should not conclude, for example, that SVM and LR (Logistic Regression) are inferior for on-line filtering. One may conclude, on the other hand, that DMC and PPM set a new standard to beat on the most realistic corpus and test available at this time." [21]

While other algorithms have been used in text classification, it appears that only DMC and PPM have been used in spam filtering. The Dynamic Markov Compression [11] algorithm models an information source with a finite state machine (FSM). It constructs two variable-order Markov models (one for spam and one for legitimate email), and classifies a message according to which of the models predicts it best. The Partial Prediction Matching algorithm is a back-off smoothing technique for finite-order Markov models, similar to back-off models used in natural language processing, and has set the standard for lossless text compression since its introduction over two decades ago according to [17]. The best of both is DMC, but PPM is better known and very competitive with it (and both superior to Machine Learning approaches at spam filtering). There are some efforts that also work at the character level (in combination with some *ad hoc* scoring function), but they are different from compression as this has been principle designed to model sequential data. In particular, the IBM's Chung-Kwei system [77] uses pattern matching techniques originally developed for DNA sequences, and the Pampapathi et al. [70] have proposed a filtering technique based on the suffix tree data structure. This reflects the trend of minimizing tokenization complexity in order to avoid one of the most relevant vulnerabilities of learning-based filters.

3.4 Comparison and Summary

Heuristic filtering relying on the manual effort of communities of experts has been quickly beaten by spammers, as they have stronger (economic) motivation and time. But automated with content-based methods, a new episode in the war against spam is being written. The impact of and success of content-based spam filtering using Machine Learning and compression is significant, as the spammers have invested much effort to avoid this kind of filtering, as we review below. The main strength of Machine Learning is that it makes the filter adapted to the actual users' email. As every user is unique, every filter is different, and it is quite hard to avoid all vendors' and users' filters simultaneously.

The processing structure of learning-based filters includes a tokenization step in which the system identifies individual features (typically strings of character, often meaningful words) on which the system relies to learn and classify. This step has been recognized as the major vulnerability of learning-based filtering, and the subject of most spammers' attacks, in the form of tokenization attacks and image spam. The answer from the research community has been fast and effective, focusing on character-level modeling techniques based on compression, that detect *implicit* patterns that not even spammers know they are using! Compression-based methods have proven top effective in competitive evaluations, and in consequence, that can be presented as the current (successful) trend in content-based spam filtering.

4. Spam Filters Evaluation

Classifier system evaluation is a critical point: no progress may be expected if there is no way to assess it. Standard metrics, collections, and procedures are required, which allow cross-comparison of research works. The Machine Learning community has well-established evaluation methods, but these have been adapted and improved when facing what is probably the main difficulty in spam filtering: the problem of asymmetric misclassification costs. The fact that a false positive (a legitimate message classified as spam) is much more damaging than a false negative (a spam classified as legitimate) implies that evaluation metrics must attribute bigger weights to worse errors, but ... which weights? We analyze this point in Section 4.3.

Scientific evaluations have well-defined procedures, consolidated metrics, and public test collections that make cross-comparison of results relatively easy. The basis of scientific experiments is that they have to be reproducible. There are a number of industrial evaluations, performed by the filter vendors themselves and typically presented in their white papers, and more interestingly, those performed by specialized computer magazines. For instance, in [53], the antispam appliances BorderWare MXtreme MX-200 and Proofpoint P800 Message Protection Appliance are compared using an *ad hoc* list of criteria, including Manageability, Performance, Ease of Use, Setup, and Value. Or in [3], 10 systems are again tested according to a different set of criteria, allowing no possible comparison. The limitations of industrial evaluations include self-defined criteria, private test collections, and self-defined performance metrics. In consequence, we focus on scientific evaluations in this chapter.

The main issues in the evaluation of spam filters are the test collections, the running procedure, and the evaluation metrics. We discuss these issues in the next subsections, with special attention to what we consider a real and accurate standard in current spam filter evaluation, the TREC spam Track competition.

4.1 Test Collections

A test collection is a set of manually classified messages that are sent to a classifier in order to measure its effectiveness, in terms of hits and mistakes. It is important that test collections are publicly available, because it allows the comparison of approaches and the improvement of the technology. On the other hand, message collections may include private email, so privacy protection is an issue. There are a number of works in which the test collections employed are kept private, as they are personal emails from the author or are donated to them with the condition of being kept private, like in early works [30], [75], and [79], or even in more recent works like [36], [47], and [69]. The privacy problem can be solved in different ways:

- Serving a processed version of the messages that does not allow rebuilding them. This approach has been followed in the SpamBase, PU1, and the 2006 ECML-PKDD Discovery Challenge public collections.
- Building the collection using only messages from public sources. This is the approach in the Lingspam and SpamAssassin Public Corpus.

- Keeping the collection private in the hands of a reputable institution that performs the testing on behalf of the researchers. The TREC spam Track competition is such an institution, and makes some test collections public, and keeps some others private.

In the next paragraphs, we present the test collections that have been publicly available, solving the privacy problem:

- SpamBase[23] is an email message collection containing 4,601 messages, being 1,813 (39%) marked as spam. The collection comes in preprocessed (not raw) form, and its instances have been represented as 58-dimensional vectors. The first 48 features are words extracted from the original messages, without stop list or stemming, and selected as the most unbalanced words for the UCE class. The next 6 features are the percentage of occurrences of the special characters ';,' '(,' '[,' '!,' '$,' and '#.' The following 3 features represent different measures of occurrences of capital letters in the text of the messages. Finally, the last feature is the class label. This collection has been used in, for example, [34], and its main problem is that it is preprocessed, and in consequence, it is not possible to define and test other features apart from those already included.
- The PU1 corpus,[24] presented in [6], consists of 1,099 messages, being 481 (43%) spam and 618 legitimate. Been received by Ion Androutsopoulos, it has been processed by removing attachments and HTML tags. To respect privacy issues, in the publicly available version of PU1, fields other than 'Subject:' have been removed, and each token (word, number, punctuation symbol, etc.) in the bodies or subjects of the messages was replaced by a unique number, the same number throughout all the messages. This hashing 'encryption' mechanism makes impossible to perform experiments with other tokenization techniques and features apart from those included by the authors. It has been used in, for example, [14] and [52].
- The ECML-PKDD 2006 Discovery Challenge collection[25] has been collected by the challenge organizers in order to test how to improve spam classification using untagged data [9]. It is available in a processed form: strings that occur fewer than four times in the corpus are eliminated, and each message is represented by a vector indicating the number of occurrences of each feature in the message. The same comments to PU1 are applicable to this collection.

---

[23] This collection has been described above, but some information is repeated for the sake of comparison.
[24] Available at: http://www.aueb.gr/users/ion/data/PU123ACorpora.tar.gz.
[25] Available at: http://www.ecmlpkdd2006.org/challenge.html.

- The Lingspam test collection,[26] presented in [4] and used in many studies (including [14], [39], and very recent ones like [21]), has been built by mixing spam messages with messages extracted from spam-free public archives of mailing lists. In particular, the legitimate messages have been extracted from the Linguist list, a moderated (hence, spam-free) list about the profession and science of linguistics. The number of legitimate messages is 2,412, and the number of spam messages is 481 (16%). The Linguist messages are, of course, more topic-specific than most users' incoming email. They are less standardized, and for instance, they contain job postings, software availability announcements, and even flame-like responses. In consequence, the conclusions obtained from experiments performed on it are limited.
- The SpamAssassin Public Corpus[27] has been collected by Justin Mason (a SpamAssassin developer) with the public contributions of many others, and consists of 6,047 messages, being 1,897 (31%) spam. The legitimate messages (named 'ham' in this collection) have been further divided into easy and hard (the ones that make use of rich HTML, colored text, spam-like words, etc.). As it is relatively big, realistic, and public, it has become the standard in spam filter evaluation. Several works make use of it (including [11], [21], [22], and [67]), and it has been routinely used as a benchmark in the TREC spam Track [19].

Apart from these collections, the TREC spam Track features several public and private test collections, like the TREC Public Corpus – trec05p-1, and the Mr. X, S.B., and T.M. Private Corpora. For instance, the S.B. corpus consists of 7,006 messages (89% ham, 11% spam) received by an individual in 2005. The majority of all ham messages stems from four mailing lists (23%, 10%, 9%, and 6% of all ham messages) and private messages received from three frequent correspondents (7%, 3%, and 2%, respectively), while the vast majority of the spam messages (80%) are traditional spam: viruses, phishing, pornography, and Viagra ads.

Most TREC collections have two very singular and interesting properties:

1. Messages are chronologically sorted, allowing testing the effect of incremental learning, what we mention below as online testing.
2. They have build by using an incremental procedure [20], in which messages are tagged using several antispam filters, and the classification is reviewed by hand when a filter disagrees.

TREC collections are also very big in comparison with the previously described ones, letting the researchers arriving at more trustable conclusions.

---

[26] Available at: http://www.aueb.gr/users/ion/data/lingspam_public.tar.gz.
[27] Available at: http://spamassassin.apache.org/publiccorpus/.

In other words, TREC spam Track has set the evaluation standard for antispam learning-based filters.

4.2 Running Test Procedure

The most frequent evaluation procedure for a spam filter is batch evaluation. The spam filter is trained on a set of messages, and applied to a different set of test messages. The test messages are labeled, and it is possible to compare the judgment of the filter and of the expert, computing hits and mistakes. It is essential that the test collection is similar to, but disjoint of, the training one, and that it reflects operational settings as close as possible. This is the test procedure employed in most evaluations using the SpamBase, PU1, and Linspam test collections.

A refinement of this evaluation is to perform N-fold cross-validation. Instead of using a separate test collection, portions of the labeled collection are sometimes used as training and as test sets. In short, the labeled collection is randomly divided into N sets or folds (preserving the class distribution), and N tests are run, using $N-1$ folds as training set, and the remaining one as test set. The results are averaged over the N runs, leading to more statistically valid figures, as the experiment does not depend on unpredictable features of the data (all of them are used for training and testing). This procedure has been sometimes followed in spam filtering evaluation, like in [39].

A major criticism to this approach is that the usual operation of spam filters allows them to learn from the mistakes they made (if the user reports them, of course). The batch evaluation does not allow the filters to learn as it classifies, and ignores chronological ordering if available. The TREC organizers have instead approached filter testing as an on-line learning task in which messages are presented to the filter, one at a time, in chronological order. For each message, the filter predicts its class (spam or legitimate) by computing a score S which is compared to a fixed but arbitrary threshold T. Immediately after the prediction, the true class is presented to the filter so that it might use this information in future predictions. This evaluation procedure is supported with a specific set of scripts, requiring a filter to implement the next command-line functions:

- *Initialize* – creates any files or servers necessary for the operation of the filter.
- *Classify message* – returns ham/spam classification and spamminess score for message.
- *Train ham message* – informs filter of correct (ham or legitimate) classification for previously classified message.
- *Train spam message* – informs filter of correct (spam) classification for previously classified message.
- *Finalize* – removes any files or servers created by the filter.

This is the standard procedure used in TREC. An open question is if both methods are equivalent, in term of the results obtained in public works. Cormack and Bratko have made in [21] a systematic comparison of a number of top-performing algorithms following both procedures, and arriving at the conclusion that the current leaders are compression methods, and that the online procedure is more suitable because it is closer to operational settings.

4.3 Evaluation Metrics

We have divided the metrics used in spam filtering test studies into three groups: the basic metrics employed in the initial works, the ROCCH method as a quite advanced one that addresses prior errors, and the TREC metrics as the current standard.

4.3.1 Basic Metrics

The effectiveness of spam filtering systems is measured in terms of the number of correct and incorrect decisions. Let us suppose that the filter classifies a given number of messages. We can summarize the relationship between the system classifications and the correct judgments in a confusion matrix, like that shown in Table II. Each entry in the table specifies the number of documents with the specified outcome. For the problem of filtering spam, we take spam as the positive class (+) and legitimate as the negative class (−). In this table, the key 'tp' means 'number of true-positive decisions' and 'tn,' 'fp,' and 'fn' refer to the number of 'true-negative,' 'false-positive,' and 'false-negative' decisions, respectively.

Most traditional TC evaluation metrics can be defined in terms of the entries of the confusion matrix. F_1 [85] is a measure that gives equal importance to recall and precision. Recall is defined as the proportion of class members assigned to a category by a classifier. Precision is defined as the proportion of correctly assigned

TABLE II
A SET OF N CLASSIFICATION DECISIONS REPRESENTED
AS A CONFUSION MATRIX

	+	−
+	tp	fp
−	fn	tn

documents to a category. Given a confusion matrix like the one shown in the table, recall (*R*), precision (*P*), and F_1 are computed using the following formulas:

$$R = \frac{tp}{tp + fn}$$

$$R = \frac{tp}{tp + fp}$$

$$F_1 = \frac{2RP}{R + P}$$

Recall and precision metrics have been used in some of the works in spam filtering (e.g., [4–6, 42, 80]). Other works make use of standard ML metrics, like accuracy and error [71, 75]. Recall that not all kinds of classification mistakes have the same importance for a final user. Intuitively, the error of classifying a legitimate message as spam (a false positive) is far more dangerous than classifying a spam message as legitimate (a false negative). This observation can be re-expressed as the cost of a false positive is greater than the cost of a false negative in the context of spam classification. Misclassification costs are usually represented as a cost matrix in which the entry C(A,B) means the cost of taking a A decision when the correct decision is B, that is the cost of A given B (cost(A|B)). For instance, C(+,−) is the cost of a false-positive decision (classifying legitimate email as spam) and C(−,+) is the cost of a false-negative decision.

The situation of unequal misclassification costs has been observed in many other ML domains, like fraud and oil spills detection [74]. The metric used for evaluating classification systems must reflect the asymmetry of misclassification costs. In the area of spam filtering, several cost-sensitive metrics have been defined, including weighted accuracy (WA), weighted error (WE), and total cost ratio (TCR) (see e.g., [5]). Given a cost matrix, the cost ratio (CR) is defined as the cost of a false positive over the cost of a false negative. Given the confusion matrix for a classifier, the WA, WE, and TCR for the classifier are defined as:

$$WA = \frac{CR \times tn + tp}{CR(tn + fp) + (tp + fn)}$$

$$WE = \frac{CR \times fp + fn}{CR(tn + fp) + (tp + fn)}$$

$$TCR = \frac{tn + fp}{CR \times fp + fn}$$

The WA and WE metrics are versions of the standard accuracy and error measures that penalize those mistakes that are not preferred. Taking the trivial rejecter that classifies every message as legitimate (equivalent to not using a filter) as a baseline, the TCR of a classifier represents to what extent is a classifier better than it. These metrics are less standard than others used in cost-sensitive classification, as Expected Cost, but to some extent they are equivalent. These metrics have been calculated for a variety of classifiers, in three scenarios corresponding to three CR values (1, 9, and 999) [4–6, 42, 80].

The main problem presented in the literature on spam cost-sensitive categorization is that the CR used does not correspond to real world conditions, which are unknown and may be highly variable. There is no evidence that a false positive is neither 9 nor 999 times worse than the opposite mistake. As class distributions, CR values may vary from user to user, from corporation to corporation, and from ISP to ISP. The evaluation methodology must take this fact into account. Fortunately, there are methods that allow evaluating classifiers effectiveness when target (class distribution and CR) conditions are not known, as in spam filtering. In the next subsection, we introduce the ROCCH method for spam filtering.

4.3.2 The ROCCH Method

The receiver operating characteristics (ROC) analysis is a method for evaluating and comparing a classifiers performance. It has been extensively used in signal detection, and introduced and extended by Provost and Fawcett in the Machine Learning community (see e.g., [74]). In ROC analysis, instead of a single value of accuracy, a pair of values is recorded for different class and cost conditions a classifier is learned. The values recorded are the false-positive (FP) rate and the true-positive (TP) rate, defined in terms of the confusion matrix as:

$$FP = \frac{fp}{fp + tn}$$

$$TP = \frac{tp}{tp + fn}$$

The TP rate is equivalent to the recall of the positive class, while the FP rate is equivalent to 1 less than the recall of the negative class. Each (FP,TP) pair is plotted as a point in the ROC space. Most ML algorithms produce different classifiers in different class and cost conditions. For these algorithms, the conditions are varied to obtain a ROC curve. We will discuss how to get ROC curves by using methods for making ML algorithms cost-sensitive.

One point on a ROC diagram dominates another if it is above and to the left, that is, has a higher TP and a lower FP. Dominance implies superior performance for a variety of common performance measures, including expected cost (and then WA and WE), recall, and others. Given a set of ROC curves for several ML algorithms, the one which is closer to the left upper corner of the ROC space represents the best algorithm.

Dominance is rarely got when comparing ROC curves. Instead, it is possible to compute a range of conditions in which one ML algorithm will produce at least better results than the other algorithms. This is done through the ROC convex hull (ROCCH) method, first presented in [74]. Concisely, given a set of (FP,TP) points, that do not lie on the upper convex hull, corresponds to suboptimal classifiers for any class and cost conditions. In consequence, given a ROC curve, only its upper convex hull can be optimal, and the rest of its points can be discarded. Also, for a set of ROC curves, only the fraction of each one that lies on the upper convex hull of them is retained, leading to a slope range in which the ML algorithm corresponding to the curve produces best performance classifiers. An example of ROC curves taken from [39] is presented in Fig. 7. As it can be seen, there is no single dominator.

The ROC analysis allows a visual comparison of the performance of a set of ML algorithms, regardless of the class and cost conditions. This way, the decision of which is the best classifier or ML algorithm can be delayed until target (real world) conditions are known, and valuable information can be obtained at the same time. In the most advantageous case, one algorithm is dominant over the entire slope range.

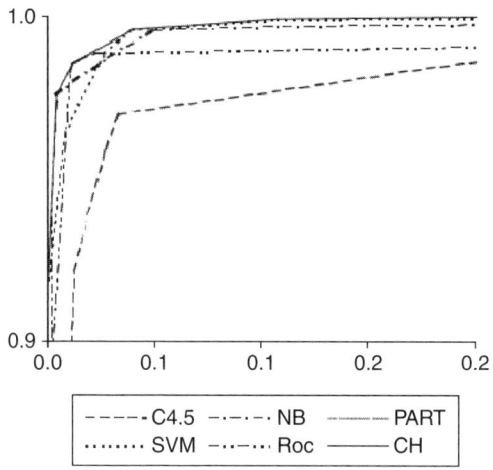

FIG. 7. A ROC curve example.

Usually, several ML algorithms will lead to classifiers that are optimal (among those tested) for different slope ranges, corresponding to different class and cost conditions. Operatively, the ROCCH method consists of the following steps:

1. For each ML algorithm, obtain a ROC curve and plot it (or only its convex hull) on the ROC space.
2. Find the convex hull of the set of ROC curves previously plotted.
3. Find the range of slopes for which each ROC curve lies on the convex hull.
4. In case the target conditions are known, compute the corresponding slope value and output the best algorithm. In other case, output all ranges and best local algorithms or classifiers.

We have made use of the ROCCH method for evaluating a variety of ML algorithms for the problem of spam filtering in [39]. This is the very first time that ROC curves have been used in spam filtering testing, but they have become a standard in TREC evaluations.

4.3.3 TREC Metrics

TREC Spam Track metrics [19] are considered also a standard in terms of spam filtering evaluation. The main improvement over the ROC method discussed above is their adaptation to the online procedure evaluation.

As online evaluation allows a filter to learn from immediately classified errors, its (TP,FP) rate is always changing and the values improving with time. The ROC graph is transformed into a single numeric figure, by computing the Area Under the ROC curve (AUC). As the evaluated filters are extremely effective, it is better to report the inverse $1-AUC$ value, which is computed over time as the learning-testing online process is active. This way, it is possible to get an idea of how quickly the systems learn, and at which levels of error the performance arrives a plateau. In Fig. 8, a ROC Learning Curve graph is also shown, taken from [19] for Mr. X text collection. It is easy to see how filters start committing a relatively high number of mistakes at the beginning of the ROC Learning Curve, and they improve their result finally achieving an average performance level.

5. Spam Filters in Practice

Implementing effective and efficient spam filters is a nontrivial work. Depending on concrete needs, it should be implemented in a certain way and the management of spam messages can vary depending on daily amount of messages, the estimated impact of possible false positives, and other peculiarities of users and companies.

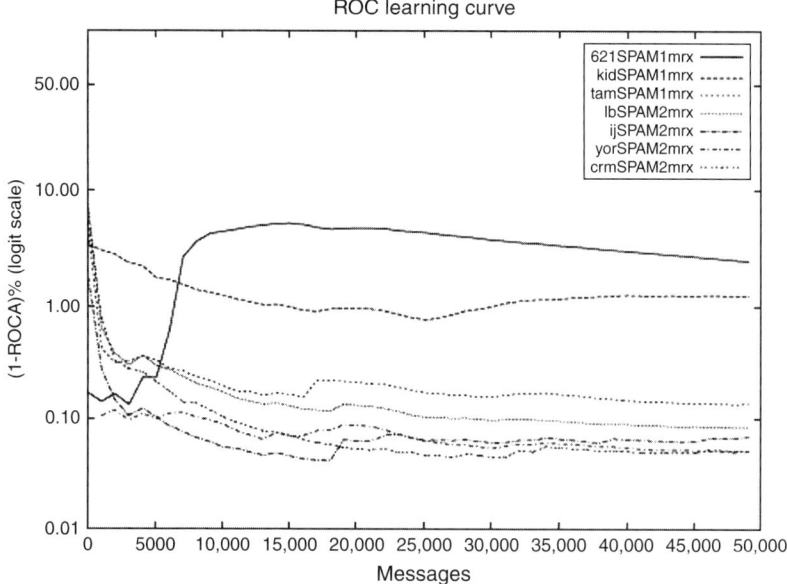

FIG. 8. A ROC learning curve from TREC.

Figure 9 shows a diagram of the mail transportation process, where spam filters can be implemented at any location in this process, although the most common and logical ones are in the receiver's SMTP Server/Mail Delivery Agent (server side) or in the receiver's email client (client side). There are also some enterprise solutions that perform the spam filtering in external servers, which is not reflected in this diagram as not being usual at the current moment.

There are important questions that we cannot obviate and we need to test carefully for their response before implementing or implanting a spam filter solution. Should we implement our spam filter in the client side or in the server side? Should we use any kind of collaborative filtering? Should the server delete the spam messages or the user should decide what to do with those undesirable messages?

5.1 Server Side Versus Client Side Filtering

The first important choice when implementing or integrating a spam filter in our IT infrastructure is the location of the filter: in the client side or in the server side. Each location has its own benefits and disadvantages and those must be measured to choose the right location depending on the user/corporation concrete needs.

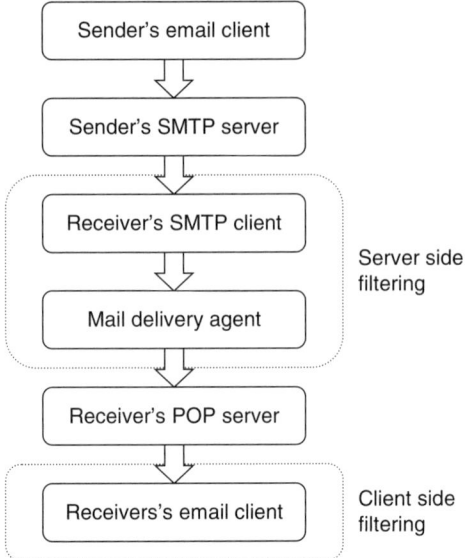

Fig. 9. Email transport process and typical location of server and client side filters. Filters can, virtually, be implemented in any step of the process.

The main features of server side spam filters are:

- They reduce the network load, as the server does not send the mails categorized as spam, which can be the biggest part of the received messages.
- They also reduce the computational load on the client side, as the mail client does not need to check each received message. This can be very helpful when most clients have a low load capacity (handhelds, mobile phones, etc.).
- They allow certain kinds of collaborative filtering or a better integration of different antispam techniques. For example, when detecting spam messages received in different users' account from the same IP, this can be added to a black list preventing other users to receive messages from the same spammer.

On the other side, client side spam filters:

- Allow a more personalized detection of management of the messages.
- Integrate the knowledge from several email accounts belonging to the same user, preventing the user to receive the same spam in different email accounts.
- Reduce the need of dedicated mail servers as the computation is distributed among all the users' computers.

As corporations and other organizations grow, their communications also need to grow and they increasingly rely on dedicated servers and appliances that are responsible for email services, including the email server, and the full email security suite of services, including the antivirus and antispam solutions. Appliances and dedicated servers are the choice for medium to big corporations, while small organizations and individual users can group the whole Internet services in one machine. In big corporations, each Internet-related service has its own machine, as the computational requirements deserve special equipment, and it is a nice idea to distribute the services in order to minimize risks.

5.2 Quarantines

Using statistical techniques is really simple to detect almost all the received spam but a really difficult, almost impossible, point is to have a 0 false-positive ratio. As common bureaucratic processes and communication are delegated to be conducted via email, false positives are being converted into a really annoying question; losing only one important mail can have important economic consequences or, even, produce a delay in some important process.

On the other hand as each user is different, what should be done with a spam message should be left to the user's choice. A clocks lover user would conserve Rolex-related spam. A key point here is that spam is sent because the messages sent are of interest to a certain set of users.

These points are difficult to solve with server spam filter solution as the spam filter is not as personalized as a spam filter implemented on the client side. Quarantine is a strategy developed to face these points. The messages detected as spam by the spam filter are stored in the server for a short period of time and the server mails a Quarantine Digest to the user reporting all the messages under quarantine. The user is given the choice of preserving those messages or deleting them.

Quarantine is a helpful technique that allows to:

- Reduce the disk space and resources the spam is using on the mail servers.
- Reduce the user's level of frustration when they receive spam.
- Keeps spam from getting into mailing lists.
- Prevent auto replies (vacation, out of office, etc.) from going back to the spammers.

In Fig. 10, we show an example of quarantine, in the Trend Micro InterScan Messaging Security Suite, a solution that includes antivirus and antispam features, and that is designed for serve side filtering. The quarantine is accessed by actual

Spam Quarantine		Log Off
Approved Senders [15] (Of Max 50 addresses)		Display: 15 per page
Delete Not Spam		1 - 15 of 33
Sender	**Subject**	**Received**
☐ baller_34_21@starproperty.co.nz	Requested info	14/10/07 9:23:01
☐ raquel@influenza.etc.br	20th October NEURÓTICA:BIOII @ Intermediae Matadero, Madrid	13/10/07 22:00:06
☐ raquel@influenza.etc.br	20th October NEURÓTICA:BIOII @ Intermediae Matadero, Madrid	13/10/07 22:00:06
☐ a-aimew@abacus.cz	Let's chat	13/10/07 20:48:17
☐ a-aimew@abacus.cz	Let's chat	13/10/07 20:48:17
☐ wkxnayn@book-keepingnetwork.com.au	Download cheap softwares in a matter of seconds	13/10/07 12:26:14
☐ wkxnayn@book-keepingnetwork.com.au	Download cheap softwares in a matter of seconds	13/10/07 12:26:14
☐ herschel.danby@we-ga.dk	Get ready for sex in 15 min	13/10/07 9:28:11

FIG. 10. An example of list of messages in the Trend Micro InterScan Messaging Security Suite.

users through a web application, where they log and screen the messages that the filter has considered spam. This quarantine in particular features the access to a white list of approved senders, a white list where the user can type in patterns that make the filter ignore messages, like those coming from the users' organization.

5.3 Proxying and Tagging

There are a number of email clients that currently implement their own content-based filters, most often based on the Graham's Bayesian algorithm. For instance, the popular open-source Mozilla Thunderbird client includes a filter that is able to learn from the users' actual email, leading to better personalization and increased effectiveness.

However, there are a number of email clients that do not feature an antispam filter at all. Although there are a number of extensions or plugins for popular email clients (like the ThunderBayes extension for Thunderbird, or the SpamBayes Microsoft

Outlook plugin – both including the SpamBayes filter controls into the email clients), the user may wish to keep the antispam software out of the email client, in order to change any of them if a better product is found. There are a number of spam filters that run as POP/SMTP/Imap proxy servers. These products download the email on behalf of the user, analyze it deciding if it is spam or legitimate, and tag it accordingly. Example of these products are POPFile (that features general email classification, apart from spam filtering), SpamBayes (proxy version), K9, or SAwin32 (Fig. 11).

As an example, the program SAwin32 is a POP proxy that includes a fully functional port of SpamAssassin for Microsoft Windows PCs. The proxy is configured to access the POP email server from where the user downloads his email and to check it using the SpamAssassin list, rule, and Bayesian filters. If found to be spam, a message is tagged with a configurable string in the subject (e.g., the default is '*** SPAM ***') and the message is forwarded to the user with an explanation in the body and the original message as an attachment. The explanation presents a digest of the message, and describes the tests that have been fired by the message, the spam

```
X-Mozilla-Status: 0001
X-Mozilla-Status2: 10000000
X-SAproxy-Timeout: 0
Return-Path: <frederick@escortcorp.com>
   Received: from localhost by mercury   with SpamAssassin (version 3.2.3);   Wed, 10 Oct 2007 06:53:56 +0200
   Message-Id: <000901c80ae7$02f2d627$9246d393@gwunfog>
X-Spam-Flag: YES
X-Spam-Checker-Version: SpamAssassin 3.2.3 (2007-08-08) on mercury
X-Spam-Level: **************
X-Spam-Status: Yes, score=16.6 required=6.3 tests=DRUGS_DIET,FB_GET_MEDS,   FB_GVR,FH_HELO_EQ_D_D_D_D,HELO_DYNAMIC_IPAI
MIME-Version: 1.0
Content-Type: multipart/mixed; boundary="----------=_470C5AE4.7E980000"

SAproxy believes that this mail is spam. The original message
has been attached intact in RFC 822 format.

Content preview: Good afternoon Bro. We want to inform you that now you have
    astonishing chance to solve your ED troubles. Huge variety of high quality
    meds at the lowest prices. Save your time and money purchase medical products
    only in our shop. C_l*!_C_K he|re. Good afternoon Bro. We want to inform
    you that now you have astonishing chance to solve your ED troubles. Huge variety
    of high quality meds at the lowest prices. Save your time and money purchase
    medical products only in our shop. C_l*!_C_K he|re. [...]

Content analysis details:   (16.6 points, 6.3 required)

 2.4 FH_HELO_EQ_D_D_D_D       Helo is d-d-d-d
 4.4 HELO_DYNAMIC_IPADDR      Relay HELO'd using suspicious hostname (IP addr
                              1)
 2.6 TVD_QUAL_MEDS            BODY: TVD_QUAL_MEDS
 0.5 FB_GVR                   BODY: Looks like generic viagra
 3.6 FB_GET_MEDS              BODY: Looks like trying to sell meds
```

FIG. 11. An example of message scanned by the SAwin32 proxy, implementing the full set of tests and techniques of SpamAssassin.

score of the message, and other administrative data. The proxy also adds some nonstandard headers, beginning with 'X-,'[28] like 'X-Spam-Status,' that includes the spam value of the message ('Yes' in this case), and the tests and scores obtained.

The user can then configure a filter in his email client, sending to a spam box (a local quarantine) all the messages that arrive tagged as spam. The user can also suppress the additional tag and rely on the spam filter headers.

Also, most proxy-type filters also include the possibility of not downloading the messages tagged as spam to the email client, and keeping a separate quarantine in the proxy program, many often accessed as a local web application.

The tagging approach can also work at the organization level. The general email server tags the messages for all users, but each one prepares a local filter based on the server-side tags if that is needed.

5.4 Best and Future Practical Spam Filtering

As stated in [27], 'the best results in spam filtering could be achieved by a combination of methods: Black lists stop known spammers, graylists eliminate spam sent by spam bots, decoy email boxes alert to new spam attacks, and Bayesian filters detect spam and virus attack emails right from the start, even before the black list is updated.' False positives are still a problem due to the important effects that loosing an important mail can produce, along with the bouncing emails created by viruses [8], an important new kind of spam.

Probably, in the future, all these techniques will be mixed with economic and legal antispam measures like computing time-based systems,[29] money-based systems [53], strong antispam laws [15], and other high-impact social measures [80]. Each day, the filters apply more and more measures to detect spam due to its high economic impact.

6. Attacking Spam Filters

6.1 Introduction

As the volume of bulk spam email increases, it becomes more and more important to apply techniques that alleviate the cost that spam implies. Spam filters evolve to better recognize spam and that has forced the spammers to find

---

[28] Non-standard headers in emails begin with 'X-.' In particular, our example includes some Mozilla Thunderbird headers, like 'X-Mozilla-Status.'

[29] See, for instance, the HashCash service: http://www.hashcash.org.

new ways to avoid the detection of their spam, ensuring the delivery of their messages. For example, as statistical spam filters began to learn that words like 'Viagra' mostly occur in spam, spammers began to obfuscate them with spaces and other symbols in order to transform spam-related words in others like 'V-i-a-g-r-a' that, while conserving all the meaning for a human being, are hardly detected by software programs.

We refer to all the techniques that try to mislead the spam filters as attacks, and the spammers employing all these methods as attackers, since their goal is to mislead the normal behavior of the spam filter, allowing a spam message to be delivered as a normal message. A good spam filter would be most robust to past, present, and future attacks, but most empirical evaluations of spam filters ignore this because the spammers' behavior is unpredictable and then, the real effectiveness of a filter cannot be known until its final release.

The attacker's point of view is of vital importance when dealing with spam filters because it gives full knowledge about the possible attacks to spam filters. This perspective also gives a schema of the way of thinking of spammers, which allows predicting possible attacks and to detect tendencies that can help to construct a robust and safe filter. Following this trend, there exist some attempts to compile spammers' attacks as the Graham Cumming's Spammer's Compendium [42].

Usually, a spam filter is part of a corporate complex IT architecture, and attackers are capable to deal with all the parts of that architecture, exploiting all the possible weak points. Attending to this, there exist direct attacks that try to exploit the spam filter vulnerabilities and indirect attacks that try to exploit other weak points of the infrastructure. Indirect attacks' relevance has been growing since 2004 as spammers shifted their attacks away from content and focuses more on the SMTP connection point [75].

Spammers usually make use of other security-related problems like virus and trojans in order to increase their infrastructure. Nowadays, a spammer uses trojan programs to control a lot of zombie-machines from which he is able to attack many sites while not being easily located as he is far from the origin of the attacks. That makes it more and more difficult to counteract the spammers' attacks and increases the side effects of the attacks.

6.2 Indirect Attacks

Mail servers automatically send mails when certain delivery problems, such as a mail over-quota problem, occur. These automatically sent emails are called bounce messages. These bounce messages can be seen, in a way, as auto replies (like the out of office auto replies) but are not sent by human decisions, and in fact are sent automatically by the mail server. All these auto replies are discussed in the RFC3834[30] that points out that it must be sent to the Return-Path established in the received email that has caused this auto reply. The return message must be sent without Return-Path in order to avoid an infinite loop of auto replies.

From these bounce messages, there exist two important ones: the NDRs (Non-Delivery Reports), which are a basic function of SMTP and inform that a certain message could not be delivered and the DSNs (Delivery Status Notifications) that can be explicitly required by means of the ESMTP (SMTP Service Extension) protocol. The NDRs implement part of the SMTP protocol that appears on the RFC2821[31]: 'If an SMTP server has accepted the task of relaying the mail and later finds that the destination is incorrect or that the mail cannot be delivered for some other reason, then it must construct an "undeliverable mail" notification message and send it to the originator of the undeliverable mail (as indicated by the reverse-path).'

The main goal of a spammer is to achieve the correct delivery of a certain nondesired mail. To achieve this goal, some time is needed to achieve a previous target. Indirect attacks are those that use characteristics outside the spam filter in order to achieve a previous target (like obtain valid email addresses) or that camouflages a certain spam mail as being a mail server notification. These indirect attacks use the bounce messages that the mail servers send in order to achieve their goals. Three typical indirect attacks are:

1. NDR (Non-Delivery Report) Attack
2. Reverse NDR Attack
3. Directory Harvest Attack

The NDR Attack consists of camouflaging the spam in an email that appears to be a NDR in order to confuse the user who can believe that he sent the initial mail that could not be delivered. The curiosity of the user may drive him to open the mail and the attachment where the spam resides. The NDR Attacks have two weak points:

1. Make intensive use of the attacker's mail server trying to send thousands or millions of NDR like messages that many times are not even opened by the recipient.
2. If the receiver's mail server uses a black list where the attacker's mail server's IP is listed, the false NDR would not ever reach their destination. Then, all the efforts made by the spammer would not be useful at all.

---

[30] Available at: http://www.rfc-editor.org/rfc/rfc3834.txt.
[31] Available at: http://www.ietf.org/rfc/rfc2821.txt.

To defeat these two weak points, the Reverse NDR Attack was devised. In a Reverse NDR Attack, the intended target's email is used as the sender, rather than the recipient. The recipient is a fictitious email address that uses the domain name for the target's company (for instance, 'example.com'), such as noexist@example.com. The mail server of Example Company cannot deliver the message and sends an NDR mail back to the sender (which is the target email). This return mail carries the NDR and the original spam message attached and the target can read the NDR and the included spam thinking they may have sent the email. As can be seen in this procedure, a reliable mail server that is not in any black list and cannot be easily filtered sends the NDR mail.

Another attack that exploits the bounce messages is the DHA (Directory Harvest Attack) which is a technique used by spammers attempting to find valid email addresses at a certain domain. The success of a Directory Harvest Attack relies on the recipient email server rejecting email sent to invalid recipient email addresses during the Simple Mail Protocol (SMTP) session. Any addresses to which email is accepted are considered valid and are added to the spammer's list. There are two main techniques for generating the addresses that a DHA will target. In the first one, the spammer creates a list of all possible combinations of letters and numbers up to a maximum length and then appends the domain name. This is a standard brute force attack and implies a lot of workload in both servers. The other technique is a standard dictionary attack and is based on the creation of a list combining common first names, surnames, and initials. This second technique usually works well in company domains where the employees email addresses are usually created using their real names and surnames and not nicknames like in free mail services like Hotmail or Gmail.

6.3 Direct Attacks

The first generation of spam filters used rules to recognize specific spam features (like the presence of the word 'Viagra') [46]. Nowadays, as spam evolves quickly, it is impossible to update the rules as fast as new spam variations are created, and a new generation of more adaptable, learning-based spam filters has been created [46].

Direct attacks are attacks to the heart of those statistical spam filters and try to transform a given spam message into a stealthy one. The effectiveness of the attacks relies heavily on the filter type, configuration, and the previous training (mails received and set by the user as spam). One of the simplest attacks is called *picospam* and consists of appending random words to a short spam message, trying that those random words would be recognized as 'good words' by the spam filter. This attack is very simple and was previously seen to be ineffective [63] but shows the general spirit of a direct attack to a spam filter.

There are many spammer techniques [25, 91], which can be grouped into four main categories [96]:

- *Tokenization*: The attacks using tokenization work against the feature selection used by the filter to extract the main features from the messages. Examples of tokenization attacks include splitting up words with spaces, dashes, and asterisks, or using HTML, JavaScript, or CSS tricks.
- *Obfuscation*: With this kind of attacks, the message's contents are obscured from the filter using different kinds of encodings, including HTML entity or URL encoding, letter substitution, Base64 printable encoding, and others.
- *Statistical*: These methods try to skew the message's statistics by adding more good tokens or using fewer bad ones. There exist some variations of this kind of attacks depending on the methods used to select the used words. Weak statistical or passive attacks, that use random words and strong statistical or active attacks, which carefully select the words that are needed to mislead the filter by means of some kind of feedback. Strong statistical attacks are more refined versions of weak attacks, being more difficult to develop and their practical applicability can be questioned.
- *Hiding the text*: Some attacks try to avoid the use of words and inserts the spam messages as images, Flash, RTF, or in other file format; some other attacks insert a link to a web page where the real spam message resides. The goal is that the user could see the message but the filter could not extract any relevant feature.

We discuss instances of these attacks in the next sections.

6.3.1 Tokenization Attacks

The statistical spam filters need a previous tokenization stage where the original message is transformed into a set of features describing the main characteristics of the email. A typical tokenization would count the occurrence of the words appearing in the email and would decompose the words by locating the spaces and other punctuation signals that separate the words. Tokenization attacks are conceived to attack this part of the filter, trying to avoid the correct recognition of spammy words by means of inserting spaces or other typical word delimiters inside the words. A typical example for avoiding the recognition of the word 'VIAGRA' would be separating the letters in this way 'V-I.A:G-R_A.' The user would easily recognize the word 'VIAGRA' but many filters would tokenize the word into multiple letters that do not represent the real content of the email.

6.3.1.1 Hypertextus Interruptus.
The tokenization attack is based on the idea of splitting the words using HTML comments, pairs of zero width tags, or bogus tags. As the mail client renders the HTML, the user would see the message as if not containing any tag or comment, but the spam filters usually separates the words according to the presence of certain tags. Some examples trying to avoid the detection of the word 'VIAGRA':

- VIA<!- -garbage- -> GRA
- VI</n>AGRA
- VIAG<xyz>R<xyz>A
- V<comment>xyz</comment>IAGRA
- VIAGR<b></b>A

A typical tokenizer would decompose each of the last lines into a different set of words:

- VIA, GRA
- VI, AGRA
- VIAG, R, A
- V, IAGRA
- VIAGR, A

This makes more difficult to learn to distinguish the VIAGRA-related spam as each of the spam messages received contains a different set of features. The only way to face this kind of attacks is to parse carefully the messages trying to avoid HTML comments or extracting the features from the output produced by an HTML rendering engine.

6.3.1.2 Slice and Dice.
Slice and Dice means to break a certain body of information down into smaller parts and then examine it from different viewpoints. Applied to spam attacks, slice and dice consists of dividing a spam message into text columns and then rewrite the message putting each text column in a column inside an HTML table. Applying the typical VIAGRA example, we could render it using a table as follows:

```
<table><tr><td>V</td><td>I</td><td>A</td><td>G</td>
<td>R</td><td>A</td></tr></table>
```

The user would see only VIAGRA but the tokenizer would extract one feature by each different letter in the message.

6.3.1.3 Lost in Space. This is the most basic tokenization attack, and consists of adding spaces or other characters between the letters that composes a word in order to make them unrecognizable to word parsers. 'V*I*A*G*R*A,' 'V I A G R A,' and 'V.I.A.G.R.A' are typical examples applying this simple and common technique. Some spam filters recognize the words by merging nonsense words separated by common characters and studying if they compose a word that could be considered as a spam feature.

6.3.2 Obfuscation Attacks

In obfuscation attacks, the message's contents are obscured to the filter using different encodings or misdirection like letter substitution, HTML entities, etc. The way these attacks affect the spam filter is very similar to the way tokenization attacks affect. The features extracted from an obfuscation attack do not correspond with the standard features of the spam previously received. Obfuscating a word like 'VIAGRA' can be done in many ways:

- 'V1AGRA'
- 'VI4GR4'
- 'VÍAGRÀ'

All the previous ways to obfuscate the word 'VIAGRA' produce a different feature that would be used by the spam filter to distinguish or learn how to distinguish a spam message related to the VIAGRA.

A prominent form of obfuscation is the utilization of *leetspeak*. Leetspeak or leet (usually written as l33t or 1337) is an argot used primarily on the Internet, but becoming very common in many online video games due to the excellent reception of this argot from the youngsters who use to obfuscate their mails or SMS trying to avoid their parents to understand what they write to other friends. The leet speech uses various combinations of alphanumeric characters to replace proper letters. Typical replacements are '4' for 'A,' '8' or '13' for 'B,' '(' for 'C,' ')' or '|)' for 'D,' '3' for 'E,' 'ph' for 'F,' '6' for 'G,' '#' for 'H,' '1' or '!' for 'I,' etc. Using these replacements, 'VIAGRA' could be written as 'V14GR4' or 'V!4G2A,' which can be understood by a human being but would be intelligible by a spam filter.

Foreign accent is an attack very similar to the leetspeak, but do not replace letters with alphanumeric characters, it uses accented letters to substitute vocals or even characters like 'ç' to substitute 'c' due to their similarity. 'VIAGRA' could be rewritten in huge set of ways like 'VÍÁGRÁ,' 'VÌÀGRÀ,' 'VÏAGRÄ,' etc.

The simplest way to affront these attacks is to undo these replacements, which can be very simple in the foreign accent attacks because there exists a univocal correspondence between an accented letter and the unaccented letter. But the leetspeak is more difficult to translate as when a certain number or character is found, it would be needed to study whether the alphanumeric is representing a certain letter or it must continue being an alphanumeric.

6.3.3 Statistical Attacks

While tokenization and obfuscation attacks are more related with the preprocessing stage of a spam filter, the statistical attacks have the main goal of attacking the heart of the statistical filter. Statistical attacks, more often called Bayesian poisoning [27, 37, 88] as most of the classifiers used to detect spam are Bayesian or Good Word Attacks [63], are based on adding random, or even carefully selected, words that are unlikely to appear in spam messages and are supposed to cause the spam filter to believe the message is not a spam (a statistical type II error). The statistical attacks have a secondary effect, a higher false-positive rate (statistical I error) because when the user trains their spam filter with spam messages containing normal words, the filter learns that these normal words are a good indication of spam.

Statistical attacks are very similar, and what most vary among them is the way the words are added into the normal message and the way the words are selected. According to the word selection, there exist active attacks and passive attacks. According to the way the words are inserted in the email, there exist some variations like Invisible Ink and MIME-based attacks. Attacks can combine approaches, and for each possible variation of including the words in the message, the attack can be active or passive.

6.3.3.1 Passive Attacks.
In passive attacks, the attacker constructs the word list to be used as the good words to be added in the spam messages without any feedback from the spam filter. Calburn [16] explains this process as 'The automata will just keep selecting random words from the legit dictionary ... When it reaches a Bayesian filtering system, [the filtering system] looks at these legitimate words and the probability that these words are associated with a spam message is really low. And the program will classify this as legitimate mail.'

The simplest passive attack consists of selecting a random set of words that would be added to all the spam messages sent by the attacker. If the same words are added to all the spam messages sent, that set of words would finish being considered as a good indicative of a spam message and the attack would convert into unproductive. Another simple yet more effective attack consists of a random selection of words per each spam mail sent or for each certain number of spam messages sent. Wittel [96] shows that the addition of random words was ineffective against the filter CRM-114 but effective against SpamBayes.

A smarter attack can be achieved by selecting common or hammy words instead of performing a random selection. Wittel [96] shows that attacks using common words are more effective against SpamBayes even when adding fewer words than when the word selection was randomizing. Instead, the work in [88] shows that by adding common words, the filter's precision decreases from 84% to 67% and from 94% to 84% and proposes to ignore common words when performing the spam classification to avoid this performance falling.

6.3.3.2 Active Attacks.
In active attacks, the attacker is allowed to receive some kind of feedback to know whether the spam filter labels a message as spam or not. Graham-Cumming [43] presents a simple way of getting this feedback by including a unique web bug at each message sent. A web bug is a graphic on a web page or inside an email that is designed to monitor who is reading a certain web or mail by counting the hits or accesses to this graphic. Having one web bug per word or per each word set allows the spammer to control what are the words that makes a spam message to look like a ham message to a certain spam filter.

Lowd [63] shows that passive attacks adding random words to spam messages is ineffective as a form of attacks and also demonstrates that adding hammy words was very effective against naïve Bayesian filters. Lowd [63] also shows in detail two active attacks that are very effective against most typical spam filters.

The best way of preventing active attacks is to close any door that allows the spammer to receive any feedback from our system such as nondelivery reports, SMTP level errors, or web bugs.

6.3.3.3 Invisible Ink.
This statistical attack consists of the addition of some real random words in the message but not letting the user to see those words. There are some variants of doing this:
- Add the random words before the HTML.
- Add an email header packer with the random words.
- Write a colored text on a background of the same color.

For avoiding this spammer practice, spam filters should work only with the data the user can see, avoiding headers and colored text over the same color background (Fig. 12).

6.3.4 Hidden Text Attacks

These kind of attacks try to show the user a certain spam message but avoiding the spam filter to capture any feature from the content. The important point here is to place the spam message inside an image or other format, as RTF or PDF, that the

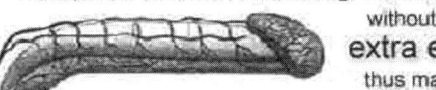

FIG. 12. An example of image-based spam.

email client will show to the user embedded into the message. We describe several kinds of attacks that try to avoid using text, or disguise it in formats hard to process effectively and efficiently.

6.3.4.1 MIME Encoding.
MIME (Multipurpose Internet Mail Extensions) is an Internet standard that allows the email to support plain text, non-text attachments, multi-part bodies, and non-ASCII header information. In this attack, the spammer sends a MIME document with the spam message in the HTML section and any normal text in the plain text section, which makes more difficult the work of the spam filter.

6.3.4.2 Script Hides the Contents.
Most email clients parse the HTML to extract the features and avoid the information inside SCRIPT tags as they usually contain no important information when we are extracting features to detect spam contents. Recently, some attackers have made use of script languages to change the contents of the message on mouse-over event or when the email is opened. Using this technique, the spam filter would read a normal mail that would be converted into a spam mail when opened by the email client.

6.3.4.3 Image-Based Spam.
The first image-based spam was reported by Graham-Cumming in 2003 and was very simple; it included only an image with the spam text inside. At that time images were loaded from Web sites using simple HTML image tags, but as email clients started to avoid the remote image load, spammers started to send the images as MIME attachments. First attempts to detect this kind of spam made the developers to use OCR (Optical Character Recognition) techniques to pull words out of the images and use them as features to classify the message as spam or ham.

But OCR is computationally expensive; its accuracy leaves much to be desired and can be easily mislead by adding noise to the images or even obfuscating the contents with the same techniques used to obfuscate the text content in text spam. According to IronPort statistics [49], 1% of spam was image based in June 2005 and one year later, it had risen to 16%. Cosoi [23] shows that the evolution of image-based spam was from 5% by March 2006 to almost 40% at the end of 2006. Samosseiko [83] asserts that more than the 40% of the spam seen in SophosLabs at the end of 2006 is image-based spam. All these numbers show the real importance of image-based spam in the present spam's world (Fig. 13).

For a more detailed review of actual image-based techniques, Cumming [45] shows the evolution of image spam along the last years, from single images containing the whole spam, combinations (more or less complicated) of different images to show the spam to the user, integrating images with tokenization and obfuscation attacks, using strange fonts, adding random pixels to avoid OCR, hashing, etc.

Interestingly, a number of researchers have devised approaches to deal with specific form of image spam. In particular, Biggio and others [10] have proposed to detect spam email by using the fact that spammers try to add noise to images in order to avoid OCRs, what it is called 'obscuring' by the authors. Also, and in a more general approach, Byun and others make use of a suite of synthesized attributes of

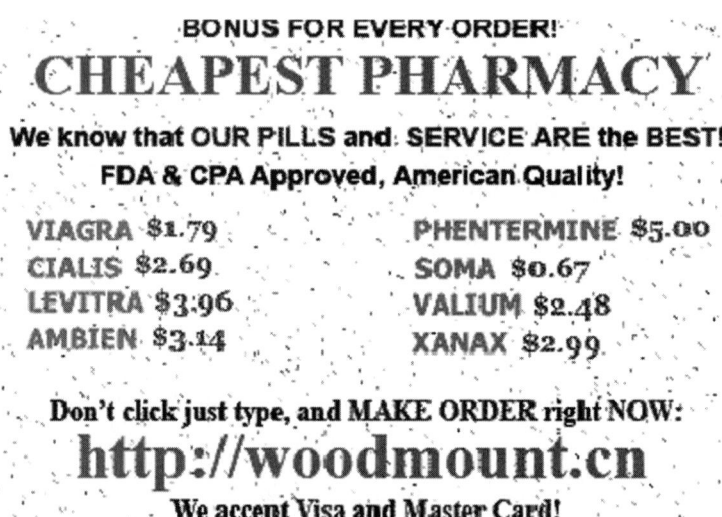

FIG. 13. An example of image-based spam with noise in order to avoid text recognition using OCR software.

spam images (color moment, color heterogeneity, conspicuousness, and self-similarity) in order to characterize a variety of image spam types [12]. These are promising lines of research, and combined with other techniques offer the possibility of high accuracy filtering.

6.3.4.4 Spam Using Other Formats. As spam filter developers increase the precision of their software, spammers develop new variation of their attacks. One simple variation of image-based attacks consists of using a compressed PDF instead of an image to place the spam content trying to avoid the OCR scanning as PDF do not use to contain spam. Another attack consists of embedding RTF files, containing the spam message, which are sniffed by Microsoft email clients.

7. Conclusions and Future Trends

Spam is an ever growing menace that can be very harmful. Its effects could be very similar to those produced by a Denial of Service Attack (DoS). Political, economical, legal, and technical measures are not enough to end the problem, and only a combination of all of them can lower the harm produced by it.

Among all those approaches, content-based filters have been the best solution, having big impact in spammers that have had to search new ways to pass those filters.

Luckily, systems based on Machine Learning algorithms allow the system to learn adapt to new treats, reacting to countermeasures used by spammers.

Recent competitions in spam filtering have shown that actual systems can filter out most of the spam, and new approaches like those based on compression can achieve high accuracy ratios. Spammers have designed new and refined attacks that hit one of the critical steps in every learning method: the tokenization process, but compression-based filters have been very resistant to this kind of attack.

REFERENCES

[1] Abadi M., Birrell A., Burrows M., Dabek F., and Wobber T., December 2003. Bankable postage for network services. In *Proceedings of the 8th Asian Computing Science Conference,* Mumbai, India.
[2] Ahn L. V., Blum M., and Langford J., February 2004. How lazy cryptographers do AI. *Communications of the ACM.*
[3] Anderson R., 2004. Taking a bit out of spam. *Network Computing,* Magazine Article, May 13.
[4] Androutsopoulos I., Koutsias J., Chandrinos K. V., Paliouras G., and Spyropoulos C. D., 2000. An evaluation of Naive Bayesian anti-spam filtering. In *Proceedings of the Workshop on Machine Learning in the New Information Age, 11th European Conference on Machine Learning (ECML),* pp. 9–17. Barcelona, Spain.

[5] Androutsopoulos I., Paliouras G., Karkaletsis V., Sakkis G., Spyropoulos C. D., and Stamatopoulos P., 2000. Learning to filter spam e-mail: A comparison of a naive Bayesian and a memory-based approach. In *Proceedings of the Workshop on Machine Learning and Textual Information Access, 4th European Conference on Principles and Practice of Knowledge Discovery in Databases (PKDD)*, pp. 1–13. Lyon, France.

[6] Androutsopoulos I., Koutsias J., Chandrinos K. V., and Spyropoulos C. D., 2000. An experimental comparison of naive Bayesian and keyword-based anti-spam filtering with encrypted personal e-mail messages. In *Proceedings of the 23rd Annual International ACM SIGIR Conference on Research and Development in Information Retrieval*, pp. 160–167. Athens, Greece, ACM Press, New York, US.

[7] Belkin N. J., and Croft W. B., 1992. Information filtering and information retrieval: Two sides of the same coin? *Communications of the ACM*, **35**(12): 29–38.

[8] Bell S., 2003. Filters causing rash of false positives: TelstraClear's new virus and spam screening service gets mixed reviews. http://computerworld.co.nz/news.nsf/news/CC256CED0016AD1ECC-256DAC000D90D4? Opendocument.

[9] Bickel S., September 2006. ECML/PKDD discovery challenge 2006 overview. In *Proceedings of the Discovery Challenge Workshop, 17th European Conference on Machine Learning (ECML) and 10th European Conference on Principles and Practice of Knowledge Discovery in Databases (PKDD)*, Berlin, Germany.

[10] Biggio B., Giorgio Fumera, Ignazio Pillai, and Fabio Roli, August 23, 2007. Image spam filtering by content obscuring detection. In *Proceedings of the Fourth Conference on Email and Anti-Spam (CEAS 2007)*, pp. 2–3. Microsoft Research Silicon Valley, Mountain View, California.

[11] Bratko A., Cormack G. V., Filipic B., Lynam T. R., and Zupan B., Dec 2006. Spam filtering using statistical data compression models. *Journal of Machine Learning Research*, **7**: 2699–2720.

[12] Byun B., Lee C.-H., Webb S., and Calton P., August 2–3, 2007. A discriminative classifier learning approach to image modeling and spam image identification. In *Proceedings of the Fourth Conference on Email and Anti-Spam (CEAS 2007)*, Microsoft Research Silicon Valley, Mountain View, California.

[13] Caropreso M. F., Matwin S., and Sebastiani F., 2001. A learner-independent evaluation of the usefulness of statistical phrases for automated text categorization. In *Text Databases and Document Management: Theory and Practice*, A. G. Chin, editor, pp. 78–102. Idea Group Publishing, Hershey, US.

[14] Carreras X., and Márquez L., 2001. Boosting trees for anti-spam email filtering. In *Proceedings of RANLP-2001, 4th International Conference on Recent Advances in Natural Language Processing*.

[15] Caruso J., December 2003. Anti-spam law just a start, panel says. *Networld World*, http://www.networkworld.com/news/2003/1218panel.html.

[16] Claburn T., 2005. Constant struggle: How spammers keep ahead of technology, Message Pipeline. http://www.informationweek.com/software/messaging/57702892.

[17] Cleary J. G., and Teahan W. J., 1997. Unbounded length contexts for PPM. *The Computer Journal*, **40**(2/3): 67–75.

[18] Cohen W. W., and Hirsh H., 1998. Joins that generalize: Text classification using WHIRL. In *Proceedings of KDD-98, 4th International Conference on Knowledge Discovery and Data Mining*, pp. 169–173. New York, NY.

[19] Cormack G. V., and Lynam T. R., 2005. TREC 2005 spam track overview. In *Proc. TREC 2005 – the Fourteenth Text REtrieval Conference*, Gaithersburg.

[20] Cormack G. V., and Lynam T. R., July 2005. Spam corpus creation for TREC. In *Proc. CEAS 2005 – The Second Conference on Email and Anti-spam*, Palo Alto.

[21] Cormack G. V., and Bratko A., July 2006. Batch and on-line spam filter evaluation. In *CEAS 2006 – Third Conference on Email and Anti-spam,* Mountain View.
[22] Cormack G., Gómez Hidalgo J. M., and Puertas Sanz E., November 6-9, 2007. Spam filtering for short messages. In *ACM Sixteenth Conference on Information and Knowledge Management (CIKM 2007),* Lisboa, Portugal.
[23] Cosoi C. A., December 2006. The medium or the message? Dealing with image spam. *Virus Bulletin,* http://www.virusbtn.com/spambulletin/archive/2006/12/sb200612-image-spam.dkb.
[24] Cranor L. F., and LaMacchia B. A., 1998. Spam! *Communications of the ACM,* **41**(8): 74–83.
[25] Dalvi N., Domingos P., Sanghai M. S., and Verma D., 2004. Adversarial classification. In *Proceedings of the Tenth International Conference on Knowledge Discovery and Data Mining,* pp. 99–108. ACM Press, Seattle, WA.
[26] Dantin U., and Paynter J., 2005. Spam in email inboxes. In *18th Annual Conference of the National Advisory Committee on Computing Qualifications,* Tauranga, New Zealand.
[27] Deerwester S., Dumais S. T., Furnas G. W., Landauer T. K., and Harshman R., 1990. Indexing by latent semantic indexing. *Journal of the American Society for Information Science,* **41**(6): 391–407.
[28] Domingos P., and Pazzani M. J., 1997. On the optimality of the simple Bayesian classifier under zero-one loss. *Machine Learning,* **29**(2–3): 103–130.
[29] Domingos P., 1999. MetaCost: A general method for making classifiers cost-sensitive. In *Proceedings of the Fifth International Conference on Knowledge Discovery and Data Mining,* pp. 155–164. San Diego, CA, ACM Press.
[30] Drucker H., Vapnik V., and Wu D., 1999. Support vector machines for spam categorization. *IEEE Transactions on Neural Networks,* **10**(5): 1048–1054.
[31] Dumais S. T., Platt J., Heckerman D., and Sahami M., 1998. Inductive learning algorithms and representations for text categorization. In *Proceedings of CIKM-98, 7th ACM International Conference on Information and Knowledge Management,* G. Gardarin, J. C. French, N. Pissinou, K. Makki, and L. Bouganim, eds. pp. 148–155. ACM Press, New York, US, Bethesda, US.
[32] Dwork C., Goldberg A., and Naor M., August 2003. On memory-bound functions for fighting spam. In *Proceedings of the 23rd Annual International Cryptology Conference (CRYPTO 2003).*
[33] Eckelberry A., 2006. What is the effect of Bayesian poisoning? *Security Pro Portal,* http://www.netscape.com/viewstory/2006/08/21/what-is-the-effect-of-bayesian-poisoning/.
[34] Fawcett T., 2003. "*In vivo*" spam filtering: A challenge problem for KDD. *SIGKDD Explorations,* **5**(2): 140–148.
[35] Fuhr N., Hartmann S., Knorz G., Lustig G., Schwantner M., and Tzeras K., 1991. AIR/X—a rule-based multistage indexing system for large subject fields. In *Proceedings of RIAO-91, 3rd International Conference "Recherche d'Information Assistee par Ordinateur,"* pp. 606–623. Barcelona, Spain.
[36] Garcia F. D., Hoepman J.-H., and van Nieuwenhuizen J., 2004. Spam filter analysis. In *Security and Protection in Information Processing Systems, IFIP TC11 19th International Information Security Conference (SEC2004),* Y. Deswarte, F. Cuppens, S. Jajodia, and L. Wang, eds. pp. 395–410. Toulouse, France.
[37] Gee K., and Cook D. J., 2003. Using latent semantic Iidexing to filter spam. *ACM Symposium on Applied Computing,* Data Mining Track.
[38] Gómez-Hidalgo J. M., Maña-López M., and Puertas-Sanz E., 2000. Combining text and heuristics for cost-sensitive spam filtering. In *Proceedings of the Fourth Computational Natural Language Learning Workshop, CoNLL-2000,* Association for Computational Linguistics, Lisbon, Portugal.
[39] Gómez-Hidalgo J. M., 2002. Evaluating cost-sensitive unsolicited bulk email categorization. In *Proceedings of SAC-02, 17th ACM Symposium on Applied Computing,* pp. 615–620. Madrid, ES.

[40] Gómez-Hidalgo J. M., Maña-López M., and Puertas-Sanz E., 2002. Evaluating cost-sensitive unsolicited bulk email categorization. In *Proceedings of JADT-02, 6th International Conference on the Statistical Analysis of Textual Data*, Madrid, ES.
[41] Goodman J., 2004. IP addresses in email clients. In *Proceedings of The First Conference on Email and Anti-Spam.*
[42] Graham-Cumming J., 2003. The Spammer's Compendium. In *MIT Spam Conference.*
[43] Graham-Cumming J., 2004. How to beat an adaptive spam filter. In *MIT Spam Conference.*
[44] Graham-Cumming J., February 2006. Does Bayesian poisoning exist? *Virus Bulletin.*
[45] Graham-Cumming J., November 2006. The rise and rise of image-based spam. *Virus Bulletin.*
[46] Graham P., 2002. A plan for spam. Reprinted in Paul Graham, *Hackers and Painters, Big Ideas from the Computer Age*, O'Really (2004). Available: http://www.paulgraham.com/spam.html.
[47] Graham P., January 2003. Better Bayesian filtering. In *Proceedings of the 2003 Spam Conference.* Available:http://www.paulgraham.com/better.html.
[48] Gray A., and Haahr M., 2004. Personalised, collaborative spam filtering. In *Proceedings of the First Conference on Email and Anti-Spam (CEAS).*
[49] Hahn J., 2006. Image-based spam makes a comeback. *Web Trends*, http://www.dmconfidential.com/blogs/column/Web_Trends/916/.
[50] Hall R. J., March 1998. How to avoid unwanted email. *Communications of the ACM.*
[51] Hird S., 2002. Technical solutions for controlling spam. In *Proceedings of AUUG2002, Melbourne.*
[52] Hovold J., 2005. Naive Bayes spam filtering using word-position-based attributes. In *Proceedings of the Second Conference on Email and Anti-spam, CEAS*, Stanford University.
[53] InfoWorld Test Center. Strong spam combatants: Brute anti-spam force takes on false-positive savvy. Issue 22 May 31, 2004.
[54] Jeffrey E., and Friedl F., August 2006. Mastering Regular Expressions. 3rd edn. O'Really.
[55] Joachims T., 1998. Text categorization with support vector machines: learning with many relevant features. In *Proceedings of ECML-98, 10th European Conference on Machine Learning*, pp. 137–142. Chemnitz, Germany.
[56] Joachims T., 1999. Transductive inference for text classification using support vector machines. In *Proceedings of ICML-99, 16th International Conference on Machine Learning*, pp. 200–209. Bled, Slovenia.
[57] Keogh E., Lonardi S., and Ratanamahatana C. A., 2004. Towards parameter-free data mining. In *Proceedings of the Tenth ACM SIGKDD International Conference on Knowledge Discovery and Data Mining*, pp. 206–215. Seattle, WA, USA, August 22–25, 2004). KDD '04. ACM, New York, NY.
[58] Kolcz A., Chowdhury A., and Alspector J., 2004. The impact of feature selection on signature-driven spam detection. In *Proceedings of the First Conference on Email and Anti-Spam (CEAS).*
[59] Larkey L. S., and Croft W. B., 1996. Combining classifiers in text categorization. In *Proceedings of SIGIR-96, 19th ACM International Conference on Research and Development in Information Retrieval*, H.-P. Frei, D. Harman, P. Schäuble, and R. Wilkinson, eds. pp. 289–297. ACM Press, New York, US, Zurich, CH.
[60] Lewis D. D., and Gale W. A., 1994. A sequential algorithm for training text classifiers. In *Proceedings of SIGIR-94, 17th ACM International Conference on Research and Development in Information Retrieval*, W. B. Croft and C. J. van Rijsbergen, eds. pp. 3–12. Springer Verlag, Heidelberg, DE, Dublin, IE.
[61] Lewis D. D., 1998. Naive (Bayes) at forty: The independence assumption in information retrieval. In *Proceedings of ECML-98, 10th European Conference on, Machine Learning*, C. Nédellec and

C. Rouveirol, eds. pp. 4–15. Springer Verlag, Heidelberg, DE, Chemnitz, DE. Lecture Notes in Computer Science, 1398.
[62] Li Y. H., and Jain A. K., 1998. Classification of text documents. *The Computer Journal*, **41**(8): 537–546.
[63] Lowd D., and Meek C., 2005. Adversarial Learning. In *Proceedings of the Eleventh ACM SIGKDD International Conference on Knowledge Discovery and Data Mining (KDD)*, ACM Press, Chicago, IL.
[64] Lucas M. W., 2006. PGP & GPG: email for the Practical Paranoid. No Starch Press.
[65] McCallum A., and Nigam K., 1998. A comparison of event models for Naive Bayes text classification. In *Proceedings of the AAAI-98 Workshop on Learning for Text Categorization*.
[66] MessageLabs, 2006. MessageLabs Intelligence: 2006 Annual Security Report. Available: http://www.messagelabs.com/mlireport/2006_annual_security_report_5.pdf.
[67] Meyer T. A., and Whateley B., 2004. Spambayes: Effective open-source, bayesian based, email classification system. In *Proceedings of the First Conference on Email and Anti-spam (CEAS)*.
[68] Mitchell T. M., 1996. Machine learning. McGraw Hill, New York, US.
[69] O'Brien C., and Vogel C., September 2003. Spam filters: Bayes vs. chi-squared; letters vs. words. In *Proceedings of the International Symposium on Information and Communication Technologies*.
[70] Pampapathi R., Mirkin B., and Levene M., 2006. A suffix tree approach to anti-spam email filtering. *Mach. Learn*, **65**(1): 309–338.
[71] Pantel P., and Lin D., 1998. Spamcop: A spam classification and organization program. In *Learning for Text Categorization: Papers from the 1998 Workshop,* Madison, Wisconsin. AAAI Technical Report WS-98-05.
[72] Platt J., 1998. Fast training of support vector machines using sequential minimal optimization. B. Schölkopf, C. Burges, and A. Smola, eds. Advances in Kernel Methods – Support Vector Learning.
[73] Postini White Paper, 2004. Why content filter is no longer enough: Fighting the battle against spam before it can reach your network. Postini Pre-emptive email protection.
[74] Provost F., and Fawcett T., 1997. Analysis and visualization of classifier performance: Comparison under imprecise class and cost distributions. In *Proceedings of the Third International Conference on Knowledge Discovery and Data Mining*.
[75] Provost J., 1999. Naive-bayes vs. rule-learning in classification of email. Technical report Department of Computer Sciences at the University of Texas at Austin.
[76] Quinlan R., 1986. Induction of decision trees. *Machine Learning*, **1**(1): 81–106.
[77] Rigoutsos I., and Huynh T., 2004. Chung-kwei: A pattern-discovery-based system for the automatic identification of unsolicited e-mail messages (spam). In *Proceedings of the First Conference on Email and Anti-Spam (CEAS)*.
[78] Saarinen J., 2003. Spammer ducks for cover as details published on the web, NZHerald. http://www.nzherald.co.nz/section/1/story.cfm?c_id=1&objectid=3518682.
[79] Sahami M., Dumais S., Heckerman D., and Horvitz E., 1998. A Bayesian approach to filtering junk e-mail. In *Proceedings of the AAAI-98 Workshop on Learning for Text Categorization*, AAAI Press, Madison, WI.
[80] Sakkis G., Androutsopoulos I., Paliouras G., Karkaletsis V., Spyropoulos C. D., and Stamatopoulos P., 2001. Stacking classifiers for anti-spam filtering of e-mail. In *Proceedings of EMNLP-01, 6th Conference on Empirical Methods in Natural Language Processing,* Pittsburgh, US, Association for Computational Linguistics, Morristown, US.
[81] Salton G., 1981. A blueprint for automatic indexing. *SIGIR Forum*, **16**(2): 22–38.
[82] Salton G., and McGill M. J., 1983. Introduction to Modern Information Retrieval. McGraw Hill, New York, US.

[83] Samosseiko D., and Thomas R., 2006. The game goes on: An analysis of modern spam techniques. In *Proceedings of the 16th Virus Bulletin International Conference*.
[84] Sculley D., and Brodley C. E., 2006. Compression and machine learning: a new perspective on feature space vectors. In *Data Compression Conference (DCC'06)*, pp. 332–341.
[85] Sebastiani F., 2002. Machine learning in automated text categorization. *ACM Computing Surveys*, **34**(1): 1–47.
[86] Seigneur J.-M., and Jensen C. D., 2004. Privacy recovery with disposable email addresses. *IEEE Security and Privacy*, **1**(6): 35–39.
[87] Sergeant M., 2003. Internet-level spam detection and SpamAssassin 2.50. In *Spam Conference*.
[88] Stern H., Mason J., and Shepherd M., A linguistics-based attack on personalized statistical e-mail classifiers. Technical report CS-2004-06, Faculty of Computer Science, Dalhousie University, Canada. March 25, 2004.
[89] Taylor B., 2006. Sender reputation in a large webmail service. In *Proceedings of the Third Conference on Email and Anti-Spam (CEAS)*, Mountain View, California.
[90] Teahan W. J., and Harper D. J., 2003. "Using compression based language models for text categorization". In *Language Modeling for Information Retrieval*, W. B. Croft and J. Laferty, eds. The Kluwer International Series on Information Retrieval, Kluwer Academic Publishers.
[91] Theo V. D., 2004. New and upcoming features in SpamAssassin v3, ApacheCon.
[92] Thomas R., and Samosseiko D., October 2006. The game goes on: An analysis of modern spam techniques. In *Virus Bulletin Conference*.
[93] Turing A. M., 1950. Computing machinery and intelligence. *Mind*, **59**: 433–460.
[94] Watson B., 2004. Beyond identity: Addressing problems that persist in an electronic mail system with reliable sender identification. In *Proceedings of the First Conference on Email and Anti-Spam (CEAS)*, Mountain View, CA.
[95] Wiener E. D., Pedersen J. O., and Weigend A. S., 1995. A neural network approach to topic spotting. In *Proceedings of SDAIR-95, 4th Annual Symposium on Document Analysis and Information Retrieval*, pp. 317–332. Las Vegas, US.
[96] Wittel G. L., and Wu F., 2004. On attacking statistical spam filters. In *Proceedings of the Conference on Email and Anti-spam (CEAS)*.
[97] Witten I. H., and Frank E., 2000. Data Mining: Practical Machine Learning Tools and Techniques with Java Implementations. Morgan Kaufmann, Los Altos, US.
[98] Yang Y., and Chute C. G., 1994. An example-based mapping method for text categorization and retrieval. *ACM Transactions on Information Systems*, **12**(3): 252–277.
[99] Yang Y., and Pedersen J. O., 1997. A comparative study on feature selection in text categorization. In *Proceedings of ICML-97, 14th International Conference on Machine Learning*. D. H. Fisher, editor.
[100] Yerazunis B., 2003. Sparse binary polynomial hash message filtering and the CRM114 discriminator. In *Proceedings of the Spam Conference*.
[101] Yerazunis B., 2004. The plateau at 99.9. In *Proceedings of the Spam Conference*.
[102] Zdziarski J., 2004. Advanced language classification using chained tokens. In *Proceedings of the Spam Conference*.
[103] Zimmermann P. R., 1995. The Official PGP User's Guide. MIT Press.

The Use of Simulation Techniques for Hybrid Software Cost Estimation and Risk Analysis

MICHAEL KLÄS

Fraunhofer Institute for Experimental Software Engineering, 67663 Kaiserslautern, Germany

ADAM TRENDOWICZ

Fraunhofer Institute for Experimental Software Engineering, 67663 Kaiserslautern, Germany

AXEL WICKENKAMP

Fraunhofer Institute for Experimental Software Engineering, 67663 Kaiserslautern, Germany

JÜRGEN MÜNCH

Fraunhofer Institute for Experimental Software Engineering, 67663 Kaiserslautern, Germany

NAHOMI KIKUCHI

Oki Electric Industry Co., Ltd., Warabi-shi, Saitama 335-8510, Japan

YASUSHI ISHIGAI

Information-Technology Promotion Agency, Software Engineering Center, Bunkyo-Ku, Tokyo 113-6591, Japan

Abstract

Cost estimation is a crucial field for companies developing software or software-intensive systems. Besides point estimates, effective project management also requires information about cost-related project risks, for example, a probability distribution of project costs. One possibility to provide such information is the application of Monte Carlo simulation. However, it is not clear whether other simulation techniques exist that are more accurate or efficient when applied in this context. We investigate this question with CoBRA®,[1] a cost estimation method that applies simulation, that is, random sampling, for cost estimation. This chapter presents an empirical study, which evaluates selected sampling techniques employed within the CoBRA® method. One result of this study is that the usage of Latin Hypercube sampling can improve average simulation accuracy by 60% and efficiency by 77%. Moreover, analytical solutions are compared with sampling methods, and related work, limitations of the study, and future research directions are described. In addition, the chapter presents a comprehensive overview and comparison of existing software effort estimation methods.

1. Introduction 117
2. Background 119
   2.1. CoBRA® Principles 119
   2.2. Simulation Techniques 122
3. Related Work 126
   3.1. Software Effort Estimation Methods 126
   3.2. Overview of Random Sampling Techniques 138
4. Problem Statement 140
5. Analytical Approaches 143
   5.1. Point Estimation 144
   5.2. Distribution Computation 144
6. Stochastic Approaches 145
   6.1. The MC Approach 145
   6.2. The LH Approach 146
   6.3. Comparison of Stochastic Algorithms 149
7. Experimental Study 149
   7.1. Experimental Planning 149
   7.2. Experimental Operation 157

---

[1] CoBRA is a registered trademark of the Fraunhofer Institute for Experimental Software Engineering (IESE), Kaiserslautern, Germany.

7.3. Experimental Results 158
7.4. Validity Discussion 162
8. Summary 166
   Acknowledgments 169
   References 169

1. Introduction

Rapid growth in the demand for high quality software and increased investment into software projects show that software development is one of the key markets worldwide [27, 28]. A fast changing market demands software products with ever more functionality, higher reliability, and higher performance. Moreover, in order to stay competitive, software providers must ensure that software products are delivered on time, within budget, and to an agreed level of quality, or even with reduced development costs and time. This illustrates the necessity for reliable software cost estimation, since many software organizations budget unrealistic software costs, work within tight schedules, and finish their projects behind schedule and budget, or do not complete them at all [100].

At the same time, software cost estimation is considered to be more difficult than cost estimation in other industries. This is mainly because software organizations typically develop products as opposed to fabricating the same product over and over again. Moreover, software development is a human-based activity with extreme uncertainties. This leads to many difficulties in cost estimation, especially in early project phases. These difficulties are related to a variety of practical issues. Examples include difficulties with project sizing, a large number of associated and unknown cost factors, applicability of cost models across different organizational contexts, or, finally, insufficient data to build a reliable estimation model on. To address these and many other issues, considerable research has been directed at gaining a better understanding of the software development processes, and at building and evaluating software cost estimation techniques, methods, and tools [12].

Traditionally, effort estimation has been used for the purpose of planning and tracking project resources. Effort estimation methods that grew upon those objectives do not, however, support systematic and reliable analysis of the causal effort dependencies when projects fail. Currently software industry requires effort estimation methods to support them in understanding their business and identifying potential sources of short-term project risks and areas of long-term process improvements. Moreover, in order to gain wider industrial acceptance, a candidate method should minimize the required overhead, for example, by utilizing a variety of already existing information sources instead of requiring extensive expert involvement and/or large project measurement databases. Yet, existing estimation methods (especially those currently used in the software industry) do not offer such comprehensive support.

An example prerequisite to accepting a certain estimation method is its applicability within a particular context, which includes its adaptability to organization-specific characteristics such as availability of required data or effort required to apply the method. Usually, the latter two issues are contradictory: The less effort a method requires to build the estimation model, the more measurement data from previous projects is needed.[2] Data-based methods focus on the latter exclusively. Contrariwise, expert-based methods require almost no measurement data, but obtaining estimates costs a lot of effort. Software organizations move between those two extremes, tempted either by low application costs or low data requirements. In fact, a great majority of organizations that actually use data-based methods do not have a sufficient amount of appropriate (valid, homogeneous, etc.) data as required by such methods. Hybrid methods offer a reasonable bias between data and effort requirements, providing at the same time reliable estimates and justifiable effort to apply the method [14, 18, 69].

Moreover, estimation methods are required to cope with the uncertainty inherent to software development itself as well as to the estimation activity. On the one hand, the preferred method should accept uncertain inputs. On the other hand, besides simple point estimates means to draw conclusions about the estimation uncertainty and effort-related project risk should be provided. The use of probabilistic simulation provides the possibility to deal with estimation uncertainty, perform cost risk analyses, and provide an add-on of important information for project planning (e.g., how probable it is not to exceed a given budget).

One of the methods that respond to current industrial objectives with respect to effort estimation is CoBRA® [14], an estimation method developed at the Fraunhofer Institute for Experimental Software Engineering (IESE). The method uses sparse measurement data (size, effort) from already completed software projects in order to model development productivity and complements it with expert evaluations in order to explicitly model the influence of various project characteristics on productivity deviations.

Random simulation is one of the core elements of the CoBRA® method; it is supposed to deal with estimation uncertainty introduced through expert evaluations. Experts first specify a causal model that describes which factors influence costs (and each other) and how. Afterwards, they quantify the impact of each factor on costs by giving the maximal, minimal, and most likely increase of costs dependent on a certain factor. The simulation component of CoBRA® processes this information into a probability distribution of project costs. This provides decision makers with a robust basis for managing software development costs, for example, planning software costs, analyzing and mitigating cost-related software risks, or benchmarking projects

---

[2] The more data is available, the less expert involvement is required and the company's effort is reduced. This does not include the effort spent on collecting the data. Above a certain maturity level, companies have to collect the data needed anyway.

with respect to software costs. Numerous practical benefits of CoBRA® have been proven in various industrial applications (e.g., [14, 97]).

Originally, CoBRA® utilized the Monte Carlo (MC) simulation technique. Yet, experience from applying the CoBRA® method in an industrial project [14] suggested that MC might not be the optimal solution for that purpose. Latin Hypercube (LH) was proposed as an alternative method that can improve the performance of the CoBRA® implementation. Yet, neither a definition of the performance nor empirical evidence is available that would support this hypothesis. This led us to the question of whether there exist techniques that deliver more accurate results in the context of software cost estimation in general, and for the CoBRA® method in particular, than simple MC sampling or performing the simulation in a more efficient way. In addition, we are interested in the extent of accuracy improvement that can be expected from the use of such a technique.

In this chapter, we describe analytical considerations as well as the results of an empirical study we conducted in order to answer these questions. Our major objective was to evaluate the magnitude of possible CoBRA® accuracy and efficiency improvement related to the selection of a certain simulation technique and its parameters. In order to achieve our goal, we derived several analytical error estimations and compared selected simulation techniques (including MC and LH) in an experiment employing various settings (i.e., parameters) within the same CoBRA® application (i.e., on the same input data).

The chapter is organized as follows: Section 2 presents the necessary theoretical foundations regarding the CoBRA® cost estimation method and the simulation techniques relevant for this chapter. Section 3 provides an overview of related work. It presents a comprehensive overview and comparison of existing estimation methods as a well as summary of common simulation techniques. In Section 4, the research questions (RQ) are described. Section 5 presents analytical results regarding one of the research questions. Section 6 sketches simulation algorithms that were selected for comparison and motivates why an empirical study is necessary for such a comparison. Section 7 presentsthe empirical study, including its results and limitations. Finally, Section 8 summarizes the findings of the study and outlines perspectives of further research work.

2. Background

2.1 CoBRA® Principles

CoBRA® is a hybrid method combining data- and expert-based cost estimation approaches [14]. The CoBRA® method is based on the idea that project costs consist of two basic components: nominal project costs (Equation 1) and a cost overhead (CO) portion (Equation 2).

$$\text{Cost} = \underbrace{\text{Nominal Productivity} \cdot \text{Size}}_{\text{Nominal Cost}} + \text{Cost Overhead} \qquad (1)$$

$$\text{Cost Overhead} = \sum_i \text{Multiplier}_i(\text{Cost Factor}_i) \\ + \sum_i \sum_j \text{Multiplier}_{ij}(\text{Cost Factor}_i, \text{Indirect Cost Factor}_j) \qquad (2)$$

Nominal cost is the cost spent only on developing a software product of a certain size in the context of a *nominal* project. A nominal project is a hypothetical 'ideal' project in a certain environment of an organization (or business unit). It is a project that runs under optimal conditions; that is, all project characteristics are the best possible ones ('perfect') at the start of the project. For instance, the project objectives are well defined and understood by all staff members and the customer and all key people in the project have appropriate skills to successfully conduct the project. CO is the additional cost spent on overcoming imperfections of a real project environment such as insufficient skills of the project team. In this case, a certain effort is required to compensate for such a situation, for example, team training has to be conducted.

In CoBRA®, CO is modeled by a so-called causal model. The causal model consists of factors affecting the costs of projects within a certain context. The causal model is obtained through expert knowledge acquisition (e.g., involving experienced project managers). An example is presented in Fig. 1. The arrows indicate direct and indirect relationships. A sign ('+' or '−') indicates the way a cost factor contributes to the overall project costs. The '+' and '−' represent a positive and negative relationship, respectively; that is, if the factor increases, the project costs will also increase ('+') or decrease ('−'). For instance, if *Requirements volatility*

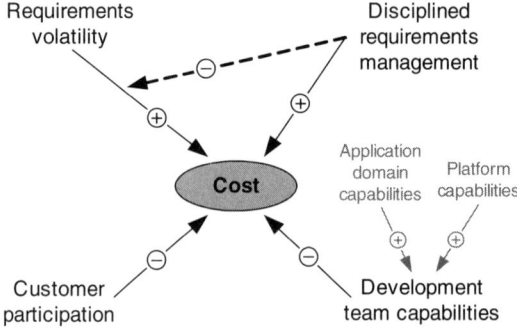

FIG. 1. Example causal cost model.

increases, costs will also increase. One arrow pointing to another one indicates an interaction effect. For example, an interaction exists between *Disciplined requirements management* and *Requirements volatility*. In this case, increased disciplined requirements management compensates for the negative influence of volatile requirements on software costs.

The CO portion resulting from indirect influences is represented by the second component of the sum shown in Equation 2. In general, CoBRA® allows for expressing indirect influences on multiple levels (e.g., influences on *Disciplined requirements management* and influences on influences thereon). However, in practice, it is not recommended for experts to rate all factors due to the increased complexity of the model and the resulting difficulties and efforts. Further details on computing the CO can be found in [14].

The influence on costs and between different factors is quantified for each factor using experts' evaluation. The influence is measured as a relative percentage increase of the costs above the nominal project. For each factor, experts are asked to give the increase of costs when the considered factor has the worst possible value (extreme case) and all other factors have their nominal values. In order to capture the uncertainty of evaluations, experts are asked to give three values: the maximal, minimal, and most likely CO for each factor (triangular distribution).

The second component of CoBRA®, the nominal project costs, is based on data from past projects that are similar with respect to certain characteristics (e.g., development and life cycle type) that are not part of the causal model. These characteristics define the context of the project. Past project data is used to determine the relationship between CO and costs (see Equation 1). Since it is a simple bivariate dependency, it does not require much measurement data. In principle, merely project size and effort are required. The size measure should reflect the overall project volume, including all produced artifacts. Common examples include lines of code or function points [60]. Past project information on identified cost factors is usually elicited from experts.

On the basis of the quantified causal model, past project data, and current project characteristics, a CO model (distribution) is generated using a simulation algorithm (e.g., MC or LH). The probability distribution obtained could be used further to support various project management activities, such as cost estimation, evaluation of cost-related project risks, or benchmarking [14]. Figure 2 illustrates two usage scenarios using the cumulative cost distribution: Calculating the project costs for a given probability level and computing the probability for exceeding given project costs.

Let us assume (scenario A) that the budget available for a project is 900 U and that this project's costs are characterized by the distribution in Fig. 2. There is roughly a 90% probability that the project will overrun this budget. If this probability

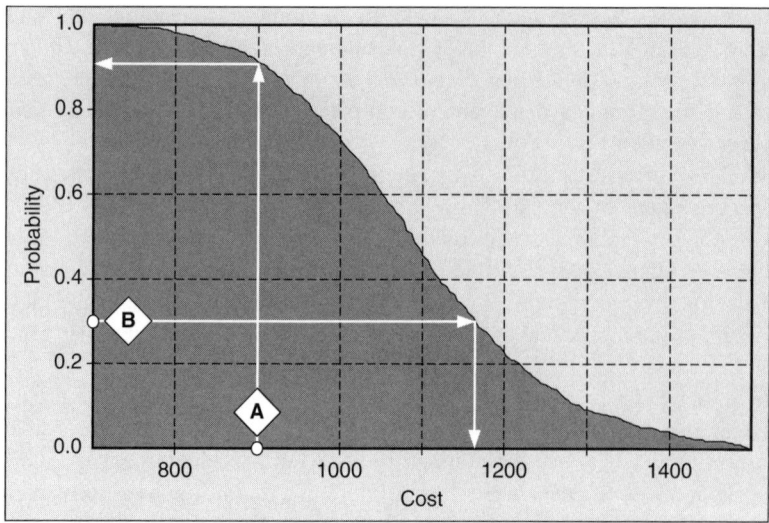

Fig. 2. Example cumulative cost distribution.

represents an acceptable risk in a particular context, the project budget may not be approved. On the other hand, let us consider (scenario B) that a project manager wants to minimize the risks of overrunning the budget. In other words, the cost of a software project should be planned so that there is minimal risk of exceeding it. If a project manager sets the maximal tolerable risk of exceeding the budget to 30%, then the planned budget for the project should not be lower than 1170 U.

The advantage of CoBRA® over many other cost estimation methods is its low requirements with respect to measurement data. Moreover, it is not restricted to certain size and cost measures. The method provides means to develop an estimation model that is tailored to a certain organization's context, thus increasing model applicability and performance (estimation accuracy, consistency, etc.) A more detailed description of the CoBRA® method can be found in [14].

2.2 Simulation Techniques

Simulation is an approach to obtaining knowledge about the behavior or properties of a real system by creating and investigating a similar system or a mathematical model of the system.

Before computer simulations were possible, formally modeled systems had to be analytically solved to enable predicting the behavior of the system from a given number of parameters and initial conditions. Stochastic computer simulation methods like MC sampling and LH sampling make it possible to handle more complex systems with larger numbers of input variables and more complex interactions between them.

2.2.1 Basic Principles of the MC Method

The *MC* method [90] is the most popular simulation approach, used in numerous science areas (see Section 3). One significant advantage of the MC method is the simplicity of its algorithmic structure. In principle, an MC algorithm consists of a process of producing random events (so-called *trials*). These events are used as input for the mathematical model and produce possible occurrences of the observed variable. This process is repeated n times (*iterations*) and the average of overall occurrences is calculated as a result.

The mathematical foundation of the MC method is built upon the *strong law of large numbers* [59]:

Let $\eta_i (i \in N)$ be a sequence of integrable, independent random variables with identical distributions and a finite expected value $\mu = E(\eta_i)$. It follows that for nearly any realization ω:

$$\lim_{n \to \infty} \frac{\sum_{i=1}^{n} \eta_i(\omega)}{n} = \mu \qquad (3)$$

This means that we can transform η into the integral, stochastically, by taking repeated samples from η and averaging $\eta_i(\omega)$.

The *expected error* for n iterations can be estimated with [90]:

$$\text{EAR} = E\left(\left|\frac{1}{n}\sum_{i=1}^{z} \eta_i(\omega) - \mu\right|\right) \leq \frac{\sigma}{\sqrt{n}} \qquad (4)$$

where σ is the standard derivation of the random variable η_i.

In this chapter, however, we also consider *sampling* as an attempt to approximate an unknown probability distribution with the help of randomly sampled values. In the context of MC, this means that we are not only interested in the expectation of the random variable (η), but also in an approximation of its distribution D_η, describable by its *probability density function* (pdf) $f(x)$ [90].

The range of random samples obtained (result of independent realizations of η) build up an interval $[x,y]$. After breaking it into a number of equal length intervals and counting the frequency of the samples that fall into each interval, we may construct a *histogram* that approximates the density distribution of the sampled random variable η (see Fig. 3).

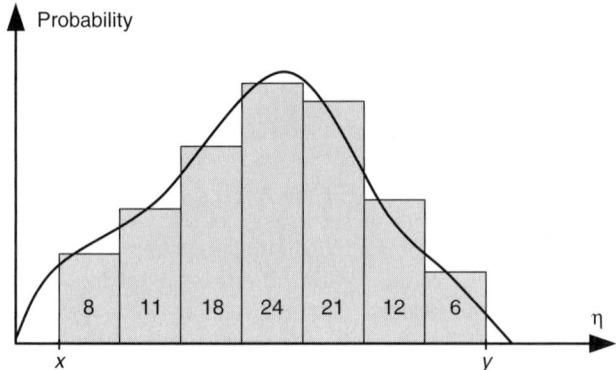

FIG. 3. Example histogram of a probability variable.

2.2.2 Stratification and LH Sampling

The expected error of the MC method depends on the number of iterations n and the variance σ of the random variable η (see Equation 4). The error can thus be reduced either by increasing the number of trials and/or reducing the variance of the random variable. Numerous variance reduction techniques (VRT) are discussed in the literature (see Section 3). One of them is stratification. In this section, we present the idea of stratification in general, as well as LH as a modification for multidimensional problems that can be used without deeper knowledge of the random variable. We explain both: how LH can be used for mean computation and how it can be used for sampling.

When we use stratification for variance reduction, we break the domain of the input distribution into s independent domains (so-called *strata*). In a one-dimensional case, this is simply a partitioning of the range of the input distribution (Fig. 4). If we use strata of equal size and consider the input variable ξ as being uniformly distributed across the half-open interval $[0,1)$, we can get random variables ξ_j for each j-th strata as:

$$\xi_j = \frac{j - \xi}{s} \tag{5}$$

So, we can simulate any random variable η with the density function $f(x)$ if its inverse cumulative density $F^{-1}(x)$ is given:

$$\xi_s = \frac{1}{m \times s} \sum_{i,j=1}^{m,s} \eta_{i,j} = \frac{1}{n} \sum_{i,j=1}^{m,s} F^{-1}(\xi_j) \text{ with } n = m \times s \tag{6}$$

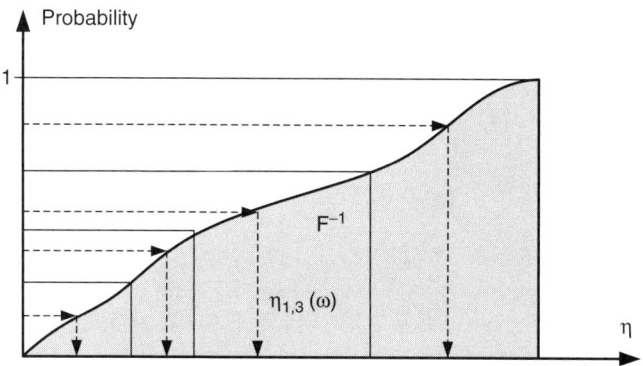

FIG. 4. Example stratification for s and $n = 4$.

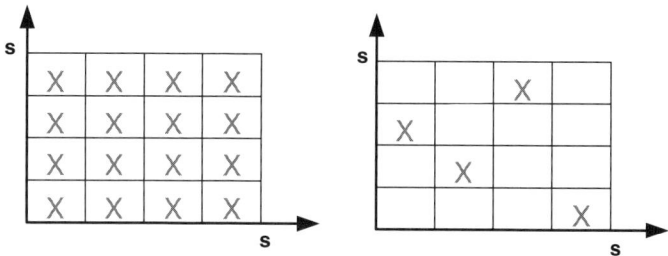

FIG. 5. Two-dimensional straightforward application of stratification requires 16 cells to get four strata (left), the LH algorithm selects only four cells (right).

Moreover, we obtain for $m = 1$, a variance reduction of:

$$\text{Var}_{\text{new}} = \text{Var}_{\text{old}} - \sum_{j=1}^{n} \frac{(E - E_j)^2}{n^2} \tag{7}$$

where E is the expectation of $F^{-1}(x)$ over [0,1] and E_j is the expectation of $F^{-1}(x)$ over strata j.

In the multidimensional case, we have the problem that a straightforward generalization is not efficient [50]. The reason is that we must sample in too many cells (subintervals). If we want s strata in any dimension d, we need s^d cells to partition the input variable(s) (Fig. 5).

The LH technique [64] overcomes this problem by sampling each stratum only once. In order to obtain a perfect stratification within a single dimension, each randomly selected stratum is taken only once and independently of the strata within other dimensions. This can be guaranteed by using d independent permutations π_k of $\{1, \ldots, s\}$ with $k = 1, \ldots, d$.

For $s=n$, we obtain the following estimator that converges against the expected value $E(\eta)$ (see also Equation 3):

$$\xi_{\text{LH}} = \frac{1}{n}\sum_{j=1}^{n} \eta_i = \frac{1}{n}\sum_{j=1}^{n} F^{-1}\left(\xi_{j,1}, \ldots, \xi_{j,d}\right) \text{ with } \xi_{j,k} = \frac{\pi_k(j) - \xi}{n} \qquad (8)$$

In order to sample a pdf using the LH approach, we can use the same procedure as for the MC approach (see Section 2.2.1). In fact, LH is an improvement (special case) of the MC approach. Therefore, the error can be estimated in the same way as for MC (see Section 2.2.1).

3. Related Work

3.1 Software Effort Estimation Methods

3.1.1 Classification of Existing Effort Estimation Methods

Software researchers have made a number of attempts to systematize software effort estimation methods [8, 9, 12, 103].

Proposed classification schemes are subjective and there is no agreement on the best one. Some of the classes, like *Price-to-Win*, cannot really be considered to be an estimation technique. Other classes are not orthogonal, for example, expert judgment can be used following a bottom-up estimation strategy. The systematization we propose in this chapter (Fig. 6) is probably also not fully satisfactory, but is designed to overview the current status of effort estimation research and evaluate it against the most recent industrial requirements.

3.1.1.1 Data-driven Methods. Data-driven (data-intensive) methods refer to methods that provide effort estimates based on an analysis of measurement project data.

Proprietary Versus Non-proprietary. Proprietary effort estimation methods refer to methods that are not fully documented in the public domain. Examples of proprietary approaches are PRICE-S [19], SPQR/100 [44], and SoftCost-R [77]. Because of the lack of sufficient documentation, we exclude these methods from further consideration.

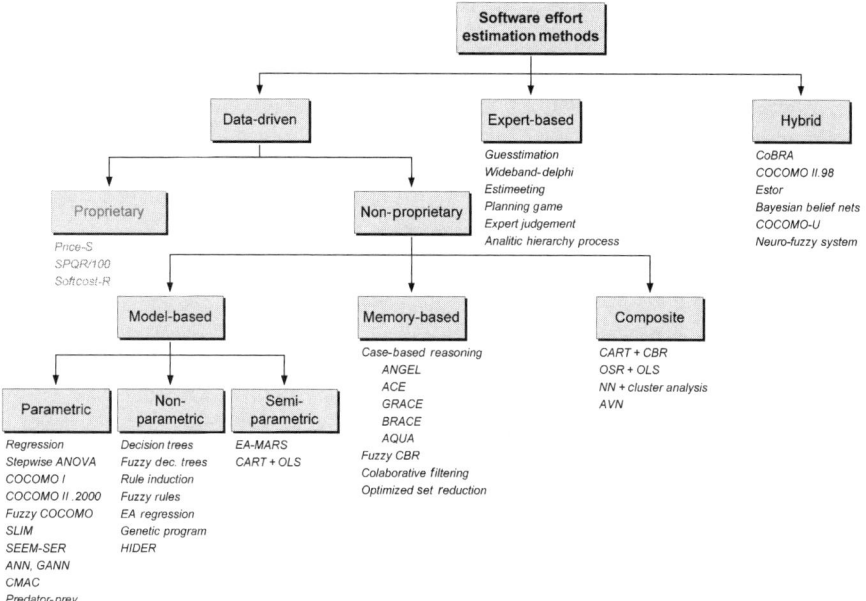

FIG. 6. Classification of software effort estimation methods.

Model-Based Methods. *Model-based methods* provide a model that relates the dependent variable (effort) to one or more independent variables, typically a size measure and one or more effort factors. The model is built, based on the historical project data and used for predictions; the data are generally not needed at the time of the prediction. An estimation model may be characterized by different parameters. They might be specified *a priori* before analyzing the project data (parametric methods) or determined completely from the data at the time of model development (nonparametric methods).

Parametric Model-Based Methods. Specify the parameters of the estimation model *a priori*. *Statistical regression* methods, for instance, require *a priori* specification of the model's functional form and assume estimation errors to follow certain parametric distributions. Typical functional forms of the *univariate regression* include [54, 61, 82]: linear, quadratic, Cobb-Douglas, log-linear, and Translog. *Multivariate regression*, on the other hand, constructs models that relate effort to many independent variables [20, 66]. Typical regression methods fit regression parameters to historical project data using ordinary least squares strategy. Yet, this technique is commonly criticized for its sensitivity to data outliers. As an alternative, robust regression has been proposed by several authors [61, 67]. Another statistical approach to handle unbalanced data is *Stepwise Analysis of Variance* (*Stepwise ANOVA*), which combines classical ANOVA with OLS regression [53]. In each step of an iterative procedure, ANOVA is applied to identify the most significant effort factor and remove its effect by computing the regression residuals; ANOVA is then applied again using the remaining variables on the residuals. Another critical point regarding classical regression is that it does not handle noncontinuous (ordinal and categorical) variables. In order to solve that problem, generalized regression methods such as *Categorical Regression* [3] and *Ordinal Regression* [82] have been proposed.

Fixed-model estimation methods such as *COCOMO* [8, 10], *SLIM* [75], and *SEER–SEM* [40] also belong to the group of parametric methods. In fact, all three models have their roots in the early *Jensen's regression models* [41, 42]. All of them actually represent regression models that were once built on multiple proprietary project data and were intended to be applied 'as is' in contexts that may significantly differ from the one they were built in. Yet, the software estimation community agrees that effort models work better when calibrated with local data [24, 66]. In order to improve the performance of fixed-model methods when applied within a specific context, various adjustment techniques have been proposed, such as periodical updates of fixed models to reflect current trends and changes in the software domain are (e.g., the family of COCOMO models [8, 10]). The frequency of official model calibrations might, however, not be sufficient to keep them up to date [66]. Therefore, additional mechanisms to fit fixed models to local context have been suggested. Already, authors of fixed models are proposing certain *local calibration* techniques as part of a model's application procedures. Moreover, several adjustment approaches were proposed by independent researchers [66, 89]. The *COCONUT* approach, for instance, calibrates effort estimation models using an exhaustive search over the space of COCOMO I calibration parameters [66].

Recently, a parametric regression method that adapts ecological *predator–prey models* to represent dynamics of software testing and maintenance effort has been proposed [17]. The basic idea is that the high population of prey allows predators to survive and reproduce. In contrast, a limited population of prey reduces the chances of predators to survive and reproduce. A limited number of predators, in turn, creates favorable conditions for the prey population to increase. Authors adopt this phenomenon to software maintenance and testing. Software corrections represent predating software defects, and associated effort is fed by defects being discovered by the user. Perfective and adaptive maintenance are both fed by user needs, and the corresponding maintenance effort adapts itself to the number of change requests.

Parametric estimation also adapts selected machine learning approaches. Examples include *Artificial Neural Networks* (*ANN*). Network parameters such as architecture or input factors must be specified *a priori*; the learning algorithm then searches for values of parameters based on project data. The simplest ANN, so-called *single-layer perceptron*, consists of a single layer of output nodes. The inputs are fed directly to the outputs via a series of weights. In this way, it can be considered the simplest kind of feed-forward network. The sum of the products of the weights and the inputs is calculated in each node, and if the value is above a certain threshold (typically 0), the neuron fires and takes the activated value (typically 1); otherwise, it takes the deactivated value (typically −1). Neurons with this kind of activation function are also called McCulloch–Pitts neurons or threshold neurons. In the area of effort estimation, the *sigmoid* and *Gaussian* functions are usually used. Since single-unit perceptrons are only capable of learning linearly separable patterns, multilayer perceptrons are used to represent nonlinear functions. Examples of such ANN are the *Multilayer feed-forward back-propagation perceptron* (*MFBP*) [32] and the *Radial Basis Function Network* (*RBFN*) [32, 87].

The *Cerebellar Model Arithmetic Computer* (*CMAC*), also known as *Albus perceptron*, is a special type of a neural network, which represents a multidimensional function approximator [81]. It discretizes values of continuous input to select so-called training points. CMAC first learns the function at training points and then interpolates to intermediary points at which it has not been trained. The CMAC method operates in a fashion similar to a lookup table, using a generalization mechanism so that a solution learned at one training point in the input space will influence solutions at neighboring points.

Finally, *Evolutionary Algorithms* (*EA*) represent an approach that can be used to provide various types of traditionally data-based estimation models. EA simulates the natural behavior of a population of individuals (chromosomes) by following an iterative procedure based on selection and recombination operations to generate new individuals (next generations). An individual (chromosome) is usually represented by a finite string of symbols called chromosome. Each member of a population encodes a possible solution in a given problem search space, which is comprised of all possible solutions to the problem. The length of the string is constant and completely dependent of the problem. The finite string of symbol alphabet can represent real-valued encodings [2, 79], tree representation [21], or software code [83]. In each simulation iteration (generation), relatively good solutions produce offspring that replace relatively worse ones retaining many features of their parents. The relative quality of a solution is based on the *fitness function*, which determines how good an individual is within the population in each generation, and what its chance of reproducing is, while others (worse) are likely to disappear. New individuals (offspring) for the next generation are typically created by using two basic operations, crossover and mutation.

Shukla, for instance, used EA to generate an optimal ANN for a specific problem [88]. The method was called a *Neuro-genetic Effort Estimation Method* (*GANN*).

The author encoded the neural network using so-called strong representation and learned it using genetic algorithm. The ANN generated in each cycle was evaluated against a fitness function, defined as the inverse of the network's prediction error when applied on the historical project data.

Nonparametric Model-Based Methods. Differ from parametric methods in that the model structure is not specified *a priori* but instead is determined from quantitative (project) data. In other words, nonparametric estimators produce their inferences free from any particular underlying functional form. The term nonparametric is not meant to imply that such methods completely lack parameters, but rather that the number and nature of the parameters are flexible and not fixed in advance.

Typical nonparametric methods originate from the machine learning domain. The most prominent examples are *decision tree* methods. The most popular one in the artificial intelligence domain, the C4.5 algorithm has not been exploited much in the software effort estimation domain due to its inability to handle numerical input data [2]. Proposed in [11], the classification and regression trees (CART) method overcame this limitation and is applicable for both categorical and numerical data. The CART algorithm was re-implemented in numerous software tools such as CART [11], CARTX [93], and GC&RT [96]. Decision trees group instances (software projects) with respect to a dependent variable. A decision tree represents a collection of rules of the form: *if (condition 1 and . . .and condition N) then Z* and basically forms a stepwise partition of the data set being used. In that sense, decision trees are equivalent to decision rules, that is, a decision tree may be represented as a set of decision rules. Successful applications in a number of different domains have motivated software researchers to apply dedicated *rule induction* (RI) methods that generate effort estimation decision rules directly from statistical data [63].

Finally, some applications of the EA can be classified as nonparametric, for instance, EA applied to generate regression equations (*EA Regression*) [16, 21]. In that case, the elements to be evolved are trees representing equations. That is, the population P of the algorithm is a set of trees to which the crossover and mutation operators are applied. A *crossover* exchanges parts of two equations, preserving the syntax of the mathematical expression, whereas a *mutation* randomly changes a term of the equation (function, variable or constant). The terminal nodes are constants or variables, and not-terminals are basic functions that are available for system definition. Each member of the population (regression equation) is evaluated with respect to the fitness function defined as the average estimation error on the historical project data obtained when using the member.

Shan et al. applied so-called *Grammar Guided Genetic Programming* (*GGGP*) to generate computer programs that estimate software development effort [83]. Generating computer programs using EA creates a separate area of machine learning called *Genetic Programming (GP)*. The objective of GP is to optimize a population of computer programs according to a fitness function determined by a program's ability to perform a given computational task (in this case accurately estimate software effort). In GGGP, generation process is guided by a specific language grammar that (1) imposes certain syntactical constraints and (2) incorporates background knowledge into the generation process. The grammar used resulted in models that are very similar to regression. The means square estimation error on the historical project data was applied as the fitness function to evaluate the model-generated GGGP process.

Finally, Aguilar-Ruiz et al. applied EA to generate a set of *Hierarchical Decision Rules (HIDER)* for the effort classification problem [2]. Members of the population are coded using a vector of real values. The vector consists of pair values representing an interval (lower and upper bound) for each effort factor. The last position in the vector represents a discrete value of effort (effort class). The EA algorithm extracts decision rules from this representation. Each member of the population (decision rule) is evaluated with respect to the fitness function that discriminates between correct and incorrect effort predictions (classifications) of historical projects using the member.

Semi-parametric Model-Based Methods. Represent methods that contain both parametric and nonparametric components. In a statistical sense, a semi-parametric method produces its inferences free from a particular functional form but within a particular class of functional forms, for example, it might handle any functional form within the class of additive models.

A typical example of the semi-parametric method where parametric and nonparametric elements were merged is the *integration of decision trees and ordinary least squares regression (CART+OLS)* proposed by Briand and colleagues [12, 13]. The method generates a CART tree and applies regression analysis to interpret projects at each terminal node of the tree.

The application of EA to generate *Multiple Regression Splines (MARS)* is another example of the semi-parametric method [79]. Instead of building a single parametric model, *EA-MARS* generates multiple parametric models, using linear regression. In this case, the search space comprises a set of cut-points (CP) in the independent variable (e.g., software size measured in function points, FP), so a different parametric estimation model can be used for the intervals that comprise such CP.

Memory-based Methods. Model-based methods, such as neural networks or statistical regression, use data to build a parameterized model (where parameters must be specified *a priori* or might be learned from data). After training, the model is used for predictions and the data are generally not needed at the time of prediction. In contrast, *memory-based* methods (or *analogy-based methods*) do not create a model but explicitly retain the available project data, and use them each time a prediction of a new project needs to be made. Each time an estimate is to be provided for a new project (target), the project data (case base) is searched for the projects (analogues) that are most similar to the target. Once the most similar projects are found, their actual efforts are used as a basis for estimation.

The *Case-based Reasoning* (*CBR*) method represents an exact implementation of the memory-based estimation paradigm. It estimates a new project (target) by adapting the effort of the most similar historical projects (analogues) [49].

In order to find project analogies, the CBR method first selects the most relevant characteristics of a software project and defines a project similarity measure upon it. The early methods, such as the *Analogy Software Tool* (*ANGEL*), required *a priori* selection of the most significant characteristics or simply used all characteristics available in the data repository [86]. Later, automatic factor selection [49] and weighting [57] techniques were applied that optimize the performance of the CBR estimator. The most commonly used similarity measure is based on the *Euclidian distance* [86]. The *Gray Relational Analysis Based Software Project Effort* (*GRACE*) method computes the *gray relational grade* [91], and the *Collaborative Filtering* (*CF*) method uses a similarity measure proposed in the field of information retrieval to evaluate the similarity between two documents [71]. The most recent *AQUA* method [57] uses a locally weighted similarity measure, where local weights are computed for different project attributes dependent on their type (nominal, ordinal, continuous, set, fuzzy, etc.) The *Analogical and Algorithmic Cost Estimator* (*ACE*) and *Bootstrap Based Analogy Cost Estimation* (*BRACE*) consider several alternative similarity measures. ACE uses average similarity computed over several distance measures [39, 102, 103], whereas BRACE analyzes each alternative similarity measure to select the one that entails optimal estimation performance [94, 95]. In addition, scaling techniques are used to transform values of project characteristics such that all have the same degree of influence, independently of the choice of units [71, 86, 91]. On the basis of the results of the similarity analysis, analogues are selected to base target estimates on. One group of methods proposes a small constant number (1–3) of nearest neighbors [13, 65, 86, 102]. Other methods, such as BRACE [94] and AQUA [57], determine the optimal number of analogies in a cross-validation analysis. The effort of selected analogues is then adapted to predict the target project. In case of a single analogue, its effort may be adopted [13] or additionally adjusted using the size of the analogue and target projects [94, 102]. Adapting several analogues includes median, mean, distance-weighted mean [65, 71, 86], and inverse rank weighted mean [49, 65]. Again, BRACE analyzes several alternative adaptation approaches to determine the optimal one [94].

Another way of implementing memory-based estimation is represented by the *Optimized Set Reduction* (*OSR*) method [13, 15, 104]. On the basis of the characteristics of the target project, OSR iteratively partition the set of project data into subsets of similar projects that assure increasing information gain. Similarly to decision trees, the final effort estimate is based on projects in the terminal subset.

Composite Methods. These integrate elements of the model- and memory-based methods. Typical examples are applications of CBR or OLS regression to adapt projects in the CART terminal node [12, 13] or OSR terminal subset [47]. Another example is the usage of *cluster analysis* to group similar projects in order to facilitate the training of the neural network [56]. The *analogy with virtual neighbor method* (*AVN*) represents a more sophisticated combination of model- and memory-based methods. It enhances the classical CBR approach in that the two nearest neighbors of a target project are identified based on the normalized effort, which is computed from the multiple-regression equation. Moreover, AVN adjusts the effort estimation by compensating for the location of the target project relative to the two analogues.

3.1.1.2 Expert-Based Methods.

Expert-based estimation methods are based on the judgment of one or more human experts. The simplest instance of the unstructured expert-based estimation is the so-called *Guesstimation* approach, where a single expert provides final estimates. An expert could just provide a guess ('*rule-of-thumb*' method) or give estimates based on more *structured reasoning*, for example, breaking the project down into tasks, estimating them separately and summing those predictions into a total estimate (bottom-up approach). An example formal way of structuring expert-based effort estimation is represented by the *Analytic Hierarchy Process* (*AHP*) method, which systematically extracts a subjective expert's predictions by means of pair-wise comparison [84].

Because of many possible causes of bias in individual experts (optimist, pessimist, desire to win, desire to please, and political), it is preferable to obtain estimates from more than one expert [68]. A group of experts may, for instance, provide their individual estimates, which are then aggregated to a final estimation, for example, by use of statistical mean or median. This is quick, but subject to adverse bias by individual extreme estimates. Alternative *group consensus approaches* try to hold group meetings in order to obtain expert agreement with regard to a single estimate. Examples of such methods are *Wide-band Delphi* [10, 68], *Estimeeting* [98], and, recently defined in the context of agile software development, *Planning Game* [7]. A more formal approach to integrating uncertain estimates of multiple human experts is *Stochastic Budgeting Simulation* (*SBS*) [22]. SBS employs random sampling to combine effort of individual effort items (work products or development activities) and project risks specified by experts in terms of triangular or Erlang distribution.

3.1.1.3 Hybrid Methods.
Hybrid methods combine data- and expert-based methods. In practice, hybrid methods are perceived as the answer to the more and more common observation that human experts, when supported by low-cost analytical techniques, might be the most accurate estimation method [43].

The *ESTOR* method, for instance, [69] provides the initial effort of a target project based on the CBR estimation and then adjusts it by applying a set of expert-based rules, which account for the remaining differences between the analog and the target project. The *Cost Estimation, Benchmarking, and Risk Analysis (CoBRA)* method applies expert-based effort causal modeling to explain the variance on the development production rate measured as effort divided by size [14, 99].

Recently, the software research community has given much attention to *Bayes Theorem* and *Bayesian Belief Networks (BBN)*. Chulani, for instance, employed Bayes theorem to combine *a priori* information judged by experts with data-driven information and to calibrate one of the first versions of the COCOMO II model, known as *COCOMO II.98* [8]. Several researchers have adapted *BBN* to build causal effort models and combine *a priori* expert judgments with a posteriori quantitative project data [23, 74]. Recently, BBN was used to construct the probabilistic effort model called *COCOMO-U*, which extends COCOMO II [8] to handle uncertainty [107].

3.1.2 Handling Uncertainty

Uncertainty is inherent to software effort estimation [46, 52]. Yet, software managers usually do not understand how to properly handle the uncertainty and risks inherent in estimates to improve current project budgeting and planning processes.

Kitchenham et al. conclude that estimation, as an assessment of a future condition, has inherent probabilistic uncertainty, and formulate four major sources of estimation uncertainty: measurement error, model error, assumption error, and scope error [52]. We claim that limited cognition of effort dependencies is another major source of estimation uncertainty. We distinguish two major sources of effort estimation uncertainty: probabilistic and possibilistic. *Probabilistic uncertainty* reflects the random character of the underlying phenomena, that is, the variable character of estimation problem parameters. Uncertainty is considered here in terms of probability, that is, by the random (unpredictable, nondeterministic) character of future software project conditions, in particular, factors influencing software development effort. *Possibilistic uncertainty (epistemological uncertainty)* reflects the subjectivity of the view on modeled phenomena due to its limited cognition (knowledge and/or experience). This may, for instance, include limited granularity of the description of the modeled phenomena, that is, a finite number of estimation parameters (e.g., effort factors and the ways they influence effort). The lack of knowledge may, for instance, stem from a partial lack of data, either because this data is impossible to collect or too expensive to collect, or because the measurement devices have limited precision. Uncertainty is considered here in terms of possibilities.

Handling Probabilistic Uncertainty. The most common approach, which actually explicitly refers to probabilistic uncertainty, is based on the analysis of the probability distribution of given input or output variables. Representing effort as a distribution of values contributes, for instance, to better understanding of estimation results [52]. Motivated by this property, a number of estimation methods that operate on probability distributions have been proposed. Recent interest in *BBN* [23, 74, 107] is one example of this trend. Another one is generally acknowledged to be the application of *random sampling* over the multiple estimation inputs represented by probability distribution [14, 22, 34, 75, 76, 94, 99]. The resulting effort distribution may then be interpreted using various analysis techniques. Confidence and prediction intervals are commonly used statistical means to reflect predictive uncertainty. Human estimators are traditionally asked to provide predicted min–max effort intervals based on given confidence levels, for example, 'almost sure' or '90 percent confident.' This approach may, however, lead to overoptimistic views regarding the level of estimation uncertainty [44]. Jørgensen proposes an alternative, so-called *pX-effort approach*, where instead of giving a number, the human estimator should give ranges and views based on probabilistic distributions [45]. The pX-view means there is an X percent probability of not exceeding the estimate.

Handling Possibilistic Uncertainty. In practice, handling possibilistic uncertainty involves mainly the application of the *fuzzy set theory* [108]. The approaches proposed in the effort estimation domain can be principally classified into two categories: (1) fuzzifying existing effort estimation methods and (2) building-specific rule-based fuzzy logic models.

The first approach simply adapts existing software effort estimation methods to handle the uncertainty of their inputs and outputs using fuzzy sets. A typical estimation process consists of three steps: (1) *fuzzification* of inputs, (2) imprecise reasoning using fuzzy rules, and (3) *de-fuzzification* of outputs. Inputs and outputs can be either linguistic or numeric. Fuzzification involves finding the membership of an input variable with a linguistic term. The membership function can, in principle, be either defined by human experts or analytically extracted from data, whereas fuzzy rules are usually defined by experts. Finally, de-fuzzification provides a crisp number from the output fuzzy set. One example implementation of such an idea is a series of fuzzy-COCOMO models [1, 38, 58, 70]. Other adaptations of traditional estimation methods include fuzzyfying similarity measures within CBR methods [37, 57] or information gain measure in decision trees [35].

Rule-based fuzzy logic models directly provide a set of fuzzy rules that can then be used to infer predictions based on the fuzzy inputs [29, 62]. Fuzzy rules may be based on human judgment [61] or learned from empirical data [106]. *Neuron-fuzzy systems (NFS)* represent a hybrid approach to learn fuzzy rules. The equivalence between neural networks (ANN) and fuzzy logos systems makes it possible to create initial fuzzy rules based on expert judgment, translate them into equivalent ANN, and learn weights for the rules from quantitative data [26]. *Neuro-Fuzzy COCOMO* employs NFS to individually model each linguistic input of the COCOMO model [36].

Another approach to handle possibilistic uncertainty is based on the *rough sets theory* [73]. A rough set is a formal approximation of a crisp set (i.e., conventional set) in terms of a pair of sets, which give the lower and the upper approximation of the original set. The lower and upper approximation sets themselves are crisp sets in the standard version of the rough sets theory. The rough sets theory was recently applied within the AQUA+ method to cover the uncertain impact of effort drivers on effort [57].

3.1.3 Evaluation and Comparison of Existing Effort Estimation Methods

Literature on software effort estimation methods is rich with studies that evaluate and compare several estimation methods [12, 46, 48, 68]. However, they do not provide a clear answer to the question 'Which method is the best one?'

The first impression one may get when looking over hundreds of empirical studies published so far, is that the only reasonable criterion to evaluate an estimation method is the accuracy of the estimates it derives. The second impression is that this criterion is probably not very helpful in selecting the best estimation method because instead of at least converging results, the reader often has to face inconsistent and sometimes even contradicting outcomes of empirical investigations [46]. The source of this apparent inconsistency is a number of factors that influence the performance of estimation methods, but which are usually not explicitly considered in the published results. Examples are: inconsistent empirical data (source, quality, preprocessing steps applied, etc.), inconsistent configuration of apparently the same method, or inconsistent design of the empirical study (e.g., evaluation strategy and measures).

Despite a plethora of published comparative studies, software practitioners have actually still hardly any basis to decide which method they should select for their specific needs and capabilities. In this chapter, we propose an evaluation framework derived directly from industrial objectives and requirements with respect to effort estimation methods.[3] In addition, the definitions of individual criteria are based on related works [1, 9, 10, 12, 16, 51].

---

[3] A series of industrial surveys were performed as part of the research presented in this section. The presentation of the detailed results is, however, beyond the scope of this chapter and will be published in a separate article.

We propose to evaluate individual criteria using the four-grade Likert scale [92]. For that purpose, we define each criterion in the form of a sentence that describes its required value from the perspective of industrial objectives and capabilities. We grouped the evaluation criteria in the two groups: criteria related to the estimation method (C01–C10) and criteria related to the estimation outputs (C11–C13).

C01. Expert Involvement: The method does not require extensive expert's involvement, that is, it requires a minimal amount of experts, limited involvement (effort) per expert, minimal expertise.

C02. Required data: The method does not require many measurement data of a specific type (i.e., measurement scale) and distribution (e.g., normal).

C03. Robustness: The method is robust to low-quality inputs, that is, incomplete (e.g., missing data), inconsistent (e.g., data outliers), redundant, and collinear data.

C04. Flexibility: The method is free from a specific estimation model and provides context-specific outputs.

C05. Complexity: The method has limited complexity, that is, it does not employ many techniques, its underlying theory is easy to understand, and it does not require specifying many sophisticated parameters.

C06. Support level: There is comprehensive support provided along with the method, that is, complete and understandable documentation as well as a useful software tool.

C07. Handling uncertainty: The method supports handling the uncertainty of the estimation (i.e., inputs and outputs).

C08. Comprehensiveness: The method can be applied to estimate different kinds of project activities (e.g., management and engineering) on various levels of granularity (e.g., project, phase, and task).

C09. Availability: The method can be applied during all stages (phases) of the software development lifecycle.

C10. Empirical evidence: There is comprehensive empirical evidence supporting theoretical and practical validity of the method.

C11. Informative power: The method provides complete and understandable information that supports the achievement of numerous estimation objectives (e.g., effective effort management). In particular, it provides context-specific information regarding relevant effort factors, their interactions, and their impact on effort.

C12. Reliability: The method provides the output that reflects the true situation in a given context. In particular, it provides accurate, precise and repeatable estimation outputs.

C13. Portability: The method provides estimation outputs that are either applicable in other contexts without any modification or are easily adaptable to other contexts.

We evaluate existing estimation methods on each criterion by rating our agreement regarding the extent to which each method fulfills a given requirement. We use the following symbols: (++) strongly agree, (+) agree, (−) disagree, and (−−) strongly disagree. Presented in the Table I, evaluation is based on individual author's experiences and critique presented in related literature. It includes both subjective and objective results.

An evaluation of existing estimation methods against industry-oriented criteria shows that none of them fully responds to industrial needs. One of the few methods that support most of the estimation objectives is the CoBRA method. Yet, it has still several significant weaknesses that may prevent its wide industrial acceptance. First is the substantial involvement of human experts. The CoBRA causal effort model (the so-called CO model) is currently built exclusively by experts, which significantly increases estimation costs and reduces the reliability of outputs. The second problem is the unclear impact of the selected simulation technique on the reliability and computational cost of estimation outputs. In this chapter, we investigate the second problem.

3.2 Overview of Random Sampling Techniques

In Section 2.2, we provided the necessary theoretical foundations of simulation techniques such as LH and MC, including literature references. Therefore, we reduce the scope of our review in this section to literature regarding comparative studies of sampling techniques (concerning their theoretical and practical efficiency/usability issues) that can be applied in the context of cost estimation with CoBRA®. We consider only methods that can be used for high problem dimensions (number of input distributions). We do not consider approaches for 'very high-dimensional' problems like Latin Supercube Sampling [72], since they are of limited practical applicability in the context of software cost estimation.

Saliby describes and compares in [80a] a sampling method called Descriptive Sampling (DS) against LH and shows that DS is an improvement over LH. He proves that DS represents the upper limit of maximal improvement over standard MC that could be achieved with LH. Yet, he also showed that the improvement of DS over LH sampling decreases with an increasing number of iterations. The experimental data provided in the chapter shows that the improvement at even relatively small numbers of iterations, such as 1,000 (in our study we used between 10,000 and 640,000 iterations) is not significant anymore (<0.4%). Moreover, 5 years later, Saliby [80b] compared the performance of six MC sampling methods: Standard Monte Carlo, Descriptive Sampling, LH, and Quasi-MC using three different numeric sequences (Halton, Sobol, Faure). Their convergence rates and precision were compared with the standard MC approach in two finance

TABLE I
EVALUATION AND COMPARISON OF EXISTING SOFTWARE EFFORT ESTIMATION METHODS

		Estim. Method	C01	C02	C03	C04	C05	C06	C07	C08	C09	C10	C11	C12	C13
Data-driven	Model-based	Uni-regression	++	-	--	-	++	++	-	++	++	++	--	+	-
		Multi-Regression	++	--	--	+	+	++	-	++	++	++	-	+	-
		CATREG	++	--	--	+	+	+	-	++	++	-	-	+	-
		Ordinal Regression	++	--	--	+	+	+	+	++	++	-	-	+	-
		S-ANOVA	++	+	+	+	+	+	-	++	++	-	-	+	-
		COCOMO I	++	-	-	--	+	++	-	-	-	++	-	-	+
		COCOMO II	++	-	-	--	+	++	-	-	+	++	-	-	+
		SLIM	++	-	-	--	-	++	-	-	+	-	-	-	+
		SEER-SEM	++	-	-	--	-	+	-	-	+	--	-	+	-
		ANN	++	-	+	+	--	++	--	++	++	+	-	+	-
		Decision Trees	++	+	+	++	-	+	-	++	++	+	++	+	+
		Rule Induction	++	+	+	++	-	+	-	++	++	-	+	+	+
		CMAC	++	-	+	+	--	-	-	++	++	-	-	-	-
		EG/GP	++	++	+	++	-	+	--	++	++	+	+	+	-
		CART+OLS	++	-	+	++	-	+	-	++	++	-	++	+	-
	Memory-based	OSR	++	+	++	++	--	+	-	++	++	+	++	-	+
		ANGEL	++	+	+	++	+	++	--	++	++	++	-	+	-
		GRACE	++	+	+	++	+	+	--	++	++	+	-	+	-
		BRACE	++	+	+	++	+	+	+	++	++	+	-	-	-
		AQUA	++	+	++	++	-	+	--	++	++	+	-	++	-
		CF	++	+	+	++	+	+	-	++	++	-	-	+	-
	Composite	CART+CBR	++	+	+	++	+	+	-	++	++	-	++	-	+
		OSR+OLS	++	+	+	++	--	+	-	++	++	-	++	-	+
		ANN+Clustering	++	-	+	+	--	+	-	++	++	-	-	+	-
		AVN	++	+	++	-	+	+	-	++	++	-	-	++	-
		ACE	++	+	++	++	+	+	--	++	++	-	-	+	-
Expert-based		Guesstimation	-	++	+	++	++	--	-	+	-	+	-	--	-
		Wideband Delphi	--	++	++	++	++	++	-	+	+	-	-	-	-
		Estimeeting	--	++	++	++	++	-	-	+	+	-	-	-	-
		AHP	-	++	++	++	+	+	-	+	+	-	-	-	+
		SBS	--	++	++	++	+	-	+	+	+	-	-	+	-
		Planning Game	--	++	++	++	++	++	+	+	+	+	-	-	+
Hybrid		COCOMO II.98	-	-	+	--	+	+	+	--	+	-	-	+	+
		CoBRA	-	+	++	++	+	++	++	++	++	+	++	++	+
		BBN	-	+	+	++	+	++	++	++	++	-	++	-	+
		ESTOR	+	-	-	++	+	-	-	++	++	-	-	-	-
		NFS	++	-	+	+	--	++	+	++	++	+	-	-	+

applications: a risk analysis and a correlated stock portfolio evaluation. In this study, the best aggregate performance index was obtained by LH sampling.

Additionally, there exist a lot of VRT that promise to improve the standard MC method. A collection of them is presented in [55]. In contrast to MC, DS, and LH, (1) these techniques require knowledge of the function to be estimated, (2) they can be used only under special circumstances, and/or (3) they require an appropriate adoption of the algorithm dependent on the function to be estimated.

Although for our context (the context of software project cost simulation), no results could be found in the literature, LH is being used successfully in other domains, for example, in the automotive industry, for robustness evaluations [109] and in the probability analysis of long-span bridges [30]. In [78], it is mentioned that standard MC often requires a larger number of samples (iterations) to achieve the same level of accuracy as an LH approach. These and other articles show us the practical applicability of LH in various domains. Therefore, we see LH sampling as a promising approach to improving the standard MC method. Nevertheless, its concrete performance improvement depends on the application area with the associated simulation equations and input distributions.

Data about LH on a more theoretical level is provided by [33]. Here, standard MC and LH sampling are compared in an experiment where two simple synthetic functions, one monotonic and one non-monotonic, are sampled with both approaches. For both functions and for 25, 50, and 100 iterations, authors showed that the samples produced by the LH approach were more stable. Additionally, Owen showed in 1997 [72] that an LH sampling with n iterations never has worse accuracy than a standard MC sampling with $n-1$ iterations.

Avramidis et al. [4] investigate various techniques to improve estimation accuracy when estimating quantiles with MC approaches. The results of this work were of special interest for us because our aim of sampling a probability function has more in common with quantile estimation (see DoP in Section 7.1.1) than with the estimation of its mean (which is the common case). Avramidis provides a theoretical proof that for the class of probability functions with a continuous and nonzero density function, the LH approach delivers at least equally accurate (or more accurate) results than the standard MC method. In a second work [5], Avramidis also provides experimental evidence that supports this conclusion in the context of the upper extreme quantiles estimation of the network completion time in a stochastic activity network.

4. Problem Statement

The problem described in the introduction is to evaluate whether more accurate and efficient techniques than standard MC exist that support software cost estimation and cost risk management. CoBRA® was selected as a representative method for this evaluation. Further, the extent of accuracy improvement above standard MC should be determined.

More precisely, we want to determine and compare the accuracy and efficiency of methods that transform the input of CoBRA® project cost simulation (i.e., the expert estimations) into the data required for the different kinds of predictions provided by CoBRA®.

In the following, fundamental research scenarios are described. Afterwards, RQ are stated based on the problem statement.

The following three scenarios cover all principle kinds of application of simulation data in the CoBRA® method. Even through they are derived from the typical application of the CoBRA® method, we see no reason why they should not be general scenarios when applying any cost estimation or cost risk analyzing methods:

AS1: The question to be answered in this scenario is: How much will the project cost? The answer should be the point estimation of the project cost, a single value that represents the average project costs. Spoken in terms of probability theory, we look for the expected value of the project costs. This value is also needed to validate the accuracy of a built CoBRA® model with the help of statistics.

AS2: The question to be answered in this scenario is: How high is the probability that the project costs stay below a given budget of X? It is possible to answer this question for any X if a probability distribution of the project costs is available. In this case, we can simply calculate P (cost < X).

AS3: The question to be answered in this scenario is: Which is a realistic budget for a given project that is not exceeded with a probability of Y? Considering the previous scenario, this is finally the inverse application of the project cost probability distribution; here, we are interested in X with P (cost < X) = Y, where Y is given.

When we compare the scenarios, we see that AS2 and AS3 make use of a variable to allow customization. Therefore, they require the knowledge of the entire cost probability distribution, whereas AS1 requires only a single characterizing value of the distribution, the expectation value. As mentioned above, the study focuses on the transformation of the simulation input (i.e., consolidated expert estimations for the different influencing factors) into the data required in the different application scenarios (AS). This is the transformation that was done till now by standard MC sampling. The *input* is a matrix $D^{(m,n)}$ of estimates of m different experts for n different cost drivers (see Fig. 7). The estimates themselves are triangular distributions that represent the expert opinion as well as the uncertainty in their answers [101]. The output for AS1 should be the (approximated) expected value; the *output* for AS2 and AS3 should be the (approximated) probability distribution for further calculations based on variable X or Y (e.g., see Fig. 8).

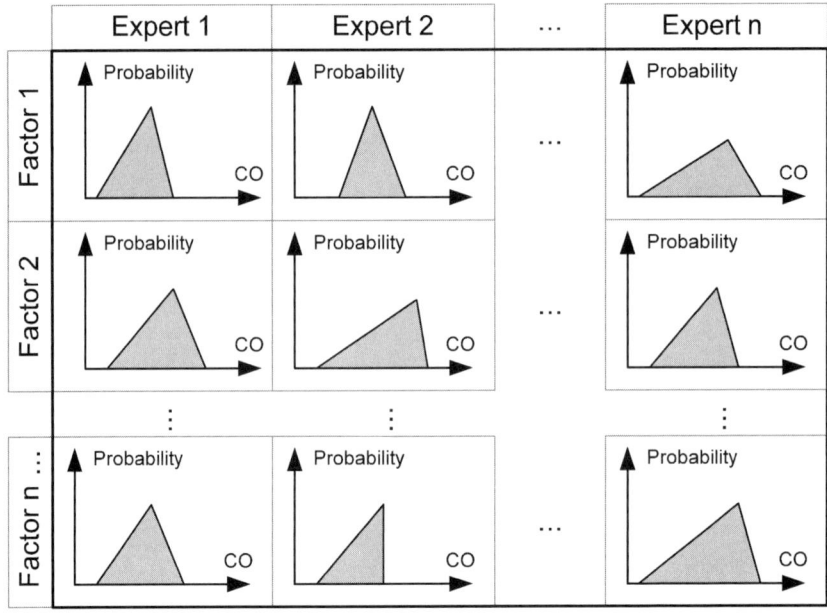

FIG. 7. Example of an expert estimation matrix $D^{(m,n)}$.

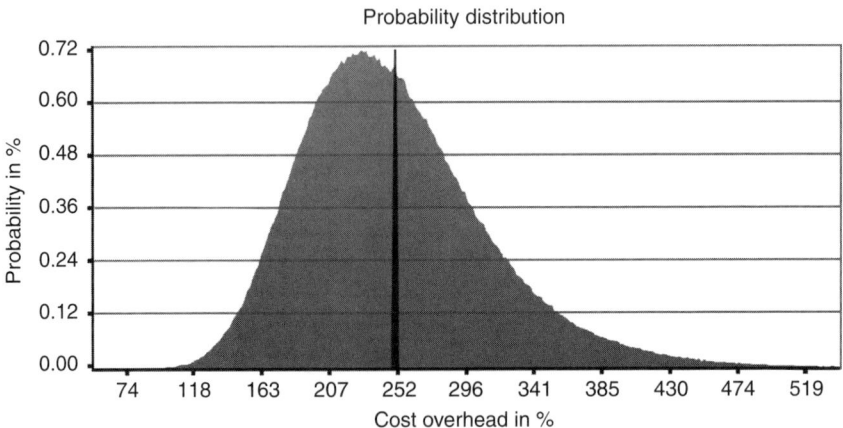

FIG. 8. Example cost overhead distribution.

On the basis of this, more detailed description of the problem and the AS that should be considered, we derived the following research questions:

RQ1: Do feasible (in terms of calculation effort) analytical solutions exist to calculate data required in AS1, AS2, and AS3?

RQ2: If data cannot be calculated analytically, do stochastic solutions exist that calculate the data used in AS1, AS2, and AS3 more accurately and more efficiently than the standard MC simulation approach?

RQ3: How high is the average accuracy and efficiency gained by choosing a certain simulation approach (e.g., LH) when compared with standard MC sampling?

In the next section, we answer RQ1 and explain why we need stochastic approaches for AS2 and AS3. Then, Section 6 presents the adoption of stochastic approaches to the context of CoBRA® and explains the need to conduct an experiment to answer RQ2 and RQ3. The planning of the experiment, its execution, and its results are described in Section 7. Finally, Section 8 summarizes the answers to all RQ (Fig. 9).

5. Analytical Approaches

In the following, we show that the data that have to be provided in application scenario AS1 (i.e., expected project costs) can be calculated analytically. However, we also show that the analytical derivation of the cost probability distribution necessary in AS2 and AS3 is not feasible with respect to the calculation effort.

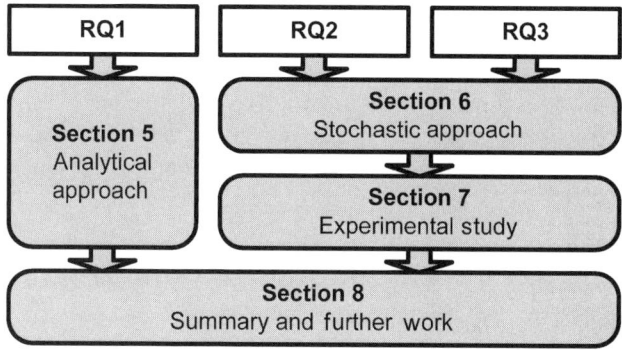

FIG. 9. Structure of chapter.

5.1 Point Estimation

To calculate the expected value analytically, in a first step, we have to derive a mathematical representation of the CO distribution D described by the $D^{(m,n)}$ matrix. Having a distribution $D_{e,v}$ that represents a distribution provided by expert e for the cost factor v, we obtain the following equation for distribution D:

$$D = \sum_{v=1}^{n} D_v \text{ with } D_v = \begin{cases} D_{1,v} \text{ with probability } \frac{1}{m} \\ D_{2,v} \text{ with probability } \frac{1}{m} \\ \cdots \\ D_{m,v} \text{ with probability } \frac{1}{m} \end{cases} \quad (9)$$

However, we are interested in the expected value (mean) E of distribution D. Assuming the linearity of the expected value and a lack of correlation between $D_1, \ldots, D_n$, we can derive the following equation:

$$E(D) = \frac{1}{m} \sum_{e=1}^{m} \sum_{v=1}^{n} E(D_{e,v}) \quad (10)$$

After considering that $D_{e,v}$'s represent the expert estimations as triangular distributions in the form $Triang\,(a_{e,v},\,b_{e,v},\,c_{e,v})$, where a, b, and c are the known minimum, most likely, and maximum cost overhead, we obtain:

$$E(D) = \frac{1}{m} \sum_{e=1}^{m} \sum_{v=1}^{n} \frac{a_{e,v} + b_{e,v} + c_{e,v}}{3} \quad (11)$$

The derivation of an algorithm that implements Equation 11 is trivial. It delivers no error, because the results are calculated analytically. The *asymptotical runtime* of the analytical mean computation is **O(m·n)**, the respective *memory usage* is **O(1)**. In practice, computation on a 12×12 test matrix consists of 432 additions, which results in non-noticeable runtime of less than 1 ms on our test machine (desktop pc).

5.2 Distribution Computation

For AS2 and AS3, the density distribution D of the expert estimation matrix $D^{(m,n)}$ is needed. In order to analytically determine a function $D(x)$ that describes the density distribution, again, we can use the mathematical representation introduced in Equation 9. The need density function of $D_v(x)$ can be computed using Equation 12, where $D_{e,v}(x)$ is the density function of the single triangular distribution.

$$D_v(x) = \frac{1}{m} \sum_{e=1}^{m} D_{e,v}(x) \qquad (12)$$

But the next step would be the calculation of a sum of nontrivial probability functions (see Equation 9). The only analytical possibility to build a sum of nontrivial probability functions is to calculate their convolution [25]. In case of probability density function D, this means to solve an $(n-1)$-multiple integral equation (Equation 13).

$$D(x) = \int_0^x D_1(x - y_1) \int_0^{y_1} D_2(y_1 - y_2) \ldots dy_1 \ldots dy_{n-1} \qquad (13)$$

This, however, is not a trivial task, especially for larger n. In principle, the integral equation for given D_v's and x might be computed using the Fast Fourier Transformation [31] or multidimensional Gaussian quadratures [97]. Yet, both approaches deliver only approximations instead of analytically proper (exact) results.

So, we cannot present an exact analytical approach to determining the probability density function for expert estimation matrix $D^{(m,n)}$. Therefore, the next logical step is to take a look at MC simulation and other competitive stochastic approaches.

6. Stochastic Approaches

This section shows the adoption of two stochastic simulation approaches to the approximation of the CO distribution as required in AS2 and AS3. The reason for choosing the first, standard MC sampling is that it has been the approach applied in CoBRA® until now. In addition, it is the simplest simulation approach known and serves as a baseline for comparison with more sophisticated simulation approaches. The second approach, LH sampling, was chosen as best competitor based on the results of the literature research (see Section 3). We implemented LH in two variants with differences in runtime and required memory. In the final subsection, we explain why an experimental approach is needed to answer RQ2 and RQ3.

6.1 The MC Approach

For AS2 and AS3, the density distribution D of the expert estimation matrix $D^{(m,n)}$ is needed. In order to obtain an approximation of D with the help of the standard MC, the algorithm follows the attempted sampling described in Section 2.2.1.

```
for i=0 to #Intervals : d[i] = 0;
for i=1 to z {
  sample = 0;
  for v=1 to n {
    sample +=
      D⁻¹[random({1,…,m})][v](random([0,1])
  }
  s = roundDown(sample);
  d[s] = d[s]+1;
}
for i=0 to #Intervals : d[i] = d[i] / z;
return d;
```

FIG. 10. MC sampling in pseudo-code.

In the algorithm, we use the unbiased estimator as defined in Equation 14 to create samples of D (Equation 9).

$$D(\omega) = \sum_{v=1}^{n} \sum_{e=1}^{m} \chi_e\left(\xi^{(m)}(\omega)\right) D_{e,v}^{-1}\left(\xi_v^{(1)}(\omega)\right) \quad (14)$$

$\xi_v^{(1)}(\omega)$ is an independent realization of the probability variable $\xi_v^{(1)}$ that is uniformly distributed on [0,1), $\xi^{(m)}(\omega)$ is a realization of $\xi^{(m)}$ that is uniformly distributed over the set $\{1,\ldots,m\}$, χ_e is the *characteristic function* of e (Equation 15), and $D_{e,v}^{-1}$ is the inverse distribution function of the triangular distribution $D_{e,v}$.

$$\chi_e(x) = \begin{Bmatrix} 1 \text{ if } x = e \\ 0 \text{ else} \end{Bmatrix} \quad (15)$$

The *asymptotical runtime* for the MC sampling is **O(z·n)** for a $D^{(m,n)}$ matrix and z iterations[4] and the respective *memory usage* for the computation is **O(#intervals)** (Fig. 10).

6.2 The LH Approach

In order to obtain the density distribution D of the expert estimation matrix $D^{(m,n)}$ with the help of LH, we make use of sampling as described in Section 2.2.2. For the purpose of calculating D (see Equation 9) with the LH approach, we first need an

---

[4] Under the assumption always valid in practice that m as well as the number of intervals is less than z.

unbiased estimator $D(\omega)$ of D. The starting point for the development of such an estimator is the realization of $D(\omega)$ presented for MC (Equation 14).

If we expect that the number of iterations (z), according to the number of strata is a multiplication of the number of experts (m), then we can optimize Equation 14 by replacing the random variable $\xi^{(m)}$ and the chi function by $1/m$ (Equation 16). This is possible because, when stratified by z, $\xi^{(m)}$ delivers (remember z is a multiplication of m) a sequence of exact values: $(z/m \cdot 1)$, $(z/m \cdot 2)$, ..., $(z/m \cdot m)$.

$$D(\omega) = \sum_{v=1}^{n}\sum_{e=1}^{m}\frac{1}{m}D_{e,v}^{-1}\left(\xi_{v}^{(1)}(\omega)\right) \qquad (16)$$

The stratification of the remaining random variable $D_{e,v}$, which is required for implementing an LH algorithm (see Section 2.2.2), is quite simple because we can construct a random number generator with the help of the inversion method for $D_{e,v}$ (triangular distribution) [101]. This means that the only random variables that must be stratified are the $\xi_{v}^{(1)}$ ($v=1\ldots n$), which are uniformly distributed on the range [0,1]. In practice, this is done by replacing $\xi_{v}^{(1)}$ by a random variable uniformly distributed between $[(\pi_{v}(i)-1)/z$ and $\pi_{v}(i)/z]$:

$$\psi_{v}(\omega,i) = \frac{\xi_{v}^{(1)}(\omega) + \pi(i) - 1}{z} \qquad (17)$$

where i is the current iteration, z the total number of iterations, and π_{v} represents random permutation over the elements 1, ..., z.

This way, we get the following estimator (Equation 18), which can be further optimized by removing $1/m$, the sum over m, and replacing e with $(i \bmod m) + 1$ (see Equation 19).

$$\xi(\omega) = \frac{1}{z}\sum_{i=1}^{z}\sum_{v=1}^{n}\sum_{e=1}^{m}\frac{1}{m}D_{e,v}^{-1}\frac{\xi_{v}^{(1)}(\omega) + \pi(i) - 1}{z} \qquad (18)$$

$$\xi(\omega) = \frac{1}{z}\sum_{i=1}^{z}\sum_{v=1}^{n}D_{(i\bmod m)+1,v}^{-1}\frac{\xi_{v}^{(1)}(\omega) + \pi(i) - 1}{z} \qquad (19)$$

The problem with implementing an estimator defined in such a way is that we need n independent permutations $p[v]$, where each $p[v][i]$ has the probability of $1/z$ of being equal to $c \in \{1, \ldots, z\}$ (*).

We have implemented two algorithms that solve this problem in different ways. The first one, LH [Fig. 11(A)], randomly chooses c from the *figures* 1, ..., z for each variable v in each iteration i. If the figure is flagged as having been chosen in an earlier iteration, a new figure is chosen; else the figure is used and flagged.

```
A
for i=0 to #Intervals : d[i] = 0;
p[*][*] = 0;
for i=1 to z {
  sample = 0;
  for v=1 to n {
    c = random(0,...,z);
    while p[v][c] == 1 do p[v][c mod z +1];
    p[v][c] == 1;
    sample +=
      D⁻¹[i mod m][v]( (random ([0,...,1])+c-1)/z)
  }
  s = roundDown(sample);
  d[s] = d[s]+1;
}
for i=0 to #Intervals : d[i] = d[i] / z;
return d
```

```
B
for i=0 to #Intervals : d[i] = 0;
for v=1 to n {
  for i=1 to z {
    p[v][i] = i;
  }
  for i=1 to z {
    c = random(1,...,z);
    d = p[v][i];
    p[v][i] = p[v][c];
    p[v][c] = d;
  }
}
for i=1 to z {
  sample = 0;
  for v=1 to n {
    sample += D⁻¹[i mod m][v](
      (random ([0,1])+p[v][i]-1)/z )
  }
  s = roundDown(sample);
  d[s] = d[s]+1;
}
for i=0 to #Intervals : d[i] = d[i] / z;
return d
```

FIG. 11. Sampling in pseudo-code: (A) LH, (B) LHRO.

For a $D^{(m,n)}$ matrix and z iterations,[5] the *asymptotical runtime* of the LH sampling algorithm is $O(z^2 \cdot n)$, and *memory usage* for the computation is $O(z \cdot n)$.

The second algorithm, LHRO [Fig. 11(B)], represents a runtime-optimized (RO) version. It creates all needed permutations $p[v]$ before the sampling part of the algorithm starts and any iteration is executed. To do this, it permutes for each variable v all numbers from $1, \ldots, z$ at random and stores the result. The proof that the permutations obtained satisfy (*) is a simple induction over z.

Computational complexities of the LHRO for a $D^{(m,n)}$ matrix and z iterations[6] are *asymptotical runtime* of $O(z \cdot n)$, and *memory usage* of $O(z \cdot n)$.

Please note that although both algorithms store $O(z \cdot n)$ values, their actual storage differs in that LH stores $z \cdot n$ flags (one bit each) and LHRO stores $z \cdot n$ integers (32 bits each).

---

[5] On the basis of the assumption always valid in practice that m as well as #*intervals* are less than z^2.
[6] On the basis of the assumption always valid in practice that m as well as #*intervals* are less than z.

6.3 Comparison of Stochastic Algorithms

When considering the MC and LH algorithms, we are unable to determine accuracy and efficiency regarding AS2 and AS3 based merely on theoretical considerations; neither can we say which one is more accurate or which one is more efficient in our context.

This is because we are not aware of any possibility to show that the errors of the algorithms based on the LH approach are lower for typical expert estimation matrices in CoBRA® application. Especially, we do not know, if the LH error is lower, to which extent it is lower compared with the error of MC.

Additionally, if considering efficiency, for example, as runtime efficiency, we have to execute the algorithms to measure the runtime. The runtime depends, among other things, on the chosen implementation language and execution environment and cannot be calculated theoretically only based merely on the pseudo-code representations.

But most notably, MC and LH are stochastic approaches, since their results suffer from a different degree of error, which not only depends on input data and number of iterations, but also on pure randomness. The same statement is true for the runtime. Therefore, in order to draw reliable conclusions regarding RQ2 and RQ3, we conduct an experiment where we compare the different approaches/algorithms in a typical CoBRA® application context.

7. Experimental Study

7.1 Experimental Planning

During experimental planning, we first define the constructs accuracy and efficiency we want to measure and use for comparing the sampling approaches. Next, we describe the independent and dependent variables relevant for our experiment and set up some hypotheses to answer RQ2. Then we present the experimental design we use to check our hypothesis and answer RQ2 and RQ3.

7.1.1 Construct: Accuracy

In order to quantify the accuracy and efficiency of the simulation techniques evaluated in the study, appropriate metrics had to be defined first. In this section, we start with the quantification of the accuracy construct. The next section deals with the accuracy-dependent efficiency construct.

Following the goal/question/metric paradigm [6], we had to consider the object of the study (i.e., the simulation methods), the purpose (i.e., characterization and comparison), the quality focus (accuracy), the perspective (CoBRA® method user), and the context. Whereas most of them are straightforward to define in our setting, we can see that two different *contexts* exist, which are described by application scenarios AS2 and AS3 (see Section 4). AS1 is not relevant at this point because it is covered by the analytical solution presented in the previous section.

But before we derived an accuracy measure for each of the two AS, we had to solve a general problem in our setting: In order to measure the accuracy of the simulation output, we need the expected output to perform some kind of comparison between actual and expected output. However, the computation of an analytic correct output distribution is not possible for a complex distribution within an acceptable period of time (see Section 5.2). So instead, we have to use an approximation with a known maximal error regarding applied measures. In the following, when we speak of *reference data*, we mean the approximation of expected output based on the approximated expected distribution (called *reference distribution*). We obtain our reference distribution by generating distributions with MC and LH algorithms and a very large number of iterations in order to get extremely high precision. The distributions obtained in this way were next compared to each other with respect to the same measures that were used in the experiment. The magnitude of obtained errors determined the upper limit of accuracy that can be measured in the experiment. It means that a derivation below the identified limit is practically not possible to observe and thus considered not to be significant.

We introduced two major types of error measure to compare the results of the sampling algorithms considered: *Derivation over Cost Overhead* (DoCO) and *Derivation over Percentiles* (DoP).

The *DoCO* measure has been designed to calculate the deviation between the sampled distribution and the reference distribution with respect to AS1, where we ask how probable it is to be below a certain cost (overhead) limit X, more formally: $P(X \leq CO)$. The DoCO value represents the derivation (absolute error) of $P_S(X \leq CO)$ calculated for the sampled distribution from the expected $P(X \leq CO)$, averaged over all meaningful CO values.

$$\text{DoCO} = \frac{1}{(\text{MaxCO} - \text{MinCO})} \times \sum_{CO=\text{MinCO}}^{\text{MaxCO}-1} |P_S(X < CO) - P_R(X < CO)| \quad (20)$$

where MaxCO = max {maximum CO of D_S with probability greater than 0, maximum CO of D_R with probability greater than 0}, and $MinCO$ = min {minimum CO of D_S with probability greater than 0, minimum CO of D_R with probability greater than 0}. $P_S(X < CO)$ returns the accumulated probability of the sampled

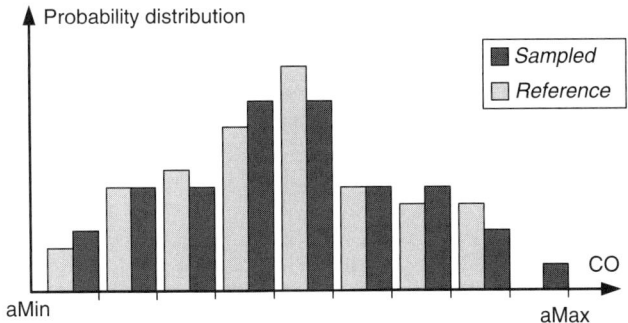

FIG. 12. Probabilities of cost overhead values of sampled and reference distribution.

distribution for all CO values less than CO, and $P_R(X < CO)$ returns the accumulated probability for the reference distribution for all CO values less than CO.

This measure is meant to give more importance to errors spread over a wider range of the distribution than to errors that have 'local' impact on a narrow range of the distribution. This property of DoCO is important because narrow local errors appear less frequently when determining the probability for a given CO interval. Consider, for example, the last two intervals in Fig. 12. The deviations between sampled and reference distributions (error) in each of the two intervals have the same magnitude but opposite signs, so that they compensate each other (total error over both intervals is equal to zero). Therefore, considering any larger interval that contains both or none of them would not be affected by derivations (errors) measured separately for each interval. Furthermore, DoCO has the advantage of being independent of errors caused by the data model because it uses the same interval boundaries as the data model, whole numbers of CO unit (short, %CO).

The *DoP* measures were derived from the practice described in AS3: For instance, we might want to estimate the software cost (or cost overhead) that has a certain probability of not being exceeded. More formally, we search for the CO value with a certain probability $P(X \leq CO) = p/100$ associated to it, where $p \in N, p \in [1,99]$. With DoP we capture the *average absolute error* over all percentiles[7] (Equation 21)

---

[7] The *Percentile* of a distribution of values is a number x_p such that a percentage p of the population values is less than or equal to x_p. For example, the 25th percentile of a variable is a value (x^p) such that 25% (p) of the values of the variable fall below that value.

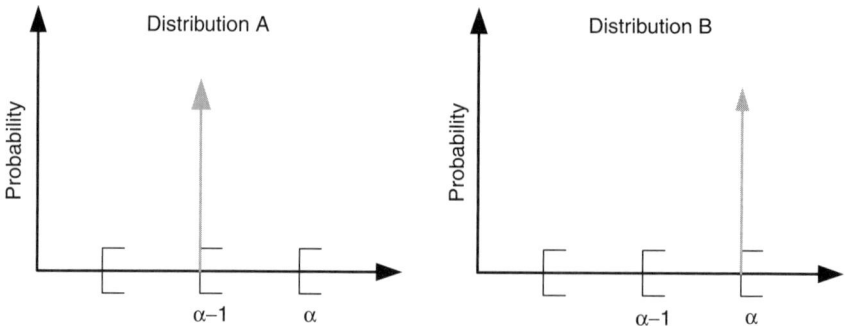

FIG. 13. Example of two distributions with the same sampling data, but different mean (D_R).

as well as the *maximum absolute error* that is reached in a single percentile (Equation 22):

$$\text{DoP}_{\text{abs}} = \frac{1}{99} \times \sum_{p=1}^{99} |\text{CO}_S(p) - \text{CO}_R(p)| \quad (21)$$

$$\text{DoP}_{\text{max}} = \max\left\{ \bigcap_{p=1}^{99} |\text{CO}_S(p) - \text{CO}_R(p)| \right\} \quad (22)$$

where $\text{CO}_S(p)$ is the CO value (CO_S) with $P_S(X \leq \text{CO}_S) = p/100$ and $\text{CO}_R(p)$ is the CO value (CO_R) with $P_R(X \leq \text{CO}_R) = p/100$, where $P_S(X \leq \text{CO}_S)$ is the accumulated probability of the sampled distribution for all CO values less than or equal to CO_S, and $P_R(X \leq \text{CO}_R)$ is the accumulated probability of the reference distribution for all CO values less than or equal to CO_R.

The disadvantage of DoP is that the data model we used to store the sampling data[8] can theoretically cause an error of 0.5% in CO (absolute error of 0.5). To explain this, let us consider two hypothetical distributions (Fig. 13):

- (A) $P_A(\alpha - 1\% \text{ CO}) = \infty$ ($\alpha \in N$) and $P_A(x) = 0$ for $x \neq (\alpha-1\% \text{ CO})$,
- (B) $P_B(\alpha - 0\% \text{ CO}) = \infty$ and $P_B(x) = 0$ for $x \neq (\alpha - 0\% \text{ CO})$.

In our data model, both distributions have the same representation: $P([0, 1\% \text{ CO}])= 0, \ldots, P([\alpha - 1\% \text{ CO}, \alpha \% \text{ CO})) = 1, P([\alpha \% \text{CO}, \alpha + 1\% \text{ CO})) = 0, \ldots$.

---

[8] To store the results of the cost overhead distribution sampling, we used a data model in which every distribution interval is represented by a numeric value. Each value represents the probability that a cost overhead value contained in the respective interval occurs. Each of the half-opened intervals covers the same range: one unit of cost overhead (1% CO). Intervals with a probability value of zero are redundant and need not be stored in the model.

Using the data from the data model, we would calculate for both distributions $CO_A(1) = CO_B(1) = \ldots = CO_A(99) = CO_B(99)$, whereas in fact, distribution A should have $CO_A(1) = \ldots = CO_A(99) = \alpha - 1\%$ CO and B $CO_B(1) = \ldots = CO_B(99) = \alpha$ %CO. Therefore, in the worst case, the minimal error of the DoP measures that can be reached is 0.5 of the CO unit (i.e., 0.5% CO).

With the understanding of derivation (error) being the opposite of accuracy, we define *accuracy* of the output as the inverse of the measured derivation regarding AS2: Accuracy=$(DoCO)^{-1}$.

We used DoCO and not DoP to calculate accuracy because DoCO is, contrary to the DoP measures, independent of the imprecision of the sampling data model and therefore allows higher accuracy concerning the calculated derivation.

7.1.2 Construct: Efficiency

From the perspective of the CoBRA® method user, who wants to perform cost estimation or cost risk analysis, the accuracy of the result and the computation effort needed are the key factors that define efficiency. Therefore, following the goal/question/metric paradigm, *efficiency* is considered as accuracy per computation effort.

On the one hand, we can define *computation effort* as a measure of computational expense, measured by the number of executed iterations;[9] on the other hand, we can define computation effort as the execution time of the simulation algorithm in a typical execution environment, measured in milliseconds. Here, the number of iterations is independent of the development language and execution environment; measuring the execution time takes into consideration that the runtime of a single iteration can be different in different simulation approaches. Therefore, we defined two efficiency measures: *accuracy per iteration* (A/I) and *accuracy per millisecond* (A/ms).

$$A/I(D_S) = \frac{(DoCO_{abs})^{-1}}{\# \text{ Iterations}} \qquad (23)$$

$$A/ms(D_S) = \frac{(DoCO_{abs})^{-1}}{\text{execution\_time}} \qquad (24)$$

---

[9] The simulation consists of many recalculation cycles, so-called *iterations*. In each iteration, cost overhead values from triangular distributions across all cost drivers are sampled and used to calculate one possible total cost overhead value (according to the CoBRA® causal model).

where #Iterations is the number of iterations performed and excecution_time is the runtime of the algorithm in milliseconds.

7.1.3 Independent and Dependent Variables

The variables that are controlled in this experiment (*independent variables*) are the chosen simulation algorithm, the number of iterations executed, the execution environment, as well as the simulation input (i.e., the matrix with the expert estimations). We control these variables, because we expect that they influence the measured *direct dependent variables* we are interested in: the simulation result (i.e., the sampled CO distribution) and the execution time of the simulation algorithm. On the basis of the sampled distribution, we can calculate the derivation regarding the reference distribution (error) with the DoCO and DoP measures, and with our accuracy definition being based on DoCO. Knowing the accuracy, we can calculate efficiency regarding the number of iterations as well as regarding the required runtime. Therefore, our *indirect dependent variables* are DoCO, DoP, accuracy, accuracy per iteration (A/I), and accuracy per millisecond (A/ms) (Fig. 14).

7.1.4 Hypotheses

On the basis of the previous definition of variables and the RQ stated in Section 4, we have the following hypotheses and corresponding 0-hypotheses (which we want to reject):

H_A: The LH approach provides more accurate results (regarding DoCO) than the standard MC approach applied with the same number of iterations ($\geq 10,000$).

H_{A0}: The LH approach provides less or equally accurate results (regarding DoCO) than the standard MC approach applied with the same number of iterations ($\geq 10,000$).

H_B: The LH approach is more efficient (per iteration performed) than the standard MC approach when applied with the same number of iterations ($\geq 10,000$).

H_{B0}: The LH approach is less efficient than or as efficient as (per performed iteration) the standard MC approach when applied with the same number of iterations ($\geq 10,000$).

H_C: The LH approach is more efficient (per millisecond of runtime) than the standard MC approach when running in the same execution environment and with the same number of iterations ($\geq 10,000$).

H_{C0}: The LH approach is less efficient than or as efficient as (per millisecond of runtime) the standard MC approach when running in the same execution environment and with the same number of iterations ($\geq 10,000$).

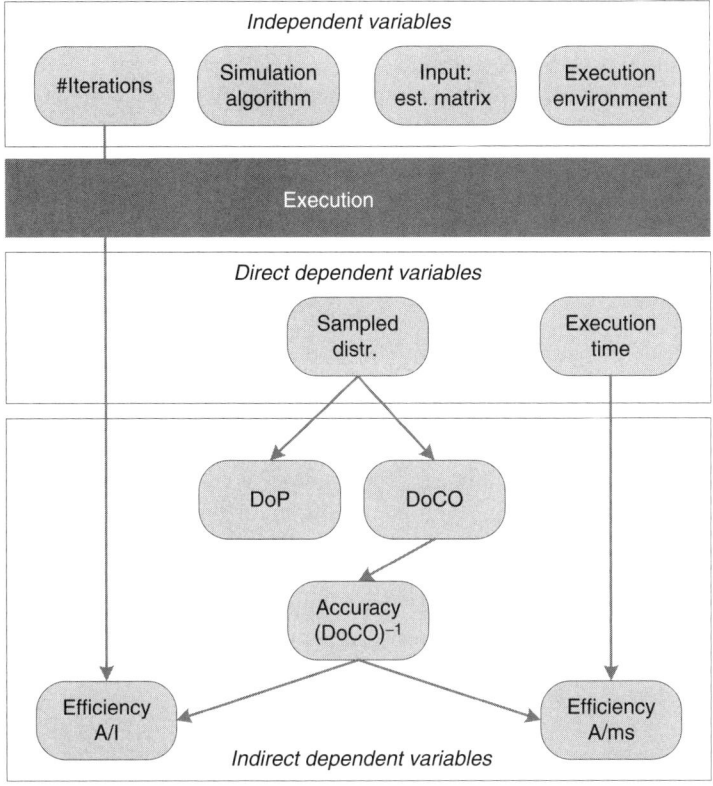

FIG. 14. Relation between (direct/indirect) dependent and independent variables.

There is a hypothesis that would be even more interesting to consider than H_C: *The LH approach is more efficient (per millisecond of runtime) than the standard MC approach when running in the same execution environment and produces a result of the same accuracy (regarding DoCO).* But this hypothesis would be hard to investigate, since receiving enough LH and MC data points with the same result accuracy would be very difficult.

7.1.5 Experimental Design

The experimental design is based on the previously stated hypotheses and on the defined dependent and independent variables. In the following, subjects and objects involved in the experiment are presented; then, the chosen experiment design and statistical tests for checking the stated hypotheses are described.

Subjects: In our setting, the methods that should be compared to each other are performed by a computer since they are algorithms; consequently, there exist no subjects in the narrow sense. But following the statistical theory used in experimentation, the randomness (ω) required in the algorithms, visible as different instantiations of a random variable, and leading to different outcomes, can be seen as a 'perfect subject.' Perfect, because it is representative (perfectly random-sampled from the totality), available without any noticeable effort (simply generated by a computer as a set of pseudo-random numbers), and really independent of any other 'subject.'

Object: In order to get a representative study object for the context of cost estimation (i.e., realistic simulation algorithms input), we decided to use data from a real CoBRA® project. For that purpose, we employed data gained during the most recent industrial application of the CoBRA® method, which took place in the context of a large Japanese software organization [97]. The input data consisted of measurement data (size, effort) and experts' assessments (cost drivers, effort multipliers) collected across 16 software projects. The resulting simulation input was a 12 × 12-triangular distribution matrix, where the 144 triangular distributions represent estimates provided by twelve experts for twelve cost drivers. This matrix size is at the upper level when compared to other CoBRA® projects known to the authors (Fig. 15).

Since the computer can generate without noticeable effort any number of pseudo-random numbers, we have as many 'subjects' as we need. Therefore, we decided to conduct a *multi-test within object study* (i.e., we have multiple subjects, but only one object), where each subject (as a concrete instantiation of used random variables) is used only in one test, which means we have a *fully randomized design*. The factor

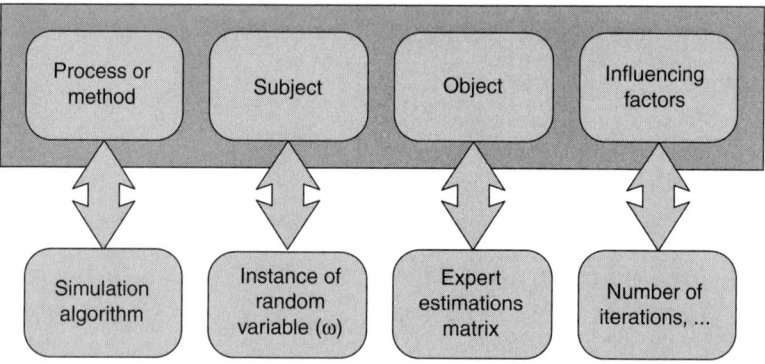

FIG. 15. Mapping of the terms: method, subject, object, and influencing factors for our experiment setting.

TABLE II
BLOCK DESIGN: DISTRIBUTION OF TESTS OVER SIMULATION ALGORITHM
AND NUMBER OF ITERATIONS

	Simulation Algorithms		
Iterations	MC	LH	LHRO
10 240 000	1, 2, 3	4, 5, 6	7, 8, 9
2 560 000	10, 11, 12	13, 14, 15	16, 17, 18
640 000	19, 20, 21	22, 23, 24	25, 26, 27
160 000	28, 29, 30	31, 32, 33	34, 35, 36
40 000	37, 38, 39	40, 41, 42	43, 44, 45
10 000	46, 47, 48	49, 50, 51	52, 53, 54

whose influence should be observed is the chosen simulation algorithm with the *three treatments*: standard MC, LH, and Latin Hypercube Runtime Optimized (LHRO). Since the number of iterations influenced the measured dependent variables accuracy and execution time, we chose a *block design* with six blocks with respect to the number of iterations performed: 10k, 40k, 160k, 640k, 2.56M, and 10.24M. Per treatment and block, three tests were performed and the median regarding the used derivation measure was selected. This was done to avoid outliers with respect to measured dependent variables (DoCO and DoP); the median was chosen because it is more robust than the mean (Table II).

To confirm our hypothesis, we conducted one-sided *Sign Tests* between the MC and both LH approaches for each of the presented hypotheses. In addition, we performed one-sided *paired t-tests* on the normalized[10] DoCO values (i.e., inverse accuracy) to provide evidence of the accuracy improvement by LH (H_A). The stricter *t*-test is possible at this point, since we consider simulation errors (which can be assumed to follow a normal distribution).

7.2 Experimental Operation

The experiment was conducted as prescribed in the experiment design. The investigated simulation algorithms were implemented in Java within the existing CoBRA® software tool. Tests were performed on a Desktop PC with Windows XP (SP2), AMD AthlonXP3000+, and 1GB DDR-RAM.

---

[10] DoCO values are normalized by multiplying them with the square of iterations performed (i.e., the expected reduction of derivation).

For each trail, we let our test system compute the approximated distributions and stored them for further analyses. In addition, the simulation runtime in milliseconds was kept. *Problems* that occurred during operation were that tests 4–6 could not be performed due to extreme computation time (we stopped after 1 h) and tests 7–9 as well as tests 16–18 were not performable due to stack overflows (memory problem).

The resulting distributions were used to calculate the corresponding DoCO and DoP values, ending up with the following tuple for each block of our test design (triple of tests):

- Median DoCO, corresponding simulation runtime
- Median DoP_{abs}, corresponding simulation runtime
- Median DoP_{max}, corresponding simulation runtime

The first kind of tuple was then used to get the needed accuracy (A) and efficiency values (A/I, A/ms) for each block.

7.3 Experimental Results

In this section, we present and discuss the results of our experiment. We check the stated hypotheses and compare the simulation approaches regarding accuracy and efficiency. The algorithms considered are the previously introduced standard MC sampling (MC), LH sampling, and the runtime optimized version of LH sampling (LHRO).

7.3.1 Accuracy of Simulation Algorithms

First, we look at the accuracy of the simulation approaches, more precisely at the DoCO as inverse accuracy. In addition, we preset the DoP results. Even if they are not used to calculate and compare the accuracy of the algorithms, they give some hints about the magnitude of the expected error when applying the different approaches (Tables III–V).

Discussion: Considering Table III, one can see that both the LH and the LHRO algorithms have comparable accuracy (regarding DoCO) for a given number of iterations. This is not surprising because both are based on the general LH sampling approach and both differ only with respect to the way they were implemented. The LH algorithms outperform the MC algorithm in *accuracy* in any given number of iterations. The relative accuracy gain with respect to MC varies for LH (mean 0.60) and LHRO (mean 0.63) around 0.6; this means a 60% improvement of accuracy of LH approaches over MC. The number of iterations seems to have no noticeable influence on this.

TABLE III
COMPARISON OF DERIVATION OVER COST OVERHEAD

Iterations	DoCO		
	MC	LH	LHRO
10 240 000	3.69E−05	–	–
2 560 000	8.13E−05	5.50E−05	–
640 000	2.19E−04	1.36E−04	1.75E−04
160 000	4.17E−04	2.66E−04	1.81E−04
40 000	8.61E−04	4.97E−04	4.92E−04
10 000	1.77E−03	1.11E−03	1.42E−03

TABLE IV
COMPARISON OF AVERAGE DERIVATION OVER PERCENTILES

Iterations	DoP Average		
	MC	LH	LHRO
10 240 000	2.34E−02	–	–
2 560 000	3.85E−02	3.31E−02	–
640 000	2.84E−01	9.42E−02	1.04E−01
160 000	1.49E−01	2.07E−01	1.03E−01
40 000	6.55E−01	2.34E−01	2.93E−01
10 000	6.73E−01	5.62E−01	8.82E−01

TABLE V
COMPARISON OF MAXIMUM DERIVATION OVER PERCENTILES

Iterations	DoP_{max}		
	MC	LH	LHRO
10 240 000	7.09E−02	–	–
2 560 000	4.03E−01	2.15E−01	–
640 000	4.90E−01	7.09E−01	8.56E−01
160 000	5.21E−01	6.28E−01	5.19E−01
40 000	1.22E+00	1.95E+00	2.28E+00
10 000	4.09E+00	2.36E+00	4.28E+00

Hypothesis testing: We check our hypothesis H_A with a sign test and can reject the null hypothesis at an α level of .05 for LH; performing a paired Student's t-test on normalized DoCO values, we can reject the null hypothesis for both implementations, LH and LHRO. Therefore, we can conclude that the LH approach delivers better results than the standard MC approach in the context of our application scenario and the range of 10 000 to 2 560 000 iterations (Table VI).

7.3.2 Efficiency of Simulation Algorithms

After we considered the accuracy of the different algorithms, we next looked at their efficiency. Efficiency is measured as accuracy per iteration as well as accuracy per millisecond of runtime (Table VII). In the following, we discuss the results obtained and check our efficiency-related hypotheses.

Discussion: The efficiency index *accuracy per iteration* shows that the LH algorithms outperform the MC algorithm. The relative accuracy per iteration gain shows the same picture as the accuracy improvement calculated for LH (mean 0.60) and LHRO (mean 0.63). This means around 60% improved A/I of the LH approaches over MC. An interesting aspect, which we can observe in Table VII, is that I/A decreases at the factor 2 when the number of iterations is quadrupled by any of the three sampling approaches. Thus, it can be concluded that the DoCO error

TABLE VI
TEST RESULTS REGARDING HYPOTHESIS H_A

Hypothesis H_A	LH	LHRO
Sign test	$p = 0.031$	$p = 0.063$
Paired Student's t-test	$p < 0.001$	$p = 0.027$

TABLE VII
COMPARISON ACCURACY: PER ITERATION (A/I) AND PER MILLISECOND (A/MS)

	A/I			A/ms		
Iterations	MC	LH	LHRO	MC	LH	LHRO
10 240 000	2.65E−03	–	–	5.50E−01	–	–
2 560 000	4.80E−03	7.10E−03	–	9.99E−01	3.48E−02	–
640 000	7.14E−03	1.15E−02	8.94E−03	1.46E+00	1.06E−01	1.07E+00
160 000	1.50E−02	2.35E−02	3.45E−02	3.12E+00	3.86E−01	4.17E+00
40 000	2.90E−02	5.03E−02	5.08E−02	6.21E+00	1.03E+00	4.20E+00
10 000	5.66E−02	9.02E−02	7.03E−02	1.21E+01	4.44E+00	1.50E+01

decreases for all algorithms at the rate of $\Theta(\text{\#iterations}^{-1/2})$, confirming our expectation based on our theoretical results (Equation 35).

The picture changes when looking at Table VII, which presents the *accuracy per millisecond* results (A/ms). The accuracy-per-iteration advantage of the LH approaches is leveled down by the additional effort for creating the needed random and independent permutations. The precision per millisecond values of the MC and the LHRO algorithms at a given number of iterations are comparable, so that the performance of both algorithms for a given number of iterations can be considered as being nearly equivalent. Both LHRO and MC outperform the LH algorithm that uses (in contrast to the LHRO) a runtime-consuming (yet memory-saving) method to create the permutations. The higher the number of iterations, the higher the gap between the A/ms value of the LH algorithm and the other algorithms. For 10 000 iterations, the A/ms efficiency of the LH algorithm is 2.7 times worse; for 2,560,000 iterations, the efficiency is worse by the factor 28.

Our conclusion that the A/ms performance of the MC and LHRO algorithm is nearly equivalent is true only if comparing them with regard to the same number of iterations. Nevertheless, for the same number of iterations, the LHRO algorithm delivers a higher accuracy than the MC algorithm. For example, the accuracy of the LHRO algorithm with 160,000 iterations is nearly equal to the accuracy of the MC algorithm with 640,000 iterations, but when comparing the corresponding A/ms values, the efficiency of the LHRO algorithm is almost twice as good. The problem is that a statistical test to prove this conclusion cannot be provided because we would need comparison data with equal accuracy for all algorithms and such data are difficult to obtain.

In order to perform a nondiscriminatory comparison between the performance of the MC and LHRO approach with respect to runtime in ms (A/ms), they have to be compared on the same level of accuracy. Because we do not have the data for MC and LHRO on the same accuracy level, the relation between the number of iterations and the achieved accuracy is considered (Fig. 16). For MC, there seems to be a linear relationship between the square root of the number of iterations and the achieved accuracy. A linear regression through the origin results in the following equation with an r^2 value of more than 0.999:

$$A_{\text{MC}} = 5.7658 \cdot \sqrt{z} \tag{25}$$

Using Equation 22, the number of iterations (z) that MC requires to reach the accuracy of the LHRO executions at 640,000, 160,000, 40,000, and 10,000 iterations can be calculated. The results are 982, 189, 918, 150, 124, 263, and 14,917 iterations. Additionally, a linear relationship between the number of iterations and the runtime of the implemented algorithm can be recognized for the implemented MC algorithm ($r^2 > 0.999$). Therefore, the runtime of MC can be calculated for 982, 189, 918, 150, 124, 263, and 14,917 iterations (i.e., for the number of iterations MC requires to

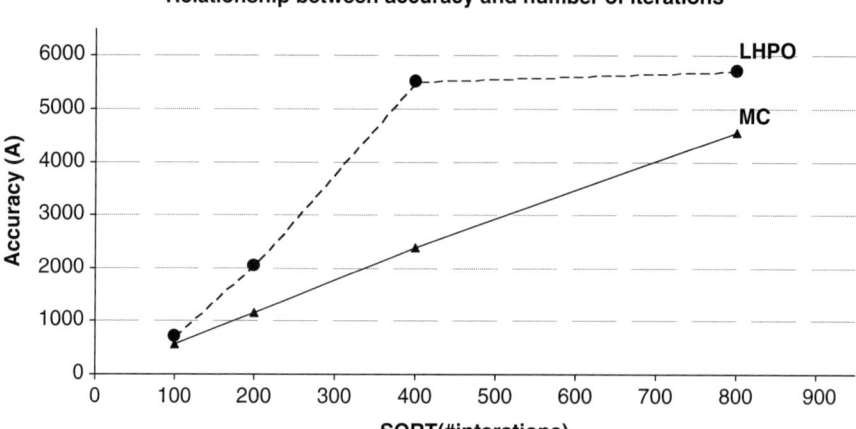

Fig. 16. Relationship between accuracy and number of iterations.

TABLE VIII
TEST RESULTS REGARDING HYPOTHESES H_B AND H_C

Hypothesis H_B	LH	LHRO
Sign test	$p = 0.031$	$p = 0.063$
Hypothesis H_C	LH	LHRO
Sign test	$p = 0.969$	$p = 0.500$

reach the same accuracy as LHPO at 640,000, 160,000, 40,000, and 10,000 iterations). Comparing these runtimes with corresponding runtimes of LHRO shows an average reduction of runtime through LHPO of ~30%, equivalent to the efficiency (A/ms) improvement of ~77% (mean).

Hypothesis testing: We checked our hypotheses H_B and H_C with a sign test at an α level of 0.05, but we could only reject H_{B0} for LH. Therefore, based on the limited number of data points, we can only conclude that the LH approach is more efficient regarding the number of iterations (A/I) (Table VIII).

7.4 Validity Discussion

First, we discuss typical threats to validity that are relevant for our experiment. Next, we present upper bound estimation for average simulation error with respect to application scenario AS2 (DoCO) and the dependent accuracy measure.

7.4.1 Threats to Validity

In this section, we consider possible threats to validity based on the experiment design and its operation. We present them in the categories proposed by Wohlin et al. [105]: First, conclusion validity, which concerns the problem of drawing a wrong conclusion from the experiment outcome; then we look at internal validity, which is threatened by uncontrolled influences distorting the results. Next, construct validity is addressed ('Do we measure what we propose to measure?'), and finally, external validity is regarded, which describes how well results can be generalized beyond our experimental setting.

Conclusion validity:

- The sign test, which is primarily used in our experiment, is one of the most robust tests. It makes no assumption beyond those stated in H_0. The paired Student's t-test is only used when comparing normalized simulation errors that can be assumed to be normally distributed.
- One problem is the low statistical power of the sign tests that results from the low number of pairs that could be compared. When testing LHRO against MC, for example, we could compare only four instead of the expected six pairs, since we were not able to execute tests 7, 8, 9, 16, 17, and 18, which result in stack overflows due to their memory requirements.

Internal validity:

- The chosen implementation in Java and the execution environment influence the runtime of the algorithms. Therefore, other implementations executed in different environments can result in different runtime and thus efficiency. To keep the results of different algorithms and iteration numbers comparable, all algorithms are implemented in the same programming language and executed on the same system. But we cannot assure that each algorithm is implemented in the most runtime-efficient way possible.
- The measures derived for AS3 (DoP) suffer from measurement inaccuracy as a result of the data model. Therefore, the algorithms are compared based on the measure derived for AS2 (DoCO), which is independent of this bias.
- The used stochastic simulation algorithms have no deterministic result. Therefore, there is a minor but not dismissible possibility of getting a very unusual result if executing a simulation. In order to avoid the presence of such outliers in analyzed data, we execute three trails (i.e., tests) for each setting and take the median.
- The reference distribution applied to measure the accuracy of the simulation results is not the analytically correct output distribution. Instead, we have to use an approximation with a known error regarding applied measures. The reason is

that computation of an analytically correct output distribution is not possible for a complex distribution within an acceptable period of time. For a more detailed discussion, see Section 7.1.

Construct validity:

- The chosen constructs that represent accuracy and efficiency are selected carefully and based on the understanding of accuracy and efficiency in typical AS of cost estimation and cost risk analysis. But we provide no empirical evidence that this is true for the majority of users.

External validity:

- The number of experts and variables as well as the characteristics of input distributions can influence the accuracy and performance of the algorithm. These input data (represented by the distribution matrix) are different in different CoBRA® AS. We performed the experiment only for one set of input data; therefore, generalizability of results (i.e., representativity of input data) is an issue. To reduce this threat as much as possible, input data was taken from an actual industrial application. The input data have a complexity in the number of experts and variables at an upper, but not untypical, level compared with other known applications.

7.4.2 Analytical Considerations

As previously mentioned, based on theoretical considerations, we cannot decide whether LH is more accurate and efficient than MC in our context. But we can give a lower boundary for the expected accuracy (regarding DoCO) that is true for both MC and LH. Such a lower boundary for accuracy allows us, on the one hand, to perform a simple validity check on the measured accuracy values; on the other hand, it can increase the external validity of our experiment results.

To derive this lower boundary, we first consider the MC algorithm. When we look at a single *individual interval* I_s, it appears that for this interval, the algorithm does nothing else than estimate the result of the integral of the density function $D(x)$ over this interval (Equation 26) with the standard MC integration [50].

$$I_s = \int_s^{s+1} D(x) dx \qquad (26)$$

Since interval size is one (measured in percentage of cost overhead), I_s delivers the average density of D on the interval $[s, s+1]$. This is interesting because the error of

SIMULATION TECHNIQUES FOR HYBRID COST ESTIMATION 165

the whole distribution can now be expressed with the help of the errors of the I_s estimators η. Considering our algorithm (Fig. 10), the estimation of I_s is calculated as follows:

$$E_s(z) = \frac{1}{z}\sum_{i=1}^{z} \eta_i, \text{ with } \eta_i = \chi_{[s,s+1]}(D(\omega_i)) \quad (27)$$

$D(\omega_i)$ is the i-th independent realization of the random variable $D(\omega)$ with distribution D and $\chi_{[s,s+1]}$ is the characteristic function for the interval $[s,s+1]$ (Equation 28).

$$\chi_{[s,s+1]}(x) = \begin{cases} 1 \text{ if } x \in [s, s+1] \\ 0 \text{ else} \end{cases} \quad (28)$$

This estimator is *unbiased*, meaning $\lim_{z \to \infty} E_s(z) = I_s$. This is clear when we recall that the expected value is the ratio of $|D_S|$ to $|D|$, and that $|D| = 1$ (D is a probability density distribution).

Now, in order to calculate the *expected error* for z iterations with Equation 4, we need the standard deviation (or variance) of our estimator. Having the variance definition (Equation 29) and $E(\eta) = E(\eta^2) = I_s$, we derive upper bounds for estimator variance (Equation 30).

$$V(\eta) = E(\eta^2) - E(\eta)^2 \quad (29)$$

$$V(\eta) = I_s - I_s^2 \leq I_s \leq 1 \quad (30)$$

The *expected error* for z iterations can be calculated with Equation 4 as follows (Equation 31):

$$\mathrm{EAR}_{I_s} \leq \frac{\sqrt{I_s}}{\sqrt{z}} \quad (31)$$

Example: If we have an average value of 0.01 over the interval $[s,s+1]$ and run 10 000 iterations, we would get an EAR_{Is} of $0.1/100 = 0.001$ (corresponding to a relative error of 10%).

In addition, Equation 31 can be used to provide a theoretical upper boundary estimation of *Deviation over Cost Overhead* (DoCO) measure (Equation 20). The term $|P_s(X < i) - P_R(X < i)|$ in the DoCO measure represents nothing else than the absolute derivation of the integral of distribution D over the interval $[0,i]$ (I_i). Expressing I_i with the intervals over $[s,s+1]$ (Equation 32) and reusing Equation 31 allows us to obtain a final estimation of the expected error estimating I_i and DoCO value (Equation 33 and Equation 34, respectively), when executing z iterations.

$$I_i = \sum_{s=0}^{i-1} \int_{s}^{s+1} D(x)\mathrm{d}x = \sum_{s=0}^{i-1} I_s \quad (32)$$

$$\mathrm{EAR}_{I_i} \leq \frac{\sqrt{\sum_{s=0}^{i-1} I_s}}{\sqrt{z}} \qquad (33)$$

$$\mathrm{DoCO} \leq \frac{1}{\sqrt{z} \cdot \#\mathrm{Intervals}} \times \sum_{i=1}^{\#\mathrm{Intervals}} \sqrt{\sum_{s=0}^{i-1} I_s} \qquad (34)$$

At a given number of iterations and intervals, DoCO becomes maximal if the double sum in Equation 34 becomes maximal. This, in turn, happens when I_1 is 1 and all other I_s are equal to 0 (consider that the sum of all I_s = |D| = 1).

$$\mathrm{DoCO} \leq \frac{1}{\sqrt{z}} \text{ and Accuracy} \geq \sqrt{z} \qquad (35)$$

Equation 35 is a weak assessment, but it is sufficient for our purpose. A stronger assessment (dependent on #Intervals) is possible regarding the exact variance of I_s.

Example: If we sample a distribution D with MC and 10,000 iterations, we get an expected DoCO value of less than 0.01.

The expected error of the LH approach can be estimated with the same equations as those used by the standard MC approach. In particular, Equation 35 holds for the expected DoCO value. The reason is that the LH approach is a special case of the standard MC approach and converges at least as fast as the MC approach (see Section 2.2.2).

8. Summary

In this chapter, we investigated the potential impact of simulation techniques on the estimation error of CoBRA®, as an exemplary cost estimation and cost risk analyzing method. This was done by deriving an analytical solution for point estimations and performing an experiment with industrial project data [97]. This experiment was necessary to compare the stochastic simulation approaches when applied to the sophisticated analysis of project costs. We could answer the question of whether there exist techniques that deliver more accurate results than simple MC sampling or performing the simulation in a more efficient way. In addition, we present results regarding the magnitude of accuracy and efficiency improvement that can be reached through the use of LH as an alternative sampling technique. The summarizing presentation of results is structured according to the three detailed RQ defined in Section 4 and the context of the three different AS; see Table IX.

TABLE IX
ANSWERED RESEARCH QUESTIONS

Application Scenario	RQ1 Analytical Feasibility	RQ2 Comparison: Accuracy and Efficiency	RQ3 Magnitude of Improvement
AS1	Yes	*Not relevant*	*Not relevant*
AS2	No	H_{A0} rejected (accuracy A)	~60% accuracy (A) and
		H_{B0} rejected (efficiency A/I)	~77% efficiency (A/ms)
		H_{C0} *not* rejected (efficiency A/ms)	gain by LHPO
AS3	No	–	–

With respect to RQ1, we showed in Section 5 that analytical mean computation (AS1) is possible and should be preferred over computation based on simulation. First of all, in contrast to stochastic approaches, the analytical approach provided repeatable and error-free results. Besides that, it has an extremely short runtime (<1 ms), that is, it has very low application costs, which makes the approach highly efficient. In that context (analytical approach is feasible and has excellent runtime), considering research questions RQ2 and RQ3 made no sense. On the basis of these results, the analytical computation of mean CO should be employed within the CoBRA® method (and implemented by a tool supporting the method). Yet, we could not find a feasible analytical approach to compute the CO distribution required in application scenarios AS2 and AS3. Thus, we had to sample the distribution with selected stochastic methods.

Since the measure derived to calculate the sampling error regarding the needs of AS3 suffers from inaccuracy based on the sampling data model, we calculated accuracy and efficiency only based on the error measure derived for AS2 (namely, DoCO). Therefore, we can answer RQ2 and RQ3 with certainty only for AS2.

Regarding RQ2, we can say that the LH algorithm and its performance-optimized version (LHRO) provide comarable accuracy for a given number of iterations. This is not surprising, since both are based on the same general sampling approach and practically differ only with respect to the way they were implemented. Yet, they outperform the MC algorithm regarding *accuracy* at any number of iterations, which we proved with a sign test at α level 0.05 (H_A).

The LH algorithms present about 60% improvement in accuracy over MC on average, with respect to DoCO (RQ3). Moreover, the number of iterations does not seem to have a noticeable influence on the improvement rate.

In case of LH, however, high accuracy is achieved at the expense of lower performance (increased runtime) as compared to MC and LHRO. Already at 10,000 iterations, efficiency measured as accuracy per millisecond (A/ms) of LH differs by a factor of 2.7 and decreases with an increasing number of iterations. It seems that LHRO has a better efficiency regarding runtime (A/ms) than MC at a given level of accuracy (~77% estimated, based on 8 data points), but we could not prove this statistically, since the required data are difficult to obtain. The data for comparing the approaches at a given number of simulation runs are easier to obtain. However, the magnitude of accuracy improvement decreases with increasing accuracy, so an improvement at a given number of simulation runs could not be proven (H_C).

With respect to the number of simulation runs (i.e., iterations), LH has a better efficiency than MC, which could be proven by rejecting the corresponding null hypothesis (H_{B0}).

Summarizing, since the results of our experiment showed that LHRO has a statistically significant higher accuracy than the MC when employed within CoBRA® and seems to also have a higher efficiency regarding runtime compared to MC at the same level of accuracy, LHRO should be the preferred simulation technique to be employed within the CoBRA® method. We showed that with 640 000 iterations on the test PC in less than 6 s, the LHRO algorithm delivers more than sufficiently accurate results.

It is, however, necessary to stress that a simulation algorithm is only responsible for transforming CoBRA® input data (expert estimations) into model output (CO distribution) and is practically independent of the quality of the input data. Although LHRO significantly improves sampling accuracy (60% increase) and efficiency (77% increase) compared to MC, its impact on the overall output of estimation strongly depends on the characteristics of the specific estimation method. For instance, the more sampling runs (e.g., for numerous input factors) are performed, the larger the gain in computational performance of the whole estimation method. Regarding the estimation accuracy, the error introduced by even the worst sampling algorithms considered here (<4.3%) is only a small fraction of the potential error caused by, for example, low quality of the input data, which in case of the CoBRA® method may easily exceed 100% [97]. Yet, considering that for valid data, CoBRA® is capable of providing estimates with less than 15% inaccuracy [97], the error introduced by the simulation technique alone may be considered significant. Therefore, we recommend LHRO as the preferable simulation technique for the CoBRA® method. Compared to MC, LHRO does not entail significant implementation overhead but may provide significant estimation precision and efficiency gains.

On the basis of the presented theoretical error estimations (especially Equation 31 and Equation 35) and confirmed by the experiment results, which reflect a typical application scenario of simulation for cost estimation purposes, the aforementioned conclusions might, in practice, be generalized on any cost estimation method employing simulation. Specific gains in accuracy and performance depend, however, on the number of simulation runs and operations performed on the sampling output (e.g., multiplying sampling outputs may multiply overall error). Moreover, the contribution of the simulation error to the overall estimation inaccuracy depends on the magnitude of error originating from other sources such as invalid estimation (and thus sampling) inputs. In practice, for instance, the quality and adequacy of input data proved, over the years and over various estimation methods, to have a large impact on prediction quality [85]. Therefore, one of the essential method acceptance criteria should be the extent to which a certain estimation method copes with potentially sparse and messy data.

Summarizing, selecting the LHRO simulation technique instead of the traditionally acknowledged MC may provide non-ignorable effort estimation accuracy and performance gains at no additional overhead. The relative estimation improvement depends, however, on the magnitude of other estimation errors. In practice, inadequate, redundant, incomplete, and inconsistent estimation inputs are still responsible for most of the loss in estimation performance (accuracy, precision, efficiency, etc.) Therefore, one of the essential method acceptance criteria should be the extent to which a certain estimation method copes with potentially sparse and messy data. In addition to efforts spent on improving the quality of data (both measurements and experts' assessments), future research should focus on a method that can handle low-quality data and support software practitioners in building appropriate, goal-oriented measurement programs. One direction of work might be, as already proposed in [97], a framework that combines expert- and data-based approaches to iteratively build a transparent effort model. The framework proposes additional quantitative methods to support experts in achieving more accurate assessments as well as to validate available measurement input data and underlying processes. Successful results of the initial validation of the proposed approach presented in [97] are very promising.

Acknowledgments

We thank OKI for providing the CoBRA® field data used to perform the study, the Japanese Information-technology Promotion Agency (IPA) for their support, Prof. Dr. Dieter Rombach and Marcus Ciolkowski from the Fraunhofer Institute for Experimental Software Engineering (IESE) for their valuable comments to this chapter, and Sonnhild Namingha from the Fraunhofer Institute for Experimental Software Engineering (IESE) for reviewing the first version of the chapter.

References

[1] Ahmed M. A., Saliu M. O., and AlGhamdi J., January 2005. Adaptive fuzzy logic-based framework for software development effort prediction. *Information and Software Technology*, **47**(1): 31–48.
[2] Aguilar-Ruiz J. S., Ramos I., Riquelme J. C., and Toro M., 2001. An evolutionary approach to estimating software development projects. *Information and Software Technology*, **43**: 875–882.

[3] Angelis L., Stamelos I., and Morisio M., 2001. Building a software cost estimation model based on categorical data. In *Proceedings of the IEEE 7th International Symposium on software Metrics*, pp. 4–15. England, UK.
[4] Avramidis A. N., and Wilson J. R., 1995. Correlation-induction techniques for estimating quantiles in simulation experiments. In *Proceedings of The Winter Simulation Conference, IEEE Press*.
[5] Avramidis A. N., and Wilson J. R., Correlation-induction techniques for estimating quantiles in simulation experiments. Technical Report 95-05, Department of Industrial Engineering. North Carolina State University, Raleigh, North Carolina.
[6] Basili V. R., and Rombach H. D., June 1988. The TAME project: towards improvement-oriented software environments. *IEEE Transactions on Software Engineering*, **14**(6): 758–773.
[7] Beck K., and Fowler M., October 2000. Planning Extreme Programming. Addison-Wesley Longman Publishing Co., Inc. Boston, MA, USA.
[8] Boehm B. W., Abts C., Brown A. W., Chulani S., Clark B. K., Horowitz E., Madachy R., Refer D., and Steece B., 2000. Software Cost Estimation with COCOMO II. Prentice Hall PTR, Upper Saddle River, NJ, USA.
[9] Boehm B., Abts C., and Chulani S., 2000. Software development cost estimation approaches – a survey. *Annals of Software Engineering*, **10**: pp. 177–205. Springer, Netherlands.
[10] Boehm B. W., 1981. Software Engineering Economics. Prentice Hall PTR, Upper Saddle River, NJ, USA.
[11] Breiman L., Friedman J., Ohlsen R., and Stone C., 1984. Classification and Regression Trees. Wadsworth & Brooks/Cole Advanced Books & Software, Monterey, CA, USA.
[12] Briand L. C., and Wieczorek I., 2002. Software resource estimation. In *Encyclopedia of Software Engineering*, J. J. Marciniak, ed. Volume 2, John Wiley & Sons, Inc., New York, NY, USA. 1160–1196.
[13] Briand L., Langley T., and Wieczorek I., 2000. A replicated assessment and comparison of common software cost modeling techniques. In *Proceedings of the 22nd International Conference on Software Engineering*, Limerick, Ireland.
[14] Briand L. C., El Emam K., and Bomarius F., 1998. CoBRA: A hybrid method for software cost estimation, benchmarking and risk assessment. In *Proceedings of the 20th International Conference on Software Engineering*, pp. 390–399.
[15] Briand L. C., Basili V. R., and Thomas W. M., November 1992. A pattern recognition approach for software engineering data analysis. *IEEE Transactions on Software Engineering*, **18**(11): 931–942.
[16] Burgess C. J., and Lefley M., 2001. Can genetic programming improve software effort estimation? A comparative evaluation. *Information and Software Technology*, **43**(14): 813–920.
[17] Calzolari F., Tonella P., and Antoniol G., July 2001. Maintenance and testing effort modeled by linear and nonlinear dynamic systems. *Information and Software Technology*, **43**(8): 477–486.
[18] Chulani S., Boehm B. W., and Steece B., July/August 1999. Bayesian analysis of empirical software engineering cost models. *IEEE Transactions on Software Engineering*, **25**(4): 573–583.
[19] Cuelenaere A. M. E., van Genuchten M. J. I. M., and Heemstra F. J., December 1987. Calibrating software cost estimation model: Why and how. *Information and Software Technology*, **29**(10): 558–567.
[20] De Lucia A., Pompella E., and Stefanucci S., 2004. Assessing effort estimation models for corrective maintenance through empirical studies. *Information and Software Technology*, **47**: 3–15.
[21] Dolado J. J., 2001. On the problem of the software cost function. *Information and Software Technology*, **43**: 61–72.
[22] Elkjaer M., April 2000. Stochastic budget simulation. *International Journal of Project Management*, **18**(2): 139–147.

[23] Fenton N., Marsh W., Neil M., Cates P., Forey S., and Tailor M., May 2004. Making resource decisions for software projects. In *Proceedings of the 26th International Conference on Software Engineering*, pp. 397–406.
[24] Ferens D. V., and Christensen D. S., April 2000. Does calibration improve predictive accuracy? *Cross-Talk: The Journal of Defense Software Engineering*, **4**: 14–17.
[25] Feller W., 1971. An Introduction to Probability – Theory and Its Application. John Wiley & Sons, NY.
[26] Garratt P. W., and Hodgkinson A. C., October 1999. A neurofuzzy cost estimator. In *Proceedings of the 3rd International Conference Software Engineering and Applications*, pp. 401–406. Arizona, USA.
[27] Gartner Inc. press releases, *Gartner Survey of 1,300 CIOs Shows IT Budgets to Increase by 2.5 Percent in 2005*, 14th January 2005, http://www.gartner.com/press_releases/pr2005.html.
[28] Gartner Inc. press releases, *Gartner Says Worldwide IT Services Revenue Grew 6.7 Percent in 2004*, 8th February 2005, http://www.gartner.com/press_releases/pr2005.html.
[29] Gray A., and MacDonell S., 1997. Applications of fuzzy logic to software metric models for development effort estimation. In *Proceedings of the Annual Meeting of the North American Fuzzy Information Processing Society*, pp. 394–399. Syracuse NY, USA.
[30] Guo Tong, Li Aiqun, and Miao Changqing, December 2005. Monte Carlo numerical simulation and its application in probability analysis of long span bridge. *Journal of Southeast University (English Edition)*, **21**(4): 469–473. China.
[31] Hayashi I., and Iwatsuki N., 1990. Universal FFT and its application. *Journal of the Japan Society of Precision Engineering*, **25**(1): 70–75.
[32] Heiat A., December 2002. Comparison of artificial neural network and regression models for estimating software development effort. *Information and Software Technology*, **44**(15): 911–922.
[33] Helton J. C., and Davis F. J., 2003. Latin hypercube sampling and the propagation of uncertainty in analyses of complex systems. *Reliability Engineering and System Safety*, **81**(1): 23–69.
[34] Hörts M., and Wohlin C., 1997. A subjective effort estimation experiment. *Information and Software Technology*, **39**(11): 755–762.
[35] Huang S.-J., Lin C.-Y., and Chiu N.-H., March 2006. Fuzzy decision tree approach for embedding risk assessment information into software cost estimation model. *Journal of Information Science and Engineering*, **22**(2): 297–313.
[36] Huang X., Capretz L. F., Ren J., and Ho D. A., 2003. Neuro-fuzzy model for software cost estimation. In *Proceedings of the 3rd International Conference on Quality Software*.
[37] Idri A., Abran A., Khoshgoftaar T., and Robert S., June 2002. Estimating software project effort by analogy based on linguistic values. In *Proceedings of the 8th International Symposium on Software Metrics, IEEE computer Society*, pp. 21–30. Ottawa, Canada.
[38] Idri A., Kjiri L., and Abran A., February–March 2000. COCOMO cost model using fuzzy logic. In *Proceeding of the 7th International Conference on Fuzzy Theory & Technology*, Atlantic City, New Jersey.
[39] Jeffery R., Ruhe M., and Wieczorek I., November 2000. A comparative study of two software development cost modeling techniques using multi-organizational and company-specific data. *Information and Software Technology*, **42**(14): 1009–1016.
[40] Jensen R. W., Putnam L. H., and Roetzheim W., February 2006. Software estimating models: Three viewpoints. *Cross-Talk, The Journal of Defense Software Engineering*, **2**: 23–29.
[41] Jensen R. W., April 1983. An improved macro-level software development resource estimation model. In *Proceedings of the 5th International Society of Parametric Analysts Conference*, pp. 88–92. St. Louis, MO.
[42] Jensen R. W., November 1980. A macro-level software development cost estimation methodology. *Proceedings of the 14th Asilomar Conference on Circuits, Systems and Computers*, Pacific Grove, CA.

[43] Johnson P. M., Moore C. A., Dane J. A., and Brwer R. S., November/December 2000. Empirically guided software effort guesstimation. *IEEE Software*, **17**(6): 51–56.
[44] Jones T. C., 1998. Estimating Software Costs. McGraw-Hill, Inc., Hightstown, NJ, USA.
[45] Jørgensen M., May/June 2005. Practical guidelines for better support of expert judgment-based software effort estimation. *IEEE Software*, **22**(3): 57–63.
[46] Jørgensen M., April 2004. Realism in effort estimation uncertainty assessments: it matters how you ask. *IEEE Transactions on Software Engineering*, **30**(4): 209–217.
[47] Jørgensen M., 1995. Experience with the accuracy of software maintenance task effort prediction models. *IEEE Transactions on Software Engineering*, **21**(8): 674–681.
[48] Jørgensen M., and Shepperd M., 2007. A systematic review of software development cost estimation studies. *IEEE Transactions on Software Engineering*, **33**(1): 33–53.
[49] Kadoda G., Cartwright M., and Shepperd M., August 2001. Issues on the effective use of CBR technology for software project prediction. In *Proceedings of the 4th International Conference on Case-Based Reasoning: Case-Based Reasoning Research and Development*, pp. 276–290.
[50] Kalos M. H., and Whitlock P. A., 1986. Monte Carlo Methods—Volume I: Basics. A Wiley-Interscience Publication, NY.
[51] Kitchenham B. A., Pickard L., Linkman S. G., and Jones P., May 2005. A framework for evaluating a software bidding model. *Information and Software Technology*, **47**(11): 747–760.
[52] Kitchenham B., Pickard L. M., Linkman S., and Jones P. W., June 2003. Modeling software bidding risks. *IEEE Transactions on Software Engineering*, **29**(6): 542–554.
[53] Kitchenham B., 1998. A procedure for analyzing unbalanced datasets. *IEEE Transactions on Software Engineering*, **24**(4): 278–301.
[54] Kitchenham B., 1992. Empirical studies of the assumptions that underline software cost-estimation models. *Information and Software Technology*, **34**(4): 211–218.
[55] L'Ecuyer P., 1994. Efficiency improvement and variance reduction. In *Proceedings of The Winter Simulation Conference*.
[56] Lee A., Cheng C. H., and Balakrishnan J., August 1998. Software development cost estimation: integrating neural network with cluster analysis. *Information and Management*, **34**(1): 1–9.
[57] Li J., and Ruhe G., 2006. A comparative study of attribute weighting heuristics for effort estimation by analogy. In *Proceedings of the International Symposium on Empirical Software Engineering*, pp. 66–74. Rio de Janeiro, Brazil.
[58] Liang T., and Noore A., March 2003. Multistage software estimation. In *Proceedings of the 35th Southeastern Symposium on System Theory*, pp. 232–236.
[59] Loève, M., 1987. Graduate texts in mathematics. 4. ed., 3.print. New York: Springer (45).
[60] Lother M., and Dumke R., 2001. Point metrics. Comparison and analysis. In *Dumke/Abran: Current Trends in Software Measurement*, pp. 228–267. Shaker Publication, Aachen, Germany.
[61] MacDonell S. G., 1997. Establishing relationships between specification size and software process effort in CASE environments. *Information and Software Technology*, **39**(1): 35–45.
[62] MacDonell S. G., and Gray A. R., 1996. Alternatives to regression models for estimating software projects. In *Proceedings of the IFPUG Fall Conference*, Dallas TX.
[63] Mair C., and Shepperd M. J., May 1999. An investigation of rule induction based prediction systems. In *Proceeding of the IEEE ICSE Workshop on Empirical Studies of Software Development and Evolution*.
[64] McKay M. D., Beckman R. J., and Conover W. J., 1979. A comparison of three methods for selecting values of input variables in the analysis of output from a computer code. *Technometrics*, **21**(2): 239–245.
[65] Mendes E., Watson I., Chris T., Nile M., and Steve C., 2003. A comparative study of cost estimation models for web hypermedia applications. *Empirical Software Engineering*, **8**(2): 163–196.

[66] Menzies T., Port D., Chen Z., Hihn J., and Stukes S., May 2005. Validation methods for calibrating software effort models. In *Proceedings of the 27th International Conference on Software Engineering*, pp. 587–595.
[67] Miyazaki Y., Terakado M., Ozaki K., and Nozaki H., 1994. Robust regression for developing software estimation models. *Journal of Systems and Software*, **27**, 3–16.
[68] Moløkken-Østvold K. J., and Jørgensen M., December 2004. Group processes in software effort estimation. *Empirical Software Engineering*, **9**(4): 315–334.
[69] Mukhopadhyay T., Vicinanza S. S., and Prietula M. J., June 1992. Examining the feasibility of a case-based reasoning model for software effort estimation. *MIS Quarterly*, **16**(2): 155–171.
[70] Musilek P., Pedrycz W., and Succi G., 2000. Software cost estimation with fuzzy models. *ACM SIGAPP Applied Computing Review*, **8**(2): 24–29.
[71] Ohsugi N., Tsunoda M., Monden A., and Matsumoto K., April 2004. Applying collaborative filtering for effort estimation with process metrics. In *Proceedings of the 5th International Conference on Product Focused Software Process Improvement*, **3009**: 274–286. Kyoto, Japan.
[72] Owen A. B., 1997. Monte Carlo variance of scrambled equidistribution quadrature. *Journal on Numerical Analysis*, **34**(5): 1884–1910.
[73] Pawlak Z., 1991. Rough Sets: Theoretical Aspects of Reasoning About Data. Kluwer Academic Publishers, Boston, MA, USA.
[74] Pendharkar P. C., Subramanian G. H., and Rodger J. A., July 2005. A probabilistic model for predicting software development effort. *IEEE Transactions on Software Engineering*, **31**(7): 615–624.
[75] Putnam L. H., and Myers W., 2003. Five Core Metrics, Dorset House, New York.
[76] Putnam L. H., and Myers W., June 2000. What we have learned. *Cross-Talk: The Journal of Defense Software Engineering*, **6**: 21–24.
[77] Reifer D. J., December 1987. SoftCost-R: User experiences and lessons learned at the age of one. *Journal of Systems and Software*, **7**(4): 279–286.
[78] *@RISK 4.5 User's Guide*. Palisade Corporation. October, 2004.
[79] Rodriguez D., Cuadrado-Gallego J. J., and Aguilar J., 2006. Using genetic algorithms to generate estimation models. In *Proceedings of the International Workshop on Software Measurement and Metrik Kongress*, Potsdam, Germany.
[80a] Saliby E., 1997. Descriptive sampling: An improvement over Latin hypercube sampling. In *Proceedings of Winter Simulation Conference*, 230–233, IEEE Press.
[80b] Saliby E., 2002. An empirical evaluation of sampling methods in risk analysis simulation: Quasi-Monte Carlo, descriptive sampling, and Latin hypercube sampling. In *Proceedings of Winter Simulation Conference*, 1606–1610.
[81] Samson B., Ellison D., and Dugard P., 1997. Software cost estimation using an Albus perceptron (CMAC). *Information and Software Technology*, **39**(1): 55–60.
[82] Sentas P., Angelis L., Stamelos I., and Bleris G. L., January 2005. Software productivity and effort prediction with ordinal regression. *Journal of Information & Software Technology*, **47**(1): 17–29.
[83] Shan Y., McKay R. I., Lokan C. J., and Essam D. L., July 2002. Software project effort estimation using genetic programming. In *Proceedings of the International Conference on Communications, Circuits and Systems*, pp. 1108–1112. Chengdu, China.
[84] Shepperd M., and Cartwright M., 2001. Predicting with sparse data. *IEEE Transactions on Software Engineering*, **27**(11): 987–998.
[85] Shepperd M., and Kadoda G., November 2001. Comparing software prediction techniques using simulation. *IEEE Transactions on Software Engineering*, **27**(11): 1014–1022.

[86] Shepperd M., and Schofield C., 1997. Estimating software project effort using analogies. *IEEE Transactions on Software Engineering*, **23**(12): 736–743.
[87] Shin M., and Goel A. L., June 2000. Empirical data modeling in software engineering using radial basis functions. *IEEE Transactions on Software Engineering*, **26**(6): 567–576.
[88] Shukla K. K., 2000. Neuro-genetic prediction of software development effort. *Information and Software Technology*, **42**: 701–713.
[89] Smith R. K., Hale J. E., and Parrish A. S., March 2001. An empirical study using task assignment patterns to improve the accuracy of software effort estimation. *IEEE Transactions on Software Engineering*, **27**(3): 264–271.
[90] Sobol I. M., 1974. The Monte Carlo Method. The University of Chicago Press, Chicago.
[91] Song Q., Shepperd M., and Mair C., 2005. Using grey relational analysis to predict software effort with small data sets. In *Proceedings of the 11th IEEE International Software Metrics Symposium*, pp. 35–45. Como, Italy.
[92] Spector P., 1992. Summated Rating Scale Construction. Sage Publications, Newbury Park, CA, USA.
[93] Srinivasan K., and Fisher D., 1995. Machine learning approaches to estimating software development effort. *IEEE Transactions on Software Engineering*, **21**(2): 126–137.
[94] Stamelos I., and Angelis L., November 2001. Managing uncertainty in project portfolio cost estimation. *Information and Software Technology*, **43**(13): 759–768.
[95] Stamelos I., Angelis L., and Sakellaris E., April 2001. BRACE: Bootstrap based analogy cost estimation. In *Proceedings of the 12th European Software Control Metrics*, pp. 17–23. London, UK.
[96] StatSoft Inc. Statistica 7 Data Miner. http://www.statsoft.com.
[97] Stroud A. H., 1971. Approximate Calculation of Multiple Integrals. Prentice-Hall, New Jersey.
[98] Taff L. M., Brochering J. W., and Hudgins W. R., 1991. Estimeetings: Development estimates and a front-end process for a large project. *IEEE Transactions on Software Engineering*, **17**(8): 839–849.
[99] Trendowicz A., Heidrich J., Münch J., Ishigai Y., Yokoyama K., and Kikuchi N., 2006. Development of a hybrid cost estimation model in an iterative manner. In *Proceeding to 28th International Conference on Software Engineering ICSE*, Shanghai, China.
[100] The Standish Group, 2005. CHAOS Chronicles. The Standish Group International, Inc., West Yarmouth, MA.
[101] Vose D., 1996. Quantitative Risk Analysis. A Guide to Monte Carlo Simulation Modelling. John Wiley & Sons, Chichester.
[102] Walkerden F., and Jeffery R., 1999. An empirical study on analogy-based software effort estimation. *Empirical Software Engineering*, **4**: 135–158.
[103] Walkerden F., and Jeffery R., 1997. Software cost estimation: A review of models, process, and practice. *Advances in Computers*, **44**: 59–125.
[104] Wieczorek I., 2002. Improved software cost estimation – a robust and interpretable modelling method and a comprehensive empirical investigation. *Empirical Software Engineering*, **7**(2): 177–180.
[105] Wohlin C., Runeson P., Höst M., Ohlsson M., Regnell B., and Wesslen A., 2000. Experimentation in Software Engineering. An Introduction. Kluwer Academic Publishers, Boston.
[106] Xu Z., and Khoshgoftaar T. M., 2004. Identification of fuzzy models of software cost estimation. *Fuzzy Sets and Systems*, **145**(1): 141–163.
[107] Yang D., Wan Y., Tang Z., Wu S., He M., and Li M., COCOMO-U: An extension of COCOMO II for cost estimation with uncertainty. In Q. Wang, et al., ed. *Software Process Change, SPW/ProSim 2006. LNCS 3966*, pp. 132–141. Springer-Verlag, Berlin, Heidelberg.
[108] Zadeh L. A., April 1965. Fuzzy sets. *Information and Control*, **8**: 338–353.
[109] Zou T., Mahadevan S., Mourelatos Z., and Meernik P., December 2002. Reliability analysis of automotive body-door subsystem. *Reliability Engineering & System Safety*, **78**(3): 315–324.

An Environment for Conducting Families of Software Engineering Experiments

LORIN HOCHSTEIN

University of Nebraska

TAIGA NAKAMURA

University of Maryland

FORREST SHULL

Fraunhofer Center Maryland

NICO ZAZWORKA

University of Maryland

VICTOR R. BASILI

University of Maryland, Fraunhofer Center, Maryland

MARVIN V. ZELKOWITZ

University of Maryland, Fraunhofer Center, Maryland

Abstract

The classroom is a valuable resource for conducting software engineering experiments. However, coordinating a family of experiments in classroom environments presents a number of challenges to researchers. Understanding how to run such experiments, developing procedures to collect accurate data, and collecting data that is consistent across multiple studies are major problems.

This paper describes an environment, the Experiment Manager that simplifies the process of collecting, managing, and sanitizing data from classroom experiments, while minimizing disruption to natural subject behavior. We have successfully used this environment to study the impact of parallel programming languages in the high-performance computing domain on programmer productivity at multiple universities across the United States.

1. Introduction 176
   1.1. Collecting Accurate Data 177
   1.2. Classroom Studies 178
2. Classroom as Software Engineering Lab 179
3. The Experiment Manager Framework 182
   3.1. Instrumentation Package 183
   3.2. Experiment Manager Roles 184
   3.3. Data Collection 186
4. Current Status 190
   4.1. Experiment Manager Effectiveness 191
   4.2. Experiment Manager Evolution 192
   4.3. Supported Analyses 193
   4.4. Evaluation 197
5. Related Work 197
6. Conclusions 198
   Acknowledgments 199
   References 199

1. Introduction

Scientific research advances by the creation of new theories and methods, followed by an experimental paradigm to either validate those theories and methods or to offer alternative models, which may be more appropriate and accurate. Computer science is not different from other sciences, and the field has been moving to adopt such experimental approaches [25]. However, in much of computer science, and in software engineering in particular, the experimental model poses difficulties possibly unique among the sciences. Software engineering is concerned about the appropriate models applicable to the development of large software systems. As such it involves the study of numerous programmers and other professionals over long periods of time. Thus, much of this research involves human behavior and in many ways is similar to research in psychology or the social sciences.

The major experimental approach accepted in scientific research is the replicated study. However, by being expensive to produce, software is not amenable to such studies. While a typical medical clinical trial may involve hundreds of subjects testing a drug or a new treatment, even one duplication of a software development is beyond the resources of most organizations. Although this approach is commonly used in various fields of research involving humans, such as clinical study in medical science, conducting many studies is still difficult and expensive, which is often a major obstacle for good software engineering research.

The problems with empirical studies in software engineering can be classified by two major problems: cost of such studies and accuracy of the data.

1.1 Collecting Accurate Data

1.1.1 Costs of Software Engineering Studies

Because of the cost of developing a piece of software, typically case studies of a single development are monitored, and after many such studies, general trends can be observed. As an example of this, we will look at the NASA Goddard Space Flight Center (GSFC) Software Engineering Laboratory (SEL), which from 1976 to 2001 conducted many such studies [6]. The experiences of the SEL are illustrative of the problems encountered in data collection. The data was collected, beginning in 1976, at GSFC from NASA developers and the main development contractor, Computer Sciences Corporation (CSC). The data was then manually reviewed at GSFC before being sent to the University of Maryland for entry into the project measures database using a UNIX-based Ingres system.

The naïve simplicity in which data was collected broke down by 1978 and a more rigorous set of processes was instituted. This could not be a part-time activity by faculty using undergraduate employees. In addition, the university researchers wanted a considerable amount of data, and soon realized that the GSFC programming staff did not have the time to comply with their requests. They had to compromise on the amount of data desired versus the amount of data that realistically could be collected. Data, which was collected on forms filled out by the programming staff, were shortened to allow for more complete collection.

The data collection process for the 20 projects then under study became more rigorous with this five-step approach:

1. Programmers and managers completed forms.
2. Forms were initially verified at CSC.
3. Forms were encoded for entry at GSFC.

4. Encoded data checked by validation program at GSFC.
5. Encoded data revalidated and entered into database at University (after several years, CSC took over total management of the database).

But, to obtain contractor cooperation, a 10% overhead cost to projects was allocated for data collection and processing activities. Eventually the overhead cost of collecting data was reduced, but the total cost of collecting, processing, and analyzing data continued to remain between 5% and 10%. However, on a $500K project, this still amounted to almost $50K just for data collection – an amount that few organizations are willing to invest. While the SEL believed that the payoff in improved software development justified this cost, it is beyond the scope of this chapter to prove that. Suffice it to say that most organizations consider the additional costs of data collection as unjustified expenses.

1.1.2 Accuracy in Collected Data

Most data collected on software development projects can be generally classified as self-reported data – the technical staff fill out effort reports on hours worked, change reports on defects found and fixed, etc. The care in which such data is reported and collected greatly affects the accuracy of the process. Unfortunately, the process is not very accurate. Self-reported measures can vary over time, due to history or maturation effects [7], and the accuracy of such measures varies across individuals. This is a particular problem when the subjects have more interest in completing the task than complying with the protocols of the study. Basili et al. [2] evaluated Software Science metrics against self-reported effort data collected from GSFC software projects. There was very little correlation between this effort and metrics known to predict effort, and there was concern that poor self-reported data was distorting the results. Perry et al. [18] analyzed previous data from project notebooks and free-form programmer diaries which were originally kept for personal use. They found that the free-form diaries were too inconsistent across subjects and sometimes lacked sufficient resolution.

1.2 Classroom Studies

The overhead in collecting accurate detailed data at the SEL was too high to maintain a data collection process. This same result has been found in other data collection studies. Instead of large replications, running studies in classrooms using students as subjects have become a favorite approach for trying out new development techniques [20]. Even though a conclusion drawn from student subjects cannot always be generalized to other environments, such experiments aid in developing approaches usable elsewhere. However, conducting larger-scale software engineering research in classroom environments can be quite complex and time consuming without proper tool support. Proper tool support is a requirement if we want to improve on the poor quality of self-reported data.

A single environment (e.g., a single class) is often insufficient for obtaining significant results. Therefore, multiple replications of a given experiment must be carried out by different faculty at different universities in order to provide sufficient data to make conclusions. This means, each experiment must be handled in a similar manner to allow for combining partial results. The complexity of providing consistent data across various experimental protocols has been overwhelming so far.

In this chapter, we describe an environment we are developing to simplify the process of conducting software engineering experiments that involve development effort and workflow and ensure consistency in data collection across experiments in classroom environments. We have used this environment to carry out research to identify the effect of parallel programming languages on novice programmer productivity in the domain of high-performance computing (e.g., MPI [12], OpenMP [11], UPC [8], and Matlab*P [10]). Although there are often issues regarding the external validity of students as subjects, this is not a major concern here since we are explicitly interested in studying student programmers.

This work was carried out in multiple universities across the United States in courses where the professors were not software engineering researchers and were, therefore, not experienced with conducting experiments that involved human subjects.

2. Classroom as Software Engineering Lab

The classroom is an appealing environment for conducting software engineering experiments, for several reasons:

- Most researchers are located at universities. Being close to your subjects is often necessary to obtain accurate results.
- Training can be integrated into the course. No extra effort is then required by the subjects since there is the assumption that the training is a valuable academic addition to the classroom syllabus.
- Required tasks can be integrated into the course.
- All subjects are performing identical programming tasks, which are not generally true in industry. This provides an easy source for replicated experiments.

In addition to the results that are obtained directly by these studies, such experiments are also useful for piloting experimental designs and protocols which can later be applied to industry subjects, an approach which has been used successfully elsewhere (e.g., [3, 4, 5]).

While there are threats to validity of such studies by using students as subjects as proxies for professional programmers (e.g., the student environment may not be representative of the ones faced by professional programmers), there are additional complexities that are specific to research in this type of environment. We encountered each of these issues when conducting research on the effects of parallel programming model on effort in high-performance computing [14]:

1. *Complexity*: Conducting an experiment in a classroom environment is a complex process that requires many different activities (e.g., planning the experimental design, identifying appropriate artifacts and treatments, enrolling students, providing for data collection, checking for process compliance, sanitizing data for privacy, and analyzing data). Each such activity identifies multiple points of failure, thus requiring a large effort to organize and run multiple studies. If the study is done at multiple universities in collaboration with other professors, these professors may have no experience in organizing and conducting such experiments.

2. *Research versus pedagogy*: When the experiment is integrated into a course, the experimentalist must take care to balance research and pedagogy [9]. Studies must have minimal interference with the course. If the students in one class are divided up into treatment groups and the task is part of an assignment, then care must be taken to ensure that the assignment is of equivalent difficulty across groups. Students who consent to participate must not have any advantage or disadvantage over students who do not consent to participate, which limits additional overhead required by the experiment. In fact, each university's Institutional Review Board (IRB), required in all United State universities performing experiments with human subjects, insists that participation (or nonparticipation) must have no effect on the student's grade in the course.

3. *Consistent replication across classes*: To build empirical knowledge with confidence, researchers replicate studies in different environments. If studies are to be replicated in different classes, then care must be taken to ensure that the artifacts and data collection protocols are consistent. This can be quite challenging because professors have their own style of giving assignments. Common projects across multiple locations often differ in crucial ways making meta-analysis of the combined results impossible [16].

4. *Participation overhead for professors*: In our experience, many professors are quite willing to integrate software engineering studies into their classroom environment. However, for professors who are unfamiliar with experimental protocols, the more effort required of them to conduct a study, the less likely it will be a success. In addition, collaborating professors who are not empirical researchers may not have the resources or the inclination to monitor the quality

of captured data to evaluate process conformance. Therefore, empirical researchers must try to minimize any additional effort required to run an empirical study in the course while ensuring that data is being captured correctly.

The required IRB approval, when attempted for the first time, seems like a formidable task. Help in understanding IRB approval would greatly aid the ability of conducting such research experiments.

5. *Participation overhead for students*: An advantage of integrating a study into a classroom environment is that the students are already required to perform the assigned task as part of the course, so the additional effort involved in participating in the study is much lower than if subjects were recruited from elsewhere. However, while the additional overhead is low, it is not zero. The motivation to conform to the data collection process is, in general, much lower than the motivation to perform the task, because process conformance cannot be graded. In addition, the study should not subvert the educational goals of the course. Putting the experiment in the context of the course syllabus is never easy.

 This can be particularly problematic when trying to collect process data from subjects (e.g., effort, activities, and defects), especially for assignments that take several weeks (e.g., we saw a reduction in process conformance over time when subjects had to fill out effort logs over the course of multiple assignments).

6. *Automatic data collection of software process*: To reduce subject overhead and increase data accuracy, it is possible to collect data automatically from the programmer's environment. Capturing data at the right level of granularity is difficult. All user-generated events can be captured (keyboard and mouse events), but this produces an enormous volume of data that may not abstract to useful information. Allowing this raw data to be used can create privacy issues, such as revealing account names, with the ability to then determine how long specific users took to build a product or how many defects they made.

 All development activities taking place within a particular development environment (e.g., Eclipse) simplifies the task of data collection, and tools exist to support such cases (e.g., Marmoset [21]). However, in many domains, development will involve a wide range of tools and possibly even multiple machines. For example, in the domain of high-performance computing, preliminary programs may be compiled on a home PC, final programs are developed on the university multiprocessor, and are ultimately run on remote supercomputers at a distant datacenter. Programmers typically use a wide variety of tools, including editors, compilers, build tools, debuggers, profilers, job submission systems, and even web browsers for viewing documentation.

7. *Data management*: Conducting multiple studies generates an enormous volume of heterogeneous data. Along with automatically collected data and

manually-reported data, additional data includes versions of the programs, pre-and post-questionnaires, and various quality outcome measures (e.g., grades, code performance, and defects). Because of privacy issues, and to conform to IRB regulations, all data must be stored with appropriate access controls, and any exported data must be appropriately sanitized. Managing this data manually is labor-intensive and error-prone, especially when conducting studies at multiple sites.

3. The Experiment Manager Framework

We evolved the Experiment Manager framework (Fig. 1) to mitigate the complexities described in the previous section. The framework is an integrated set of tools to support software engineering experiments in HPC classroom environments. While aspects of the framework have been studied by others, the integration of all features allows for a uniform environment that has been used in over 25 classroom studies over the past 4 years. The framework supports the following.

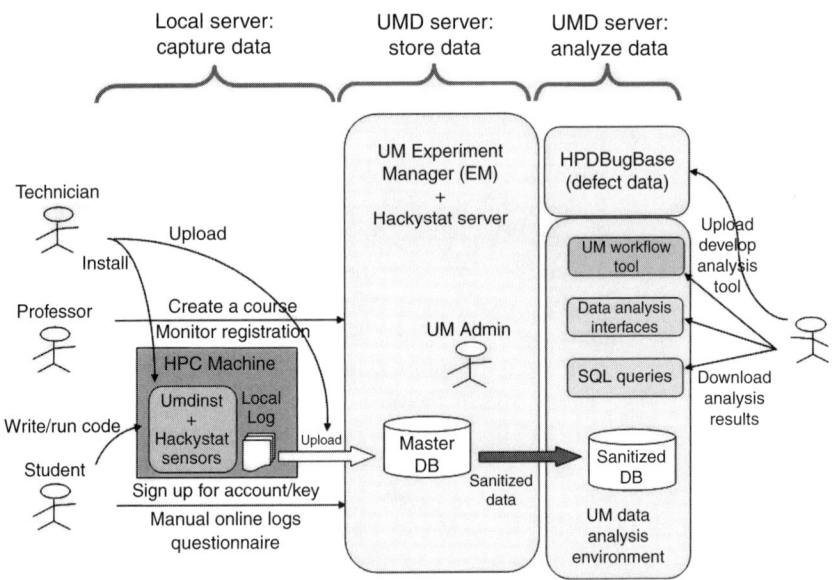

FIG. 1. Experiment Manager structure.

1. *Minimal disruption of the typical programming process*: Study participants solve programming tasks under investigation using their typical work habits, spreading out programming tasks over several days. The only additional activity required is filling out some online forms. Since we do not require them to complete the task in an alien environment or work for a fixed, uninterrupted length of time, we minimize any negative impact on pedagogy or subject overhead.
2. *Consistent instruments and artifacts*: Use of the framework ensures that the same type of data will be collected and the same type of problems will be solved, which increases confidence in meta-analysis across studies at different universities.
3. *Centralized data repository with web interface*: The framework provides a simple, consistent interface to the experimental data for experimentalists, subjects, and collaborating professors. This reduces overhead for all stakeholders and ensures that data is consistently collected across studies.
4. *Sanitization of sensitive data*: The framework provides external researcher with access to the data sets that have been stripped of any information that could identify subjects, to preserve anonymity and comply with the protocols of human subject research as set out by IRBs at American universities.

3.1 Instrumentation Package

Our instrumentation package, called *UMDinst*, supports automatic collection of software process data in a Unix-based, command-line development environment, which is commonly used in high-performance computing. The package is designed to be installed in a master account on the local server and then enabled in the accounts of each subject by executing a set up script, or be installed in the account of individual subjects. The appropriate installation mode depends on the need of a specific experiment and the configuration of the machine to be instrumented. In either case, the package can be used without the intervention of system administrators.

UMDinst package instrument programs that are involved in the software development process by replacing each command that invokes a tool (e.g., compiler) with a script that first collects the desired information and then calls the original tool. It is used for instrumenting compilers, although it is also designed to support job schedulers (common in high-performance computing environments), debuggers, and profilers. For each compile, the following data is captured:

- a time stamp when the command is executed
- contents of the source file that were compiled
- contents of local header files referenced in the source file

- the command used to invoke the compiler
- the return code of the compiler
- the time to compile

The UMDinst package includes Hackystat sensors [15] to instrument supported editors such as Emacs and vi, and to capture shell commands and time stamps. The collected data is used in studies to estimate total effort as well as to infer development activities (e.g., debugging, parallelizing, and tuning). Hackystat is a system developed by Johnson at the University of Hawaii that captures low-level events from a set of instrumented tools. Thus, while UMDinst captures data at the command-line level, Hackystat captures time stamps and events from editors and related tools that have been instrumented. The pair of tools provides a complete history of user interaction in developing a program.

3.1.1 Web Portal

The heart of the Experiment Manager framework is the web portal, which serves as a front-end to the database that contains all of the raw data, along with metadata about individual experiments. Multiple stakeholders use the web interface: experimenters, subjects, and data analysts. For example, experimenters specify treatments (in our case, parallel programming models), assignment problem, participation rate, and grades. They also upload data captured automatically from UMDinst. Subjects fill in questionnaires, and report on process data such as time worked on different activities and defects. Analysts would export data of interest, such as total effort [13] for hypothesis testing, or a stream of time stamped events for workflow modeling.

3.2 Experiment Manager Roles

We have divided the functionality of the Experiment Manager into four roles. For each role, we developed several use cases that describe its functionality, thus simplifying the design of the software.

1. *Technician*: The technician sets up the environment on the local server, usually not at the University of Maryland. This will be someone at a university with access to the machine the students use for the class. Often it is the Teaching Assistant in the course the software will be used in. The tasks for the technician are to install UMDinst so that students can use it. At the end of the semester, the technician also sends the collected data to the University of Maryland server in case it was collected locally.

2. *Professor*: A database provides the professor with sample IRB questionnaires for submittal. This cannot be fully automated since each university has its own guidelines for submitting the IRB approval form. But experience with many universities over the last 4 years allows us to help in answering the most common questions on these forms.

   The instructor first registers each class with the Experiment Manager to set up a classroom experiment. For each such class, the professor can assign several programming projects from our collected database of assignments or assign one of his own. During the semester, the system allows the professor to see if students have completed their assignments, but does not allow access to any of the collected data until the grades for the assignment are completed. In reality, the Teaching Assistant may be the person to actually perform this task, but conceptually is acting in the role of the professor.
3. *Student*: A student who takes part in the experiment provides data on HPC development. This requires the student to:
   1. Register with the Experiment Manager by filling out a background questionnaire on courses taken and experiences in computer science in general and HPC programming in particular. Although this registration process can take up to 15 min, it is required only once during the semester.
   2. Run the script to set up the wrappers for the commands that edit, compile and run programs. Once an assignment is underway, the data collection process is mostly automatic and data is collected mostly painlessly.
4. *Analyst*: An analyst accesses the collected data for evaluating some hypothesis about HPC development. At the present time, the analysis tools are relatively simple. Analysts can see the total effort and defects made by each student and collect workflow data.

Many tools exist in prototypes to support the various types of studies. For example, the HPC community is developing concepts of what productivity means in the HPC environment [26] and we have been looking at developing workflow models (e.g., how much time is spent in various activities, such as developing code, testing, parallelizing the code, and tuning the code for better performance). To support the study of workflows and productivity, we have developed a tool to allow the experimenter to apply various heuristics to the base data to see if we can automatically deduce the developer's programming activity, for example, testing and debugging versus development. This tool takes raw data, collected from online tools such as Hackystat and manual logs generated by students, and we have been developing algorithms for automatically inferring the workflow cycle [23]. Results of this work are described in Section 4. Current activities are looking at extending these tools.

3.3 Data Collection

The actual data collection activity was designed to present minimal complexity to the student (Figs. 2 and 3). Within the Experiment Manager, the student has two options. If data was not collected automatically, the student can enter a set of activities, with the times each activity started and ended (Fig. 3) (e.g., self-reported data, which we discussed earlier to be less reliable). However, the effort tool simplifies the process greatly. If the student clicks to start the tool (small oval near bottom of Fig. 3), then a small window opens on the top left corner of the screen (large oval in upper left in Fig. 3). Each time the student starts a different activity, the student only needs to pull down the menu in the effort tool and set the new activity type (Fig. 2). The time between clicks is recorded as the time of the previous activity. Thus, while the data is not totally automatic, we believe we have minimized the overhead of collecting such data.

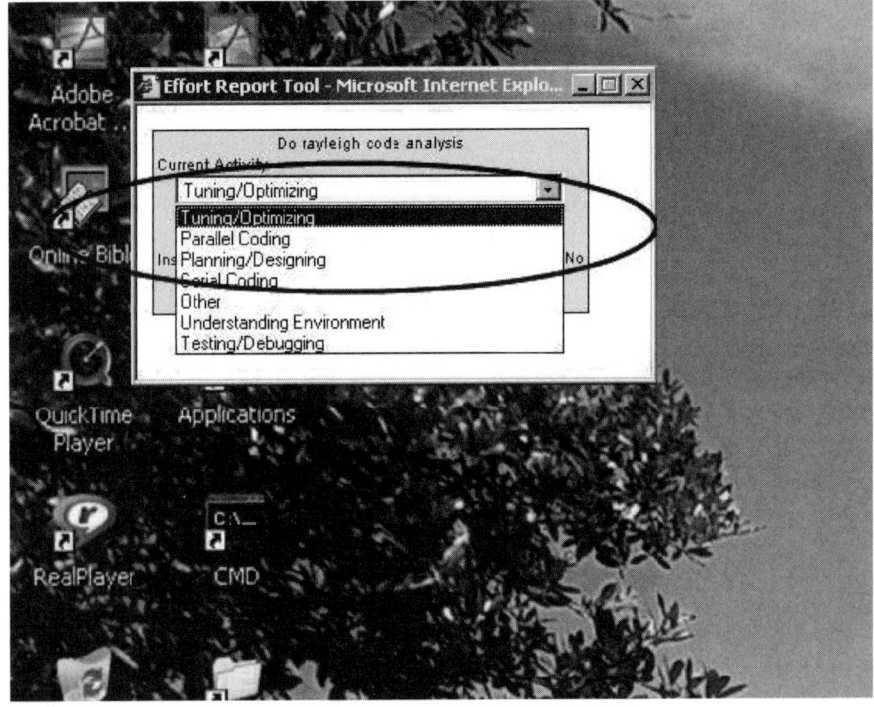

FIG. 2. Effort capture tool.

SOFTWARE ENGINEERING EXPERIMENTS 187

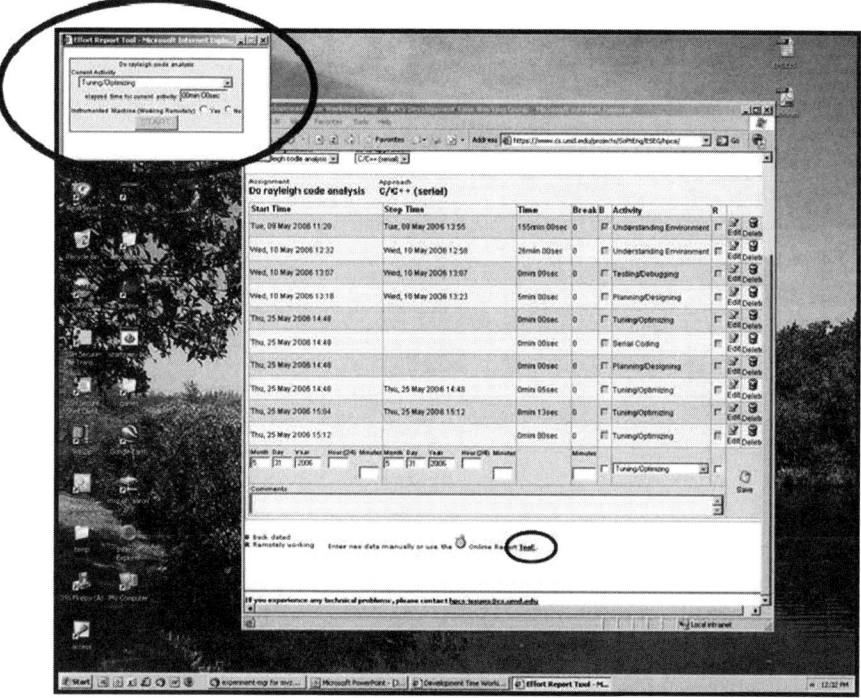

FIG. 3. Effort collection screen.

The tool automatically computes elapsed time between events and saves the data in the database. If the student stops for a period of time (e.g., goes to lunch and surfs the web), there is a *stop* button on the tool. Upon returning, the user simply clicks on *start* to resume timing.

For most HPC development, the student simply has to:

1. Log into Experiment Manager to go to effort page (Fig. 4), then click on effort tool (Fig. 3).
2. Develop program as usual.
3. Each time a new activity starts, click on the new activity in the effort tool (Fig. 2).
4. If any errors are found, the student records that defect by invoking the defect tool (Fig. 4) to explain the defect on a separate page (Fig. 5).

Only steps 3 and 4 involve any separate activity for participating in these experiments, and such activity is minimal.

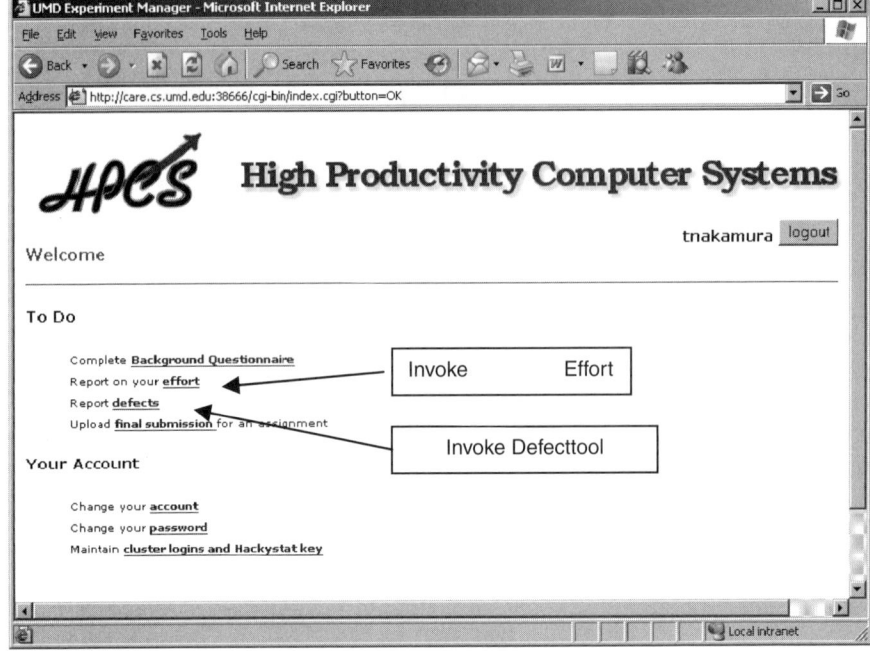

FIG. 4. Student view of Experiment Manager.

3.3.1 Data Sanitization

While personal data collected by the experiments must be kept private, we would like to provide as much data as possible to the community as part of our analysis. The sanitization process exports 'safe' data into a database that can be made accessible to other researchers, running on a separate machine.

The sanitization process is briefly described in Fig. 6. Each data object we obtained in an experiment is classified as one of:

1. *Prohibited*: Data contains personal data we cannot reveal (e.g., name or other personal identifiers).
2. *Clean*: Data we can reveal (e.g., effort data for development of another clean object).
3. *Modified*: Data we can modify to make it clean (e.g., removing all personal identification in the source program).

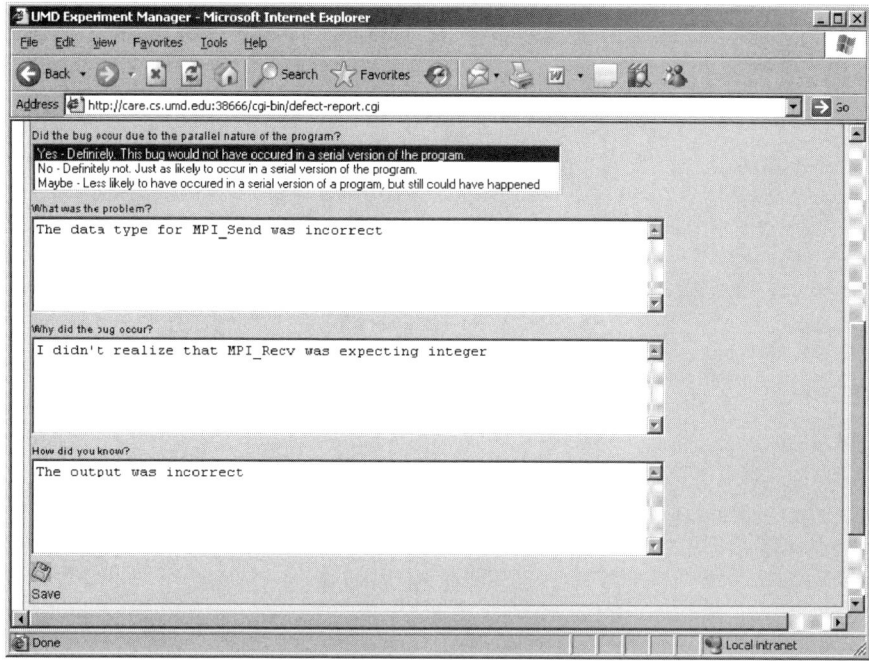

FIG. 5. Defect reporting tool.

Clean data can be moved to the analysis server and modified data can also be moved. Only prohibited data cannot be exported to others desiring to look at our collected database. Our sanitization process on data consists of the following four functions:

1. *Normalization* – Normalize the time stamps for each class on a common basis. By making each time stamp relative to 0 from the beginning of that experiment, information about in which semester it was collected (and hence from which school the data was collected) is hidden.
2. *Discretization* – Since grades are considered private data, we define a mapping table that maps grades on a small set, such as {good,bad}. Converting other interval or ratio data into less specific ordinal sets, while it loses granularity, it helps to present anonymity.
3. *Summarization* – With some of the universities, we can give out source code if we remove all personal identifiers of the students who wrote the code. But in

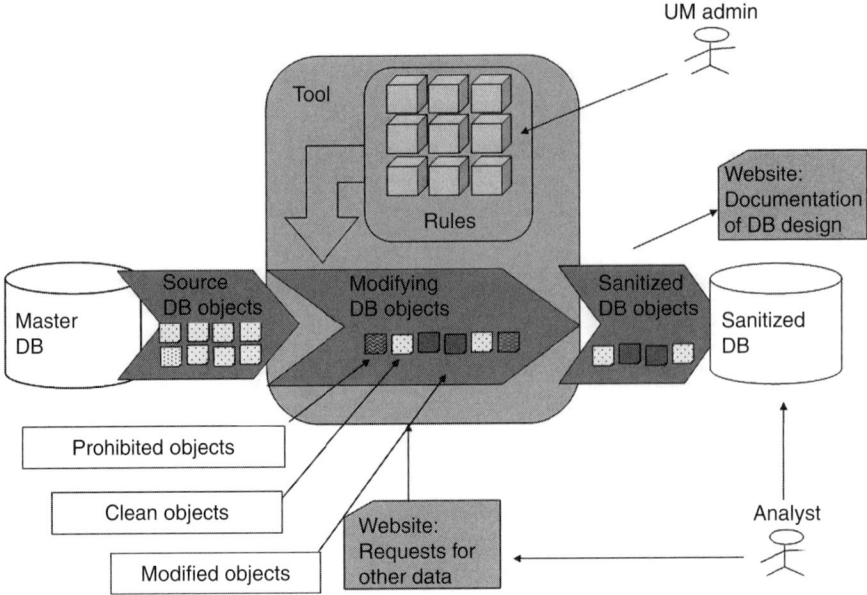

FIG. 6. Basic sanitization process.

some cases, we are prohibited from doing even that. If we cannot give out source code, we can collect patterns and counts of their occurrence in the source code. For example, we can count lines of code, or provide analyses of 'diffs' of successive versions of code.

4. *Anonymization* – We can create hash values for dates, school names, and other personal identifiers.

4. Current Status

Our Experiment Manager framework currently contains data from 25 classroom experiments conducted at various universities in the United States (Fig. 7). While some of the experiments preceded the Experiment Manager (and motivated its development) and their data was imported into the system; perhaps half the experiments used parts or all of the system.

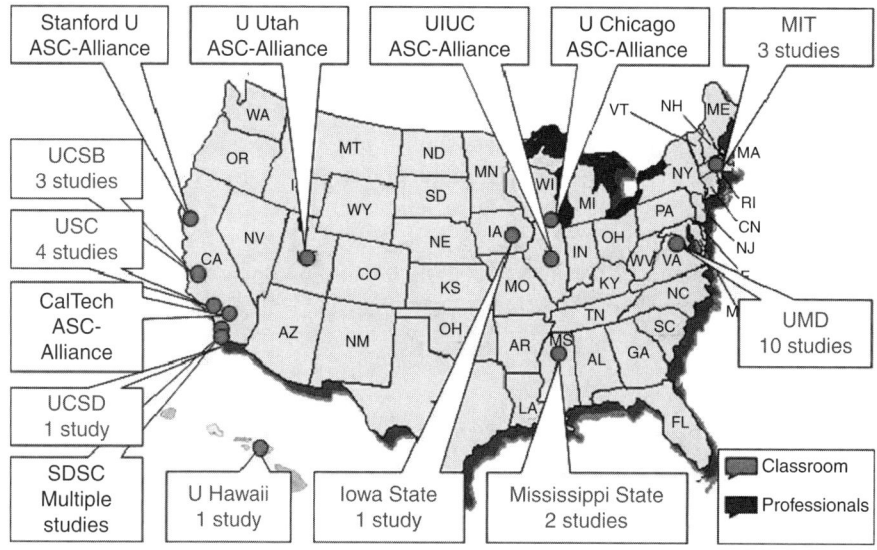

Fig. 7. Completed studies.

4.1 Experiment Manager Effectiveness

Before using the experiment manager, we discovered many discrepancies between successive classroom projects which prevented merging the results. Some of these were:

1. *It was not often obvious for how many processors the final programs were written.* Since a major goal of HPC programming is to divide an algorithm to run on multiple processors, this speedup (i.e., relative decrease in execution time by using multiple processors) is a critical measure of performance. Without knowing the initial goals for each student assignment, it was unclear how to measure performance goals for each class.
2. Related to the previous problem, *the projects all had differing requirements for final program complexity* (e.g., the number of replicated cells needing to be computed). How big a grid (e.g., number of replicated cells) were required in which to compute an answer and measure performance? This affected student programming goals.
3. *Grading requirements differed.* Was performance on an HPC machine important? Sometimes just getting a valid solution mattered. Maximum speedup, or the decrease in execution time of an HPC machine over a serial implementation, was sometimes the major goal.

By using our collected database of potential assignments, as well as our checklist of project attributes, this problem has lessened across multiple classes recently, allowing for the combination of results across different universities.

The IRB process seems like a formidable roadblock the first time any new professor encounters it. Often, in contacting faculty at a new university we would lose a semester's activity simply because the IRB approval process was too onerous the first time it was attempted. With our experience of IRB issues, and our collection of IRB forms required by various universities, this no longer is a major problem.

A related problem was the installation of software on the host computer for the collection of data. Again, this often meant the delay by a semester since the installation process was too complex. This was a major driving force to host much of this software as a web server at the University of Maryland, with a relatively simple UMDinst package that needed to be installed at each university's site.

The effort tool (pictured earlier as Figs. 2 and 3) also solved some of our data collection problems. We can collect effort data by three ways (Hackystat at the level of editor and shell event time stamps, manual data via programmer filled-in forms, and compiler time stamps via UMDinst). All give different results [13]. The use of the effort tool greatly eases the data collection problem, which we believe increases the reliability of such data.

Most of our results, so far, are anecdotal. But we have been able to address new universities and additional classes in a more methodical manner at present and believe the Experiment Manager software is a major part of this improvement.

4.2 Experiment Manager Evolution

The system is continuing to evolve. Current efforts focus on the following tasks:

- We are evolving the user interface to the Experiment Manager web-based tool. The goal is to minimize the workload of various stakeholders (i.e., roles) for setting up the experiment environment, registering with the system, entering the data, and conducting an analysis. We would like to develop small native applications that provide a more integrated interface to the operating system, making it less disruptive to users.

- We want to continue our experimentation with organizations that are often behind firewalls. Although we are currently studying professionals in an open environment, we want to use the Experiment Manager in this environment. Although the UMDinst instrumentation package can be set up in secure environments, the collected data cannot be directly uploaded to the University of Maryland servers. We have planned extensions to the Experiment Manager architecture to better

support the experimentations with these organizations. Working in these environments is necessary to see how professionals compare to the students.
- We will continue to evolve our analysis tools. For example, our prototype experience bases for evolving hypotheses and high end computing defects (e.g., www.hpcbugbase.org) will continue to evolve both in content and usability.
- We will evolve problem-specific harnesses that automatically capture information about correctness and performance of intermediate versions of the code during development to ensure that the quality of the solutions (specifically, correctness and performance) is measured consistently across all subjects.

This also requires us to evolve our experience bases to generate performance measures for each program submitted in order to have a consistent performance and speedup measure for use in our workflow and time to solution studies.

Our long-range goal is to allow the community access to our collected data. This requires additional work on a sanitized database that removes personal indicators and fulfills legal privacy requirements for use of such data.

4.3 Supported Analyses

The Experiment Manager was designed to ease data analysis, in order to support the investigation of a range of different research questions. Some of these analyses are focused in detail on a single developer being studied, while others aggregate data over several classes, allowing us to look across experimental data sets to discover influencing factors on effective HPCS development.

4.3.1 Views of a Single Subject

One view of a subject's work patterns is provided directly by the instrumentation. We refer to this view as the *physical level view* since it objectively reports incontrovertible occurrences at the operating system level, such as the time stamp of each compilation.

Figure 8 shows such a physical view for a 9 hour segment of work done by a given subject. The x-axis represents time and each dot on the graph represents a compile event. Although we have the tools to measure physical activities with a high degree of accuracy, this type of analysis does not yield much insight. For example, although we know how often the compiler was invoked, we do not know why: We cannot distinguish compilations which add new functionality from compilations which correct problems or defects that could have been avoided. This is an important distinction to make, if we want to know the cause of unnecessary rework so it can be avoided.

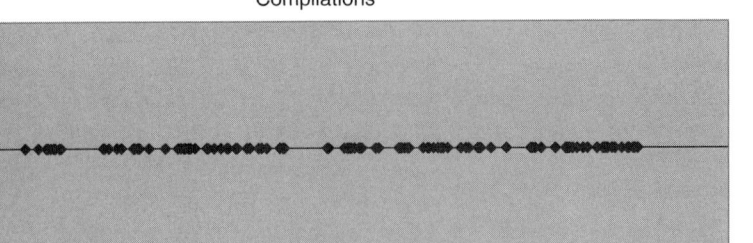

FIG. 8. Compilation events for one subject.

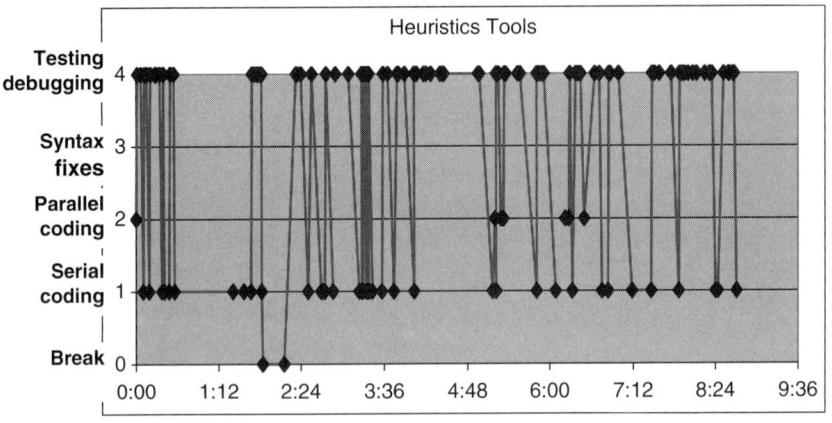

FIG. 9. Types of development activities for one subject.

A second approach is to use the logged compile times along with a snapshot of the code at each such compile and then apply a set of heuristics to guess at the semantics of the activities requiring that compilation. We have built such a tool that allows us to use various algorithms to manipulate these heuristics. The purpose of determining the semantic activities is to build baselines for predicting and evaluating the impact of new tools and languages.

Figure 9 uses the same data as Fig. 8 to illustrate this view. By evaluating the changes in the code for each compile, we can infer an activity being performed by the programmer. Again, each dot represents one compile but in this case the activity preceding each compile has been classified as one of:

- *Serial coding*: The developer is primarily focused on adding functionality through serial code. This is inferred since most of the changes since the previous compiler was in new code being added.
- *Parallel coding*: The developer is adding code to take advantage of multiple processors, not just adding function calls to the parallel library. We decided to separate out this activity since the amount of effort spent in this activity is indicative of how difficult it is to take advantage of the parallel architecture for solving the problem. This is inferred since parallel execution calls (such as to the MPI library) were added to the program.
- *Syntax fixes*: The developer is fixing errors from a previous compile. We can determine this since the previous compile failed, and the source program is changed with no intervening execution.
- *Testing and debugging*: The developer is focused on finding and fixing a problem, not adding new functionality. This activity can be identified via some typical and recognizable testing strategies, such as when a high percentage of the code added before a compile were output statements (so that variable values can be checked at runtime); creating/modifying test data files instead of the main code block; or removing test code back out of the system at the end of a debugging session. Our hypothesis is that effort spent on these activities can come from misunderstanding of the problem or the proposed solution and so could be addressed with more sophisticated aides given to developers.

Such a view helps us understand better the approach used by the subject, and how much of his/her time was spent on rework as opposed to adding new functionality. The duration data associated with each activity also helps us identify interesting events during the development: For example, when a large amount of time is spent debugging, analysts can focus on events preceding the debugging activity to understand what type of bug entered the system and how it was detected. This type of information can be used to understand how hard or easy it is for the developer to program in a given environment, and allows us to reason about what could be changed to improve the situation.

4.3.2 Validation of Workflow Heuristics

The heuristics to date have been developed by having researchers examine the collected data in detail (e.g., examining successive changes in source code versions). However, the accuracy of these heuristics is not generally known. We have developed a tool that provides a programmer with information about the inferred activities in real time. Using the tool, the developer then provides feedback about whether the heuristic has correctly classified the activity. This allows us to evaluate how well the heuristics agree with the programmer's belief about the current development activity.

4.3.3 Views of Multiple Subjects Across Several Classes

From the same data, total effort can be calculated for each developer and examined across problems and across classes to understand the range of variation and whether causal factors can be related to changes in other measures of outcome, such as the performance of the code produced.

We note that for our experimental paradigm to support effective analyses, subjects in different classes who tackle the same problem using the same HPC approach should exhibit similar results regarding effort and the performance achieved. Moreover, we must be able to find measurable differences between subjects who, for example, applied different approaches to the same problem. In a previous paper [19], we presented some initial analyses of the data showing that both conditions hold.

Such analyses have been instrumental in developing an understanding of the effectiveness of different HPC approaches in different contexts. For example, Fig. 10 shows a comparison of effort data for two HPC approaches, 'OpenMP'

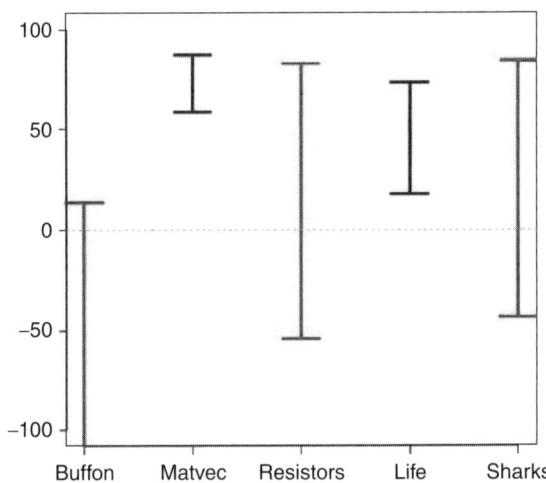

FIG. 10. Percentage effort reduction for OpenMP over MPI.

and 'MPI.' The percentage of effort saved by using OpenMP instead of MPI is shown for each of five parallel programming problems ('Buffon,' 'Matvec,' etc.) representing different classes of parallel programs. Thus, 50 on the y-axis represents 50% less effort for OpenMP; -50% would indicate that OpenMP required 50% more effort. The height of each bar represents the range of values from across an entire dataset of subjects. As can be seen, in two cases OpenMP yielded better results than MPI as all subjects required less effort; for two other cases although there were some who required less effort for MPI, the majority of data points indicated an effort savings associated with OpenMP. In only one case, for the Buffon problem, did MPI appear to give most subjects a savings in effort. As we gather more datasets using the Experiment Manager tool suite, we will continue this type of analysis to understand what other problems are in the set for which MPI requires less effort, and what it is about these situations that sets them apart from the ones where OpenMP was the less effort-intensive approach. (Interested readers can find a description of these approaches and programming problems in other publications [19].)

4.4 Evaluation

We have been performing classroom experiments in the HPC domain since early 2003. While we have not performed a careful controlled experiment of its effectiveness, we have observed anecdotally that the Experiment Manager avoids many of the problems others (including ourselves) have observed in running experiments. Many of these problems have already been reported in this paper. We have been able to run the same experiment across multiple classes in multiple universities and combine the results. Data is collected reliably and consistently across multiple development platforms. We have been able to obtain data to install into our database effortlessly without the need for students to perform any post-development activity. Faculty, who are not experimental researchers, have been able to run their own experiments with only minimal help from us. And finally, the ability to sanitize data allows us to provide copies of datasets to others wanting to perform their own analysis without running into IRB and privacy restrictions.

5. Related Work

There are various other projects that either support software engineering experiments, or support automatic data collection during development, but not both.

The SESE system [1] has many similarities: it is web-based and supports features such as managing subjects, supporting multiple roles, administering questionnaires, capturing time spent during the experiment, collection of work products, and monitoring of subject activity. By comparison, Experiment Manager supports additional data capture (e.g., intermediate source files and defects) and data analysis (e.g., sanitization and workflow analysis).

PLUM, back in 1976, was one of the first systems to automatically collect development data [24]. It, along with Hackystat [15], Ginger2 [22], Marmoset [21], and Mylyn [17] are examples of systems which are designed to collect data during the development process, but do not have data management facilities that are specifically oriented towards running multiple experiments. Hackystat, which we are using in the Experiment Manager, can collect data from several different types of applications (e.g., vi, Emacs, Eclipse, jUnit, and Microsoft Word) via sensors. It was originally designed for project monitoring rather than running experiments. We have adopted the use of some of the Hackystat sensors into our data collection system. Ginger2 is an environment for collecting an enormous amount of low-level detail during software development, including eye-tracking and skin resistance. Marmoset is an Eclipse-specific system which captures source code snapshots at each compile, and is designed for computer science education research. Mylyn (originally called Mylar) is also an Eclipse-specific system. Mylyn provides support for task-focused code development and includes a framework for capturing and reporting on information about Eclipse usage.

6. Conclusions

The classroom provides an excellent opportunity for conducting software engineering experiments, but the complexities inherent in this environment makes such research difficult to perform across multiple classes and at multiple sites. The Experiment Manager framework supports the end-to-end process of conducting software engineering experiments in the classroom environment. This allows many others to run such experiments on their own in a way that allows for the appropriate controls of the experiment so that results across classes and organization at geographically diverse locations can be compared. The Experiment Manager significantly reduces the effort on behalf of the experimentalists who are managing the family of studies, and on the subjects themselves, by applying heuristics to infer programmer activities.

We have successfully applied the Experiment Manager framework and with each application are learning and improving the interface, simplifying the use by students, making its use of value in shrinking the overall problem solving process by students, for example, various forms of harnesses, the support for analysis, in order to get a thorough understanding of the HPC development model.

Acknowledgments

This research was supported in part by Department of Energy contract DE-FG02-04ER25633 and Air Force grant FA8750-05-1-0100 to the University of Maryland. Several students worked on the Experiment Manager including Patrick R. Borek, Thiago Escudeiro Craveiro, and Martin Voelp.

References

[1] Arisholm E., Sjoberg D. I. K., Carelius G. J., and Lindsjom Y., September 2002. A web-based support environment for software engineering experiments. *Nordic Journal of Computing*, **9**(3): 231–247.
[2] Basili V. R., Selby W. R., and Phillips T.-Y., November 1983. Metric analysis and data validation across Fortran projects. *IEEE Transactions on Software Engineering, SE-9*, **6**: 652–663.
[3] Basili V., and Green S., July 1994. Software process evolution at the SEL. *IEEE Software*, **11**(4): 58–66.
[4] Basili V., July 1997. Evolving and packaging reading technologies. *Journal of Systems and Software*, **38**(1): 3–12.
[5] Basili V., Shull F., and Lanubile F., July 1999. Building knowledge through families of experiments. *IEEE Transactions on Software Engineering*, **25**(4): 456–473.
[6] Basili V., McGarry F., Pajerski R., and Zelkowitz M., May 2002. Lessons learned from 25 years of process improvement: The rise and fall of the NASA Software Engineering Laboratory. In *IEEE Computer Society and ACM International Conference on Software Engineering*, pp. 69–79. Orlando, FL.
[7] Campbell D. T., and Stanley J. C., 1963. Experimental and Quasi-Experimental Designs for Research. Houghton-Mifflin, Chicago.
[8] Carlson W., Culler D., Yellick K., Brooks E., and Warren K., 1999. Introduction to UPC and language specification (CCS-TR-99–157). Technical report, Center for Computing Sciences.
[9] Carver J., Jaccheri L., Morasca S., and Shull F., 2003. Issues in using students in empirical studies in software engineering education. In *International Symposium on Software Metrics*, pp. 239–249. Sydney, Australia.
[10] Choy R., and Edelman A., 2003. MATLAB*P 2.0: A unified parallel MATLAB. Singapore MIT Alliance Symposium.
[11] Dagum L., and Memon R., January 1998. OpenMP: An industry-standard API for shared-memory programming. *IEEE Computational Science & Engineering*, **5**(1): 46–55.
[12] Dongarra J. J., Otta S. W., Snir M., and Walker D., July 1996. A message passing standard for MPP and workstations. *Communications of the ACM*, **39**(7): 84–90.
[13] Hochstein L., Basili V., Zelkowitz M., Hollingsworth J., and Carver J., September 2005. Combining self-reported and automatic data to improve effort measurement. In *Joint 10th European Software Engineering Conference and 13th ACM SIGSOFT Symposium on the Foundations of Software Engineering*, pp. 356–365. Lisbon, Portugal.
[14] Hochstein L., Carver J., Shull F., Asgari S., Basili V., Hollingsworth J. K., and Zelkowitz M., November 2005. HPC Programmer Productivity: A Case Study of Novice HPC Programmers, Supercomputing 2005. Assoc. for Computing Machinery and Institute for Electronic and Electrical Engineers, Seattle, WA.
[15] Johnson P. M., Kou H., Agustin J. M., Zhang Q., Kagawa A., and Yamashita T., August 2004. Practical automated process and product metric collection and analysis in a classroom setting: Lessons

learned from Hackystat-UH. In *International Symposium on Empirical Software Engineering*, Los Angeles, California.
[16] Miller J., September 2000. Applying meta-analytical procedures to software engineering experiments. *Journal of Systems and Software*, **54**(1): 29–39.
[17] Murphy G. C., Kersten M., and Findlater L., July/August 2006. How are Java Software Developers using the eclipse IDE? *IEEE Software*, **23**(4): 76–83.
[18] Perry D. E., Staudenmayer N. A., and Votta L. G., 1995. Understanding and improving time usage in software development. Volume 5 of Trends in Software: Software Process John Wiley & Sons, New York.
[19] Shull F., Carver J., Hochstein L., and Basili V., 2005. Empirical study design in the area of high performance computing (HPC). In *International Symposium on Empirical Software Engineering*, Noosa Heads, Australia.
[20] Sjoberg D., Hannay J., Hansen O., Kampenes V., Karahasanovic A., Liborg N., and Rekdal A. C., September 2005. A survey of controlled experiments in software engineering. *IEEE Transactions on Software Engineering*, **31**(9): 733–753.
[21] Spacco J., Strecker J., Hovemeyer D., and Pugh W., 2005. Software repository mining with Marmoset: an automated programming project snapshot and testing system. In *Proceedings of the International Workshop on Mining Software Repositories*, pp. 1–5. St. Louis, Missouri.
[22] Torii K., Mastumoto K., Nakakoji K., Takada Y., Takada S., and Shima K., July 1999. Ginger2: An environment for computer-aided empirical software engineering. *IEEE Transactions on Software Engineering*, **25**(4): 474–492.
[23] Voelp M., August 2006. Diploma Thesis, Computer Science. University of Applied Sciences, Mannheim, Germany.
[24] Zelkowitz M. V., October 1976. Automatic program analysis and evaluation. In *International Conference on Software Engineering*, pp. 158–163. San Francisco, CA.
[25] Zelkowitz M. V., and Wallace D., May 1998. Experimental models for validating computer technology. *IEEE Computer*, **31**(5): 23–31.
[26] Zelkowitz M. V., Basili V., Asgari S., Hochstein L., Hollingsworth J., and Nakamura T., September 2005. Productivity measures for high performance computers. In *International Symposium on Software Metrics*, Como, Italy.

Global Software Development: Origins, Practices, and Directions

JAMES J. CUSICK

Wolters Kluwer

ALPANA PRASAD

Wolters Kluwer

WILLIAM M. TEPFENHART

Monmouth University

Abstract

The global software industry emerged in the wake of the first computers over 60 years ago. Computing was a global industry from its earliest days initially in the US and the UK. Today the industry touches all aspects of our modern lives in all corners of the globe. Increasingly this global industry also produces its products using globally dispersed and culturally diverse teams of scientists, engineers, technicians, and managers. This chapter explores the roots of Global Software Development (GSD), provides a detailed practice approach to conducting GSD especially with Indian suppliers, and examines current trends and their implications for the future. An exploration of the roots of this economic and technical phenomenon through the presentation of the earliest global software teams, their experiences, and how they laid the foundation for today's practitioners sets the stage. This review will place into context today's practices in GSD. Building on this foundation a detailed examination of current practices in GSD will lead to the introduction of a systematic and practical approach to conducting cross-shore development that is based on the experiences of one company which provides lessons for the industry at large. This practice approach builds on the history of GSD as well as specific adaptations of both engineering and managerial approaches to distributed development. This model for offshore development represents a tactical approach to modeling an offshore process for companies currently pursuing or planning to expand into offshore development.

Key practices that lead to success in this environment are documented and traps that can limit project effectiveness are pointed out in detail. Finally, the chapter discusses current trends in GSD and the implications for the industry in coming years. Among the topics considered include the likelihood of an acceleration in GSD, its limitations for expansion and adoption, new models of organization for effective leverage of global teams, and technical evolutions occurring due to the cross pollination of the industry and the emergence of offshore research and newly established centers of innovation. In summary, this chapter will start at the beginning of the GSD experience, provide detailed and deployable methods to conducting GSD, and point to the probable future of the field and its impacts.

1. Introduction 203
2. IT Sourcing Landscape 204
3. Global Software Development 210
   3.1. GSD as an Industry 210
   3.2. Origins of Global Development 211
   3.3. Strengths of Indian IT Industry 214
   3.4. Other Countries 216
4. Current GSD Practice 216
   4.1. Practice Introduced 217
   4.2. Practice Background 218
   4.3. Business Drivers 220
   4.4. The Supplier Selection Process 222
   4.5. Our Model for Cross-Shore Development 226
   4.6. Distributed Approach Details 228
   4.7. A Micro Engineering Process 230
   4.8. Production Support 234
   4.9. Knowledge Management 241
   4.10. Critical Loose Ends 242
   4.11. Things You Have to Live With 242
   4.12. Risks 244
   4.13. Collaborating with Vendors 244
   4.14. Results 244
   4.15. Future Direction in the Program 245
   4.16. Practice Model Concluded 245
5. A Virtual Roundtable on Outsourcing 246
   5.1. The Roundtable Mechanics 246
   5.2. The Roundtable Responses 246
   5.3. Roundtable Discussed 251

6. Future Directions in Offshoring 251
  6.1. Political Factors Affecting Offshoring 251
  6.2. Business Factors Affecting the Future of Offshoring 253
  6.3. Technology Factors Affecting Offshoring 256
  6.4. A Future Target for GSD 259
7. Conclusions 261
  Acknowledgments 262
  Appendix 1: Interview with K. (Paddy) Padmanabhan of Tata Consultancy
  Services 3/9/07 262
  Appendix 2: List of Acronyms 265
  References 267

1. Introduction

All around us there is software. In the contemporary world of technology, software plays a vital role in running everything from kitchen appliances to aircraft. The creation of all this software is now a truly global industry. Computers were born as an international industry some 60 years ago and today it is a global business which is growing in size, scope, and geography. Increasingly this global industry also produces its products using globally dispersed and culturally diverse teams of scientists, engineers, technicians, and managers. This chapter will explore the roots of Global Software Development (GSD), provide a detailed practice approach to conducting GSD, and examine current trends and their implications for the future.

The first step in this examination will be to explore the nature of software acquisition or sourcing models. A range of options will be discussed and the implications of each will be presented. The sourcing models available run the range from fully in-house to dominantly outsourced or offshored. This will lead to a discussion on Global Software Development and its variants. A special emphasis will be placed on understanding how this phenomenon developed and on defining its key characteristics. As software is such a global business, there is great interplay between companies, countries, and cultures. These aspects will be outlined and the pattern of technology transfer between the countries engaged in Global Software Development will be examined. These exchanges have led to the development of generalized practices used in running GSD and offshore engagements. A detailed discussion of such practices provides a practical tutorial on managing offshore projects and can be applied by anyone facing this challenge. This discussion is based on the in-depth experience of the authors' company and while it does not represent a broad set of experiences it does provide depth that can be exported to others interested in replicating successful offshore methods. Special focus is placed on the technical and engineering aspects of running projects in a distributed fashion. Finally, we will look ahead from today's environment to the near future. What trends do we see emerging, what impacts will there be from these endeavors, and what new opportunities will be presented to those working in this field.

The authors have dozens of years of combined software development experience in a wide variety of settings from start-ups to Fortune 10 R&D to academia. They also have experience teaching Software Engineering and developing new ideas and processes in the field. In addition, they have managed numerous offshore projects over the past 10 years working with a variety of partners in China, the UK, India, and Japan. It is from this base of practical and academic experience that the approach and recommendations presented here flow.

During the course of our careers, the importance of software has grown and the dependence on distributed sources has become endemic. The software field has changed and is changing the industries it touches. Virtually no product or service is produced today without the involvement of IT (Information Technology) with the exception of craft level production.

- Product design is done with CAD (Computer-Aided Design) tools,
- Focus groups are managed with databases,
- Advertising is planned using statistical models and demographic analysis on computers,
- Current product usage is determined by tracking the online transactions of users by scanners and POS (Point of Sale) devices linked to databases,
- POS systems report on every item moving through retail shelves [also the use of RFID (Radio Frequency ID) is growing], and
- Packages are tracked by GPS (Global Positioning System).

Thus, IT is indispensable and acquiring IT services is a buyer's game. Virtually every business needs IT and the choices available to acquire IT services are varied. From a business perspective, it boils down to cost with quality a close second. IT costs are primarily driven by labor, and so the sourcing of that labor is critical. The models for such sourcing can be laid out and examined by looking at current trends in software business arrangements.

2. IT Sourcing Landscape

To develop software today requires one or more physical locations with adequate facilities, including computing, networking, telecommunications, and tools. Most importantly, it requires a staff of trained, talented, and dedicated engineers, managers, testers, and other specialists such as business analysts and human factors engineers. Finally, capital is required. Assuming financing is available, there are a number of options available for acquiring such facilities and staff. At a minimum they include [11] the following:

- In-house development
- On-site contractors
- Onshore outsourcing
- Near Shore outsourcing
- Far Shore outsourcing
- In-house offshoring [Global R&D (Research & Development) or captive model]
- Purchased applications
- Hybrid sourcing

Each of these options has advantages and disadvantages worth exploring. In-house development offers the most control but in developed countries may have the highest cost. For some industries, such as defense, this may be the only option. A contracting option allows for scaling up and scaling down of resources more readily but also exposes the organization to a higher risk of turnover. Onshore outsourcing allows some work to be done offsite on a contract basis and can provide cost advantages as well as leverage the skills of the professional services firm. Near shore and far shore outsourcing vary only in the geographic locations. Near shore is considered to be in an aligned time zone proximity, while far shore may be around the world where day and night could be opposite. Cost is the primary advantage to both of these approaches while communications and coordination provide challenges as will be discussed in some detail below in the practices section. In-house offshoring is a model which large-scale international firms can employ effectively by setting up development centers around the world (e.g., captive model). This removes the middle man from the sourcing equation and leads to better company loyalty as the offshore staff are employees of the parent company. This taxes the corporation, however, in terms of overhead and administration. Also, software can be sourced through purchased applications, which changes the approach altogether. In this case, the company needs people skilled in the vendor technology in order to deploy and integrate with other infrastructure. The disadvantage here is that there is little flexibility in functionality once adopted.

Finally, the hybrid sourcing solution picks two or more of these approaches. Typically, companies have an in-house staff supplemented by some contractors on site and in many cases an offshore relationship as well. Thus, the benefits from all of these models can be combined. This assumes a layer of employees managing an onsite consulting crew who directly manage the offshore team. Naturally, all the difficulties inherent in these approaches also manifest themselves and require additional administration and management to keep things running smoothly.

A valuable question to ask is whether the existence of these choices is a benefit or not. If you are a US- or Europe-based business executive, today's choices are an opportunity [11]:

- It is possible to save 25–75% of your IT budget and get the same (better/worse) results.
- For a typical company spending 5% of revenue on IT, this could be a 2–4% increase in profit year over year (for large companies, this could be up to a billion dollars a year).

However, if you are a US- or Europe-based IT professional, today's choices for IT acquisition could be a problem:

- Average salary levels in the US are between 1 and 4 times greater than those of comparably skilled workers abroad.
- Current and future job prospects will be under severe price attack by foreign service suppliers.

These trends have reshaped the corporate organizations of the past. In Fig. 1, a typical corporate structure is presented. In it, there are several core functions run by a central corporate headquarters function. Everything lies within the corporation except suppliers who straddle the organizational boundary by being outside the company but closely aligned especially for parts suppliers or strategic partners. This model lasted for most of the 20th century [11].

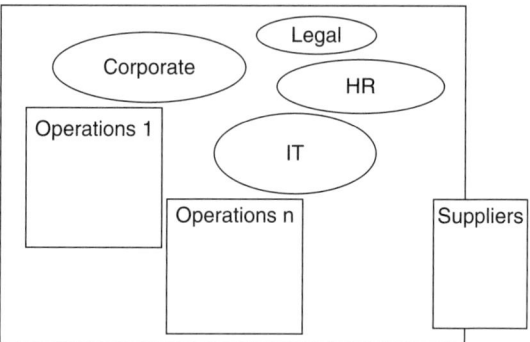

FIG. 1. The old corporate model.

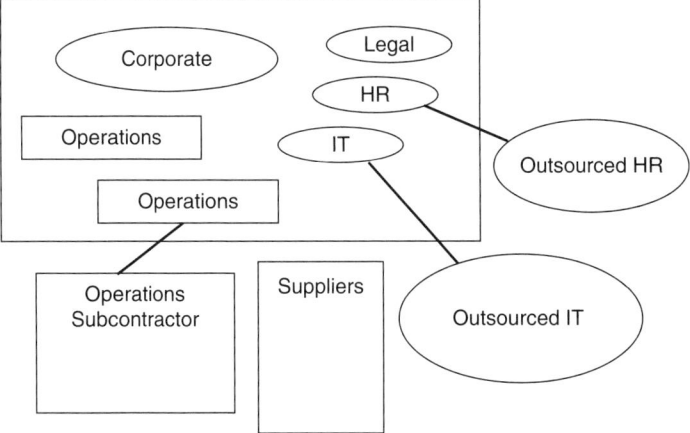

FIG. 2. The new corporate model.

Now take a look at Fig. 2. In this model, we see a major shift of corporate functions outside the traditional boundaries of the company. Here there are subcontractors of many types taking over key functions for the company and the former groups within the company have shrunk. These groups now become contract and service oversight for the outsourced functions. This has happened for most corporate functions through BPO (Business Process Outsourcing) including the following:

- IT (various capabilities)
- Human Resources (mostly routine benefits administration)
- Medical and legal transcription
- Accounting
- Helpdesk
- Customer service (account inquiry)
- Technical support
- And more

Within IT, almost anything can be outsourced either onshore or offshore. This includes requirements development, design and architecture, coding, testing, support, technical writing, network design and maintenance, and more. However, there are some functions which companies are keeping in-house. It is generally necessary to maintain a PMO (Project Management Office) function to run the outsourced projects and services. The PMO manages project initiation, staffing to one or more sources, budgeting, status, and project closure. Additionally, firms must decide what level of engineering expertise to keep in-house. It is beneficial to maintain requirements and architecture development to own the solution direction and retain core domain knowledge. On the other end of the life cycle, some degree of quality assurance will typically be retained. For example, test planning and test verification (verification that outsourced testing was completed effectively) [10]. Finally, some things need to be done onsite but can be done with subcontract personnel like deskside support and cabling.

Naturally, there are job implications from this new model. For traditional workers of typical US firms, there is job pressure and ample downsizing as a result of these new relationships. Some reports claim that 1 million US jobs have moved to India thus far [6]. However, some of this effect is mitigated by 'In-sourcing,' which is the creation of jobs by foreign corporations creating jobs in the US, for example. In Fig. 3, this offset shows an adjusted job gap that is much lower than the total outsourced figure.

These job-level changes can be put into a bigger perspective. Daniel Drezner [15] of Foreign Affairs states that:

> As for the jobs that can be sent offshore, even if the most dire-sounding forecasts come true, the impact on the economy will be negligible. [A] ... Forrester's prediction of 3.3 million lost jobs, for example, is spread across 15 years. That would mean 220,000 jobs displaced per year by offshore outsourcing. This number sounds impressive until one considers that total employment in the United States is roughly 130 million, and that ... [millions of] new jobs are expected to be added between now and 2010. Annually, outsourcing would affect less than 0.2 percent of employed Americans.

Of course for people losing jobs due to outsourcing, these statistics are of little solace. And the typical leadership response that more education is the answer

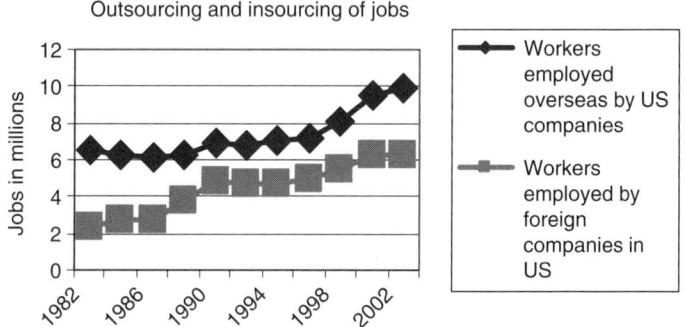

FIG. 3. Offsetting job migrations [2].

does not really apply for a master's level educated software engineer. They are already educated. The issue gets back to the earlier comment about what jobs may or may not be outsourced. For those writing code, their jobs have become commodity elements in the global job marketplace made possible by GSD and offshoring. Other sectors are yet to be affected. Pharmaceutical research is done using global teams but largely in the developed world. Life science education in India and China may soon produce quality graduates that will allow outsourcing of basic drug research and other medical research to take place in the developing world. This may then threaten the high skill and high wage jobs of thousands of US and European researchers and engineers just as we have seen with IT and other sectors.

In addition to the job impacts, it should be pointed out that not all sourcing deals work out well for one party or another. For example, there has been enough experience to date with outsourcing and offshoring that some deals are being undone. *CIO Magazine* states that '... we are starting to see the second phase of [outsourcing] – CIOs [Chief Information Officers] are renegotiating terms, contracts are expiring. Some deals just aren't working out at all ...' [7]. They go on to state that some 78% of executives who have outsourced an IT function have ended up terminating the deal early. It is common to hear in the news about functions that were moved overseas which were subsequently moved back onshore. Dell is a good example of this, recently pulling some of its technical support functions back onshore [46]. Additionally, Chase canceled a $5 billion outsourcing deal with IBM [37].

To summarize this discussion on sourcing, which lays the foundation for our exploration of Global Software Development, there are a few conclusions we can draw. First of all, the need for software is prevalent and the options for acquiring it are diverse. The outsourcing trend is deep and growing because of the low costs. Conducting outsourcing requires management changes and can be challenging due to cross-cultural issues and geographic dispersion. Outsourcing also allows one to focus on new core competencies like product innovation and also sets the stage for the ascendancy of certain higher value-added skills like requirements engineering and project management at the expense of programming. Other skills become required like the ability to develop RFPs (Request for Proposals), SLAs (Service Level Agreements), and other contract-oriented documents. Finally, this type of work requires a clear architecture and the ability to project a vision across company lines.

In terms of jobs we know that IT is critical to the economy but approaching commodity status. Sourcing decisions can both add and subtract jobs in the US and overall global IT jobs are growing with the world economy. However, for IT in developed countries, continued pressure on local jobs will be a mainstay for years to come. High-end and onsite skills will remain in the US, sometimes at a premium, but getting a foot in the door may be difficult especially for entry level positions as they will have mostly migrated overseas.

3. Global Software Development

3.1 GSD as an Industry

With this background on sourcing, we can better understand Global Software Development. Sahay defines GSD as 'software work undertaken at geographically separated locations across national boundaries in a coordinated fashion involving real time or asynchronous interaction' [48]. GSD is broadly used. Nearly half of the Fortune 500 currently engage in GSD and up to 50 countries are doing so and projections for GSD reach the $159 billion mark [39]. According to Duke University and Booze Allen, up to 45% of US companies engage in IT offshoring as of 2006 [36]. In a recent McKinsey study, this trend was summarized as follows:

> Any job that is not confined to a particular location has the potential to be globally resourced, or performed anywhere in the world. Broadly speaking, this includes any task that requires no physical or complex interaction between an employee and customers or colleagues, and little or no local knowledge [18].

The McKinsey description aptly fits many IT jobs, thereby driving the use of GSD up further. As we have noted, the implications for software is that many software functions can be offshored. According to Sahay, development in global settings remains empirically largely unexamined. New organizational patterns are developing and new social and political issues are being encountered with rapid economic growth in previously stagnant areas which still have underdeveloped populations living side by side with the newly prosperous.

Within this context, GSD has emerged as a potent force economically and technologically. Major outsourcing customers are the US, the UK, Australia, Western Europe, and expanding to Japan and Korea. US consumption reached $5.5 billion in 2000 and approximately $17.6 billion by 2005. It is estimated for India alone to reach $90 billion by 2010 [36]. Major 'Technopoles' have emerged in Ireland, India, and Israel, with emerging sources in Russia, the Philippines, and China [48]. Relative market sizes of outsourcing service sectors are provided in Table I.

The scope of this trend is staggering. In India alone, the offshore industry is projected to hit $50 billion by 2008. Ireland produces 60% of packaged software for Europe and Russia's industry is growing at 50% [48]. One recent McKinsey study concluded that up to 50% of IT jobs can be done offshore [5] while 11% of all global service jobs could be done anywhere in the world [18]. The rate of expansion in these supplier countries speaks to the high demand for low cost technical services. Some of this work is done in a pure outsource model and some is done in a joint

TABLE I
OFFSHORE MARKET SIZES

2003 Offshore Services Market Size Includes BPO and IT (in Billions)

Country	Market Size
India	$12.2
Ireland	$8.6
Canada	$3.8
Israel	$3.6
China	$3.4
Other Asia	$2.3
Latin America	$1.8
Philippines	$1.7
East Europe	$0.6
Mexico	$0.5
Australia	$0.4
Russia	$0.3
South Africa	$0.1
Thailand	$0.1
TOTAL	$39.40

Source: McKinsey [18].

cross-border fashion. We will focus on this later. However, it is worth pointing out that while impressive in scope and size, the 'entire Indian IT services industry represents less than a quarter of IBM's Global Services division' [33] or ∼3% of total software services worldwide [6].

3.2 Origins of Global Development

Understanding the current hype around GSD, it is instructive to take a step back and look at how technology development has occurred over the longer term and to look specifically at where GSD began and how it has matured. We should remind ourselves that global trade has been with us for thousands of years. Global Software Development differs from the ancient form of trade in that it is conducted in a semi-connected state through instantaneous communications, which was not possible in centuries past. However, the nature of trade between countries which GSD represents has been with us for ages. In fact, while communications were not instantaneous, there was a 'dialogue' between countries on technical matters. A thousand years ago Chinese papermaking spread Westward while Iranian windmills became known in China. Cotton textiles were first developed in India but the spinning wheel was brought from Iran into India. These interchanges of technology represented not one time exchanges but continuous dialogue around agricultural, mechanical, and civil engineering technologies [45]. It is this pattern of technical dialogue that has continued from those ancient years into today's world. The entire Indian IT capability was transferred from the West largely in one piece over a relatively short period of time. Today this technology dialogue is continuing with new innovations being invented in the developing world and being brought into the technical culture of the advanced countries. It will be instructive to look at these origins of outsourcing and the supporting technology transfer making it possible.

In looking at the origins of outsourcing we can look at the early examples cited above and we can also find references to outsourcing in early economic writings. In discussing the origins of outsourcing, Blinder [6] quotes Adam Smith from *The Wealth of Nations* in 1776 as stating that:

> It is the maxim of every prudent master of a family, never to attempt to make at home what it will cost him more to make than to buy ... If a foreign country can supply us with a commodity cheaper than we ourselves can make it, better buy it of them with some part of the produce of our own industry, employed in a way in which we have some advantage.

Clearly this is a call to outsourcing at this early date. In fact in the early days of US history, such items as covered wagons and ship's sails were outsourced to Scotland using raw materials from India [31].

3.2.1 Early Computing Sourcing

To trace the development of Global Software Development within the IT industry, one must start at the beginning of the computing age. There were numerous forerunners to the first computing machines dating back centuries. These were mostly calculating machines of various types like Schickard's in 1623 and Pascal's in 1642. Software itself has been dated as far back as 1812 with Jacquard's paper tape-driven loom [21]. From the very beginning, these calculating devices were built across Europe and then in the US. In the 1930s, several electronic calculating devices from IBM and Bell Labs began forming the technical base for the first digital computers. Arguably, the first such computer was in fact built in the UK. Colossus, built by the British in 1943, was a cipher breaking machine and while not a true multipurpose computer it was the direct forerunner of modern digital computers [55].

Harvard's Mark I went online in 1944 and influenced the further development of computers using punched card programming. It is well known that ENIAC (Electronic Numerical Integrator And Computer) was the first general purpose programmable computer built in 1946 in the US at the University of Pennsylvania's Moore School of Electrical Engineering. In the USSR, the first computer arrived in 1950. It was developed at the Kiev Institute of Electrotechnology in the Ukraine [27]. Commercial computers soon followed in 1951. The first was the UNIVAC I (Universal Automatic Computer). This computer used magnetic tape for input, output, and storage but could handle both numerical and alphabetic information. Shortly thereafter, IBM's 701 was released in 1953 [19]. These machines ushered in the modern computing era and a new international business sector.

The development of the first computers thus brought into existence programmers and software. Initially there was no possibility of physical separation of programmers from their machines. Programming had to be done through card readers or consoles attached to the machines. Development was always done on-site. These early machines were primarily located in the US and Europe and the labor markets for computer engineers and computer programmers grew up around these centers. Eventually outsourcing began to take place. Some of the earliest outsourcing occurred in the 1970s with the outsourcing of electronic payroll services. This expanded in the 1980s with accounting services, word processing, and other data processing services moving to specialty vendors [31] most of whom were onshore for their US customers but provided services to international customers as well as exporters of IT services.

3.2.2 Emergence of India as an IT Supplier

Since the 1980s outsourcing has been 'increasing ... across national and cultural borders, a phenomenon which is known as "global software outsourcing,"' and cost is a major driver as production occurs in lower wage countries [48]. This followed a pattern set up by the relocation of manufacturing to lower cost areas. Factories were moved from the NorthEast to the South and to the SouthWest in the 1980s to take advantage of lower costs and incentives [18]. Also, the government has been contracting essentially all of its IT work for 50 years. For many in industry this is new; the government has been successful by developing elaborate process controls around subcontracting. For industry developing these types of processes will be required [51].

This outsourcing was not restricted to onshore vendors. As early as 1974, there were Indian vendors providing outsource IT services [14]. The first recorded project was a venture between Tata Consultancy Services (TCS) and Burroughs. TCS (and other leading companies like Wipro and Infosys) developed their businesses despite government restrictions. (See Appendix I for a detailed interview on the origins of one vendor.) Soon regulatory changes began helping the industry grow. Importantly, in 1984 Prime Minister Rajiv Gandhi's government instituted the New Computer Policy which broadly reduced import tariffs on computing products and services [14]. During the 1990s, telecommunications bandwidth increased and costs declined.

Overall risks (perceived or real) in operating across borders were reduced [18]. This aided the development of offshore services in many countries.

In the case of India, early protectionist legislation put barriers in the path of the outsourcing industry. Thus, the early years were dominated by body shopping services where Indian programmers traveled to onsite locations outside of India. Advances in programming technologies and platforms including the PC and the Internet soon better enabled offshore development and this coincided with the easing of Indian regulations, allowing for more rapid growth of the business [14]. This growth was also fueled by investments made by multinational companies. The combination of foreign investment, relaxed policies, local ventures, and skilled labor supply led to the emergence of the IT industry with 21 companies of $4 million or more by 1980. As Dossani writes,

> The implantation of a technically sophisticated industry like software into a less-developed host country has typically been explained by the access of transnational corporations to local resources facilitated by policy reform ... [14].

The development of the Indian software services industry can be viewed in historical jumps based on market, industry, and technological evolutions, which were driven both by multinational corporations and by indigenous investment by local firms. From 1960 to 1970, the emergence of Independent Software Vendors (ISVs) allowed for the initial ventures in the field and were driven by the rise of the minicomputer. During the 1970s, wider development of custom applications emerged and a separation of hardware and software became standard practice. In India, the export of programmers and technical workers dominated. The 1980s saw a growth of complexity in software applications and lowered import tariffs in India allowing for lower cost services based in India. Finally, from the 1990s to today, managed services emerged arising on top of open systems and the Internet [14]. This has now laid the foundation for wider R&D and product-based innovation.

3.3 Strengths of Indian IT Industry

3.3.1 Large Human Resource

Every year, ~19 million students are enrolled in high schools and 10 million students in pregraduate degree courses across India. Moreover, 2.1 million graduates and 0.3 million postgraduates pass out of India's nonengineering colleges.

While some find jobs in other fields or pursue further studies abroad, the rest opt for employment in the IT industry. If the flow from high schools to graduate courses increases even marginally, there will be a massive increase in the number of skilled workers available to the industry. Even at current rates, there will approximately be 17 million people available to the IT industry by 2008.

3.3.2 Indian Education System

The Indian education system places strong emphasis on mathematics and science, resulting in a large number of science and engineering graduates. Mastery over quantitative concepts coupled with English proficiency has resulted in a skill set that has enabled the country to take advantage of the current international demand for IT.

3.3.3 Quality Manpower

Indian programmers are known for their strong technical skills and their eagerness to accommodate clients. In some cases, clients outsource work to get access to more specialized engineering talent, particularly in the area of telecommunications. India also has one of the largest pools of English-speaking professionals.

3.3.4 Strengths at a Glance

- Varied accomplishments in software development
- English language proficiency
- Government support and policies
- Cost advantage
- Strong tertiary education
- Process quality focus
- Skilled workforce
- Expertise in new technologies
- Entrepreneurship
- Reasonable technical innovations
- Reverse brain drain
- Existing long-term relationships
- Creation of global brands
- BPO (Business Process Outsourcing) and call center offerings
- Expansion of existing relationships
- Chinese domestic and export market
- Leverage relationships in West to access overseas markets
- Indian domestic-market growth

Source: www.nasscom.org.

3.4 Other Countries

With China, the situation is different. According to Zhang, 'The software outsourcing business in China is still in its infancy' [59]. China's economy did not open up until 1979 and there was little industry to build on. The IT services business started only in the 1990s. One of the authors worked with an initial R&D site in Beijing established as part of a US multinational in the late 1990s. The best engineers and graduates were recruited and they were delighted to be working for a foreign company. A number of these individuals would subsequently leave to join or form Chinese ventures in the technology services arena. The industry has grown to over $3 billion in 2004 with more than 8,000 firms, many of them under 50 employees each [59].

As has been pointed out, other key countries also developed IT services capabilities including Israel and Ireland. Each country leveraged a mix of characteristics to build its industry including incentives, education, and language skills. With the creation of these supplier sources, it is important to have concrete approaches to manage global software development across borders or in a cross-shore manner. The next section presents a complete model for the technical and managerial leadership of such globally distributed efforts. This approach has been used on numerous projects and continues to be applied today. It is offered as a guide to the sometimes challenging aspects of Global Software Development.

4. Current GSD Practice

Global Software Development has emerged as a research and practical area of experience over the past 15 years. As we have shown, the earliest computing projects were localized projects while the industry was global in scope. By the 1970s, large distributed teams were working across time zones. IBM was a pioneer in this form of work [9]. This practice became routine by the early 1990s and researchers and practitioners began documenting experiences and approaches for application in these scenarios.

The primary aspect to a global software team is its decentralized nature and its work across national boundaries [9]. These teams are distributed, rely on electronic communication, come from different organizations and cultures, and manage work via computerized processes. Typical problems encountered are around the difficulty of communications across great distance and offset time zones as well as a loss of 'teamness' and difficulty in coordination.

Numerous approaches to managing these problems have been discussed. Battin and colleagues [1] present a set of issues in GSD including distance, time zones, domain expertise, integration, government policies, and process. For each of these issues, they present tactics to counteract. These tactics include using liaisons, multiple communications channels, incremental integration, common tools, and others. These techniques influenced our approach detailed in this section.

Recent articles on GSD have been popular and widespread. Special issues of *IEEE Software* have sought to define the general success factors around global development. These issues (March 2001 and September 2006) report on requirements, best practices, knowledge management, use of componentization, educational models, managing outsource relationships, and the effects of distribution on teams lead to an established set of approaches to the challenges of GSD. A recurring conference dedicated to GSD also produces useful and far-ranging treatments of these key topics. ICGSE (International Conference on Global Software Engineering) is easily found on the Internet. Its 2007 conference was to be held in Munich (see http://www.inf.pucrs.br/icgse/).

The materials produced by these venues provide a strong base for discussing specific GSD practices. What we propose to do in this section focuses on the core challenges of selecting vendors and managing offshore projects and how to ensure technical success with distributed teams. We begin with a set of recommendations and then detail the approach which led to these recommendations. We cover staffing implications of our model and introduce the concept of an interim deliverable in order to control the quality of deliverables. Finally, we discuss those aspects of GSD which pose special problems and which managers especially need to be mindful of in order to succeed. This approach is largely reusable to any GSD environment but is based on our experience in working with multiple vendors onshore and in India.

4.1 Practice Introduced

As we have shown, global development of software is more the norm today than ever before [26]. For firms developing or maintaining software products, the impact and effects of global software development cannot be ignored. Our experiences in leading web development efforts with global teams led us to a set of concrete approaches which minimize risk and maximize effectiveness.[1] Managing multiple simultaneous projects and production support with a global composition of staff over the last several years has yielded both successful approaches and problematic traps to avoid. After providing a brief introduction on business drivers and supplier selection, we discuss in detail our processes for managing offshore teams and

---

[1] Based on [12]. © [2006] IEEE.

highlight key enablers and approaches that work [12]. We also outline those aspects of offshore development we have to live with. We believe that our findings can be applied broadly by technical managers, project leaders, and project managers to help ensure consistency, institute a common approach, and develop a best practice around global collaboration.

This section offers the reader a practitioner's view of our proven approach in an offshore model. We present our model for offshore development and insights into our management and engineering techniques which can be replicated in other environments. This work adds to the existing literature by providing a structural framework and guidelines necessary to ensure ongoing quality of offshore engagements.

4.2 Practice Background

The practical approaches described here have been developed and proven mostly at Wolters Kluwer and previous employers. Wolters Kluwer is a Netherlands-based international publisher and information services provider with operations around the world. The experience documented here focuses on global development teams managed from the New York-based Corporate Legal Services Division. The practices described are grounded on numerous projects performed primarily in the US (onshore) and India (offshore). Much of our traditionally onshore work has migrated offshore over the years through outsourcing deals made with multiple preferred vendors. This move mimics the market at large where 90% of US executive boards have discussed global delivery options [34].

Our offshore engagements today are in a mature state. We are in the fourth phase of our outsourcing agreement in the model described by Berry [4], which consists of (1) Strategy Development, (2) Selection Process, (3) Relationship Building, and (4) Sustained Management. This section focuses on the second, third, and fourth stages including sustained management and provides a complete management and engineering approach to working with our offshore collaborators. On the basis of our experiences, we have also derived numerous recommendations which we list in Table II.

The projects within our scope of experience have been as small as 1 or 2 developers and as large as 100 developers. All of these projects have been of the multitier web-based Application Service Provider architecture chiefly written in Microsoft's C# and ASP. Cycle times for these projects varied from 3 to 9 months. The processes used in developing these projects adhered to a CMMI (Capability Maturity Model Integrated) Level 2 framework.

TABLE II
SUMMARY OF RECOMMENDATIONS

Issue Category	Recommendation
Organizing for offshoring	Set clear criteria for offshoring, analyze whether a project is an offshore candidate or not. Break large projects into medium-size bundles for offshoring. Interim deliveries and reviews are key. Limit durations to keep control, shorter phases are easier to track and manage. Document requirements and baseline them. Formal document review and signoff should be required in order to move into development.
Communications and management	Both structured communication and unstructured communications are required. Structured communications provide regular means for status updates while unstructured communication encourages team bonding. Track all issues assiduously. Higher communication overhead forces more management attention, more frequent communications with formal tracking required. Distributed virtual teams require better planning; this drives both documentation and communications rigor. Onsite staff can be skittish when work moves offshore, must communicate career paths.
Managing staff	Maintain subset of teams leads release to release to build domain expertise, continuity of offshore staff should be goal. Retain domain expertise onshore and offshore, protect senior staff and reward them. Manage vendor experience level – rigorously vette candidates, take only the well qualified. Select leads carefully, leads can ensure success, take only ones with lead-level experience. Forecast resource needs early, lead time in getting appropriately skilled staff is increasing. Maintenance staff require application experience, rotate leads onto support releases.
Infrastructure issues	Infrastructure is a necessity to begin a project, work the network and facilities issues early. Develop core code infrastructure in advance, our model depends on a core framework. Mirrored computing environments require specification and investment, start early.

(*continued*)

TABLE II (Continued)

Issue Category	Recommendation
Managing development	Require adherence to best practices and standards, this drives in quality. Prepared standard ramp-up guide and enforce. Require interim deliverables, another key aspect to our model ensuring quality deliveries. Page development must achieve performance goals, this is a key technical requirement. Mange support through bundles of defects. All defect repairs reviewed onsite.
Quality and data privacy	Quality needs to be enforced through coding standards and verification. Protect data moving offshore through confidentiality agreements.

4.3 Business Drivers

Outsourcing is a critical factor in our business strategy. By 2001, it was apparent that the growing demands of the business community could not be met by the limited pool of resources onsite.

While cost was an important factor, our primary driver for outsourcing was not cost cutting or workforce reduction. The objective was to increase scalability by enabling better utilization of scarce resources onsite. This model provided us better return on investment while enabling us to support a larger number of projects.

Based on the business drivers defined, the outsourcing strategy was to retain domain expertise onsite at all times. We embarked on a workforce restructuring program where key roles of Program Management, Architecture, and Quality Verification were retained onsite. Day-to-day labor-intensive jobs of construction, test execution, etc. were mostly transferred offshore. The onsite developers were envisioned to grow up the value chain into Technical Leads who were responsible for end-to-end delivery of projects utilizing a team of offshore resources. A large population of Technical Leads, Architects, and Management staff was retained onsite to support an even larger population of developers offshore (Fig. 4).

The 2003 Restructuring Program announced by WK (Wolters Kluwer) was a catalyst for transforming the offshore program. By that time, individual business units already had limited offshore development programs on a selective basis. These were managed by the individual business units on an ad-hoc basis. There was limited process or supporting infrastructure to manage the process effectively.

WK IT Outsourcing Principles were defined and agreed to by the North American CTOs at the start of the CLS baseline project. These principles continue to guide the overall outsourcing program

WK IT Outsourcing Principles

- WK will maintain full control over IP of products
  - Comprehensive control and ownership over all software produced
  - Retain knowledge of design and implementation
- Management of software projects is lead by US based staff
- Software architecture is driven by US based staff
- Acceptance Test specification is the responsipility of US staff
- Verification of the configuration management process will be US based
- Standards around new product development should adhere to WK practices where appropriate
- Domain expertise must exist within WK for applications that are under consideration for offshoring
- All resources will be assessed on a case-by-case basis for applications that are considered to be either in legacy state or in pure maintenance mode

Resource Implications

- Program and Project Management is a US based responsibility.
- Architects and Senior Designers are US based staff positions
- The function of translating business requirements into technical specficatins will transition offshore overtime
- Coding, testing and low level design resources are in play on an application by application review
- Management and support staff has to be right sized over time based on the results of the offshoring program
- Domain expertise must remain/be built within WK

WoltersKluwer

FIG. 4. WK outsourcing principles.

The offshoring program was announced to the analyst community by the CEO (Chief Executive Officer) of WK with targeted cost savings in 2003. The direction provided by the CEO provided the structure for all CTOs (Chief Technology Officers) to work together and formulate a development strategy for individual business units to meet the ROI (Return on Investment) targets.

In order to institutionalize the process across all business units, the WK Shared Services Offshore development team was formed. The deliverable of the team was to work with the CTOs of the business units and define the offshore development process. Additionally, the team was required to develop a way to measure whether the program was truly successful or not.

The program was launched by identifying the application offshore opportunities across all units. This led to a comprehensive RFP (Request for Proposal) process in partnership with all the CTOs in each business unit.

India was the obvious choice for offshore development owing to its standing in the global marketplace as the optimal choice for offshore development and as described above. This was helped by the following factors:

- Large number of established companies in the arena
- Experience level of Indian software engineers

- Large software development population
- Infrastructure availability
- Government property laws
- Senior management familiarity with Indian processes and culture
- Fluency in English
- Workable time zone differential

4.4 The Supplier Selection Process

In order to gain maximum leverage, we engaged a third party firm to assist in the definition and selection process.

Our framework for selection was based on two primary dimensions – quality and price. Key stakeholders from individual business units provided weights for the quality criteria used in the RFP. Some special information like ERP (Enterprise Resource Planning) knowledge was also included.

The quality framework followed a 3-tier approach. The first tier comprised four primary categories – Company Background, Delivery, Processes, and Personnel. Each of these categories was further subdivided into subcategories and individual metrics at the lowest level. Weights were assigned to different criteria, subcriteria, and metrics based on detailed discussions with individual stakeholders. This provided a weighted index of qualitative metrics provided by all stakeholders. Some special information like ERP knowledge was also included. The evaluation components for reviewing vendors consisted of the following breakdown of factors including four main categories, 26 subcategories, and 150 individual metrics. The first two levels are listed below (Fig. 5).

The process for evaluation based on this set of categories included generating metric-level information based on supplier responses. Metrics examples include revenue growth, CMM certification, number of business continuation sites, etc. Next, comparisons of the metrics for each supplier were made using a weighted value. These were rolled up to arrive at an overall quality score. Price evaluation was based on the rates offered by the partners for the roles [Software Engineer, Technical Lead, QA (Quality Assurance) Tester, etc.] which they needed to fill. These included onsite, offsite, and offshore rates for each of the roles (Fig. 6).

The selection process comprised two phases. In the first phase, the RFP was floated to a wide selection of software development firms in India. Metric-level information was generated based on the responses provided by vendors. This was matched against the predetermined weights provided by key stakeholders across business units to generate scores for each subcriteria. These scores were then rolled

- Company background (19.25%)
  - Customer reference (5.0%)
  - Financial Information (4.5%)
  - Engagement structure (2.5%)
  - Geographical coverage (2.5%)
  - Key personnel (1.75%)
  - Quality assessment (1.5%)
  - Industry coverage (1.0%)
  - Type of company (0.5%)
- Delivery (39.75%)
  - Performance metrics (12.0%)
  - Subject matter expertise (9.0%)
  - Business continuity (8.0%)
  - Delivery model (3.0%)
  - Pricing practices (3.0%)
  - Relationship management (2.25%)
  - Subcontracting (1.5%)
  - Data backup (1.0%)
- Processes (18.00%)
  - Planning and estimating (6.0%)
  - Knowledge transfer (4.5%)
  - Software development lifecycle (4.0%)
  - Quality metrics (2.5%)
  - Release management (1.0%)
- Personnel (23.00%)
  - Capacity (11.0%)
  - Turnover (4.5%)
  - Training (4.0%)
  - Recruiting (3.0%)
  - Compensation practices (0.5%)

FIG. 5. Evaluation criteria.

up to generate the 'Overall Quality Score' of each supplier. The overall quality scores were reviewed by all the key stakeholders along with the recommendations of the team. On the basis of the findings, the stakeholders unanimously agreed on the top six companies which would be considered for the second round of evaluation (Fig. 7).

The second phase comprised detailed presentations by the vendors on topics of interest evinced by the key stakeholders of individual business units. A three-day marathon offsite meeting was organized where the second round participants made presentations on the topics of interest along with supporting materials. At the end of each presentation, the stakeholders provided their ratings of the vendor. At the end of all sessions, the scores were tabulated. Three companies were chosen from the six as the final partners to begin pricing negotiations (Fig. 8).

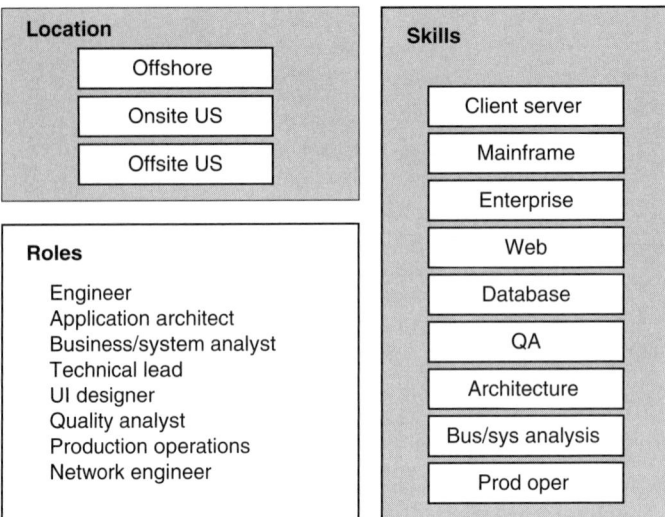

FIG. 6. Pricing evaluation criteria.

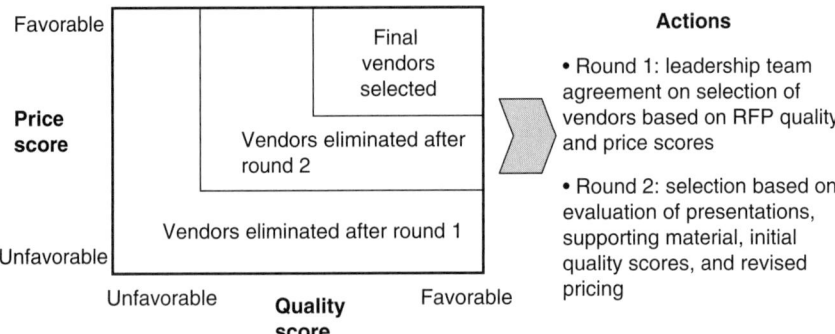

FIG. 7. Scoring matrix.

GSD: ORIGINS, PRACTICES, AND DIRECTIONS

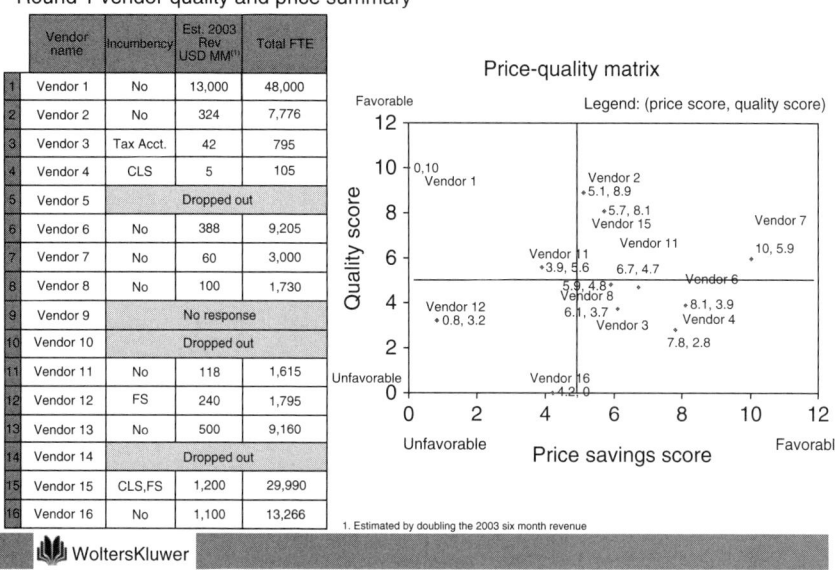

FIG. 8. Vendor quality and price summary sheet.

In order to avoid providing unlimited leverage to one partner, the plan was to work with at least two partners on an ongoing basis. However, three partners were selected as the finalists to allow for contingencies in case one of the partners dropped off during the price negotiations.

We then entered into pricing negotiations with the three finalists. Each finalist was provided a pricing percentage to adhere to for each role (Software Engineer, Sr. Software Engineer, QA Analyst, QA Tester, etc.). During the negotiation, volume was not tightly linked with pricing. This was an important criterion for negotiating rates in order to avoid creating issues related to volume-based pricing at a later stage of the relationship. Eventually, we formulated a Master Service Agreement (MSA) and a Rate Card across WK with all the finalists.

Key Success Criteria

- Senior executive backing is a must
- Involve necessary stakeholders at all levels in the decision-making process

- Send summarized information to executives
- Use mathematical model for comparisons
- Get all stakeholders together for final decision
- Gather scores immediately after presentation

Over time our offshore program has grown exponentially. Most of our customer units are utilizing preferred vendors. The number of offshore suppliers for IT initiatives has been optimized. All business units are leveraging the terms and conditions agreed upon by the MSA. Our model now enables us to deliver large-scale enterprise critical applications with a high degree of predictability. This model continues to grow and evolve to meet the ever growing needs of the business.

4.5 Our Model for Cross-Shore Development

Our model for cross-shore development follows the 'implementation' model suggested by Nissen [40]. This model has the client retaining key requirements and design functions and the vendor carrying out detailed implementation within a defined framework supervised by onsite leads. Our model requires that an onsite team lead is retained for each offshore initiative throughout the project life cycle. In this model, the offshore team is treated as an extension of the development team and not as a replacement.

There are some key success factors that underlie our model which we found early on and include the following:

- Careful setup and planning
- Knowledge transfer/training
- Use of a proven Web Delivery Foundation (WDF)
- Established policies and procedures
- Focus on communication and checkpoints

We follow a modified waterfall approach where the Concept, Analysis, and Design phases are primarily implemented onsite. Construction and Testing are primarily offshore. The resource breakdown and responsibilities are documented in Table III. This table indicates the staffing levels we typically deploy across the life cycle along with the key deliverables and who is responsible for each.

During the concept phase, we ramp up both the onshore and offshore technical leads and give them major tasks including high level requirements, architecture, and technical approach. We bring the offshore lead onsite to participate in the concept, analysis, and design phases as suggested in the literature [20, 41]. Any knowledge

TABLE III
LIFE CYCLE MODEL

Phase	Concept	Analysis	Design	Construction	QA
Location	Onsite	Onsite	Mostly onsite	Offshore	Onsite and offshore
Number of resources	2–4 IT resources	2–4 IT resources	2–4 IT resources	10–20 IT resources	10–20 IT resources
Key deliverables	HLR Estimates Assumptions/ risks New infrastructure	Detailed requirements	Class model Data model Interface definitions Sequence diagrams	Code Unit tests Code review	Defect fixing
Actors	Onsite and offshore lead	Onsite and offshore lead	Onsite and offshore lead	Onsite and offshore lead with development team	Onsite and offshore lead

transfer required is handled as a planned activity with clear deliverables. The onsite and offshore lead jointly sign off on the estimates and resource planning. During the construction phase, the offshore team takes primary responsibility for artifact development (code and tests) and delivery in accordance with defined coding standards and best practices.

The onsite team developed our core Web Delivery Framework (WDF) early on comprising infrastructure code, coding standards, best practices, and value-added tools. The WDF established the technical foundation to support offshore development and is explained in more detail below.

As each project matures, we ramp up the remaining offshore team as we get into the detailed design and construction phases. We also use an established set of process procedures, a standard estimation model, and a resource planning sheet to forecast staffing needs and plan specific responsibilities.

On average, our current onshore and offshore teams are distributed in the following manner:

Onsite Team Corporate Legal Services (CLS) (40%)
- Business Stakeholder
- Project Manager
- Technical Manager

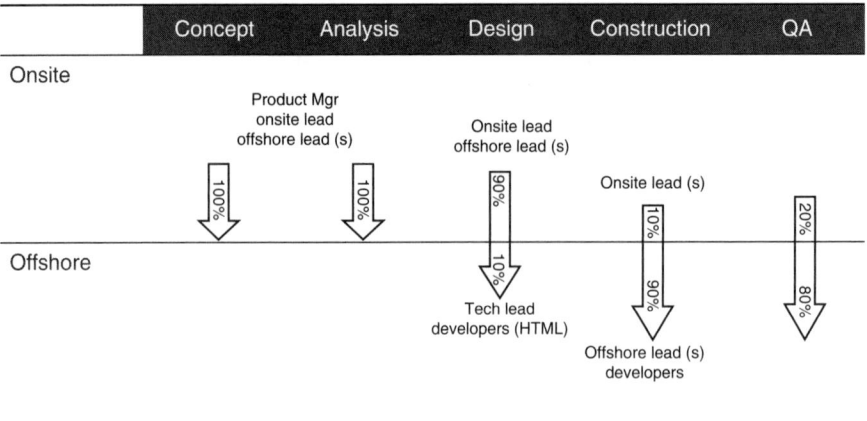

FIG. 9. Onsite and offshore balance.

- Technical Project Lead
- Business Analyst
- QA Lead

Offshore Partner (60%)
- Technical Project Lead
- Development Team
- QA Lead
- QA Team

Figure 9 represents a breakup of tasks between onsite and offshore during a project life cycle. The offshore participation in the project increases during the later phases of the project. Onsite oversight and control by CLS is maintained at all times.

4.6 Distributed Approach Details

Our model comprises management and engineering guidelines complementing each other to provide a comprehensive offshore development and management process. This process starts with a decision approach to determine which projects can be offshored. The process then covers tools, communications, planning, and the technical framework to support the offshoring. Research has shown that it is difficult to achieve iterative and incremental development in distributed development [44].

To counteract these difficulties, the use of design and code reviews, communications for fast iterations, a behavior pattern of 'immediate escalation of issues,' and frequent deliveries can be used. Our model incorporates all of these proven methods in a combined manner.

4.6.1 Key Management Guidelines

1. *Is the project offshorable*

Not all projects are well suited to an offshore model. Some key characteristics we consider in making a determination on whether to utilize offshore development on a project includes the following [24]:

- Business Process
- Interaction Requirements
- Complexity
- Current Cost
- Control Requirements
- Risk of Failure

In particular, we have found that new development projects of medium size on existing frameworks are most manageable in an offshore approach. We also consider whether there are touch points with other applications that may or may not be supported offshore. In our business, we interface with government agencies and other external organizations and if there are significant touch points with these external bodies our offshore strategy may vary. Unless the entire infrastructure footprint can be replicated offshore, numerous issues may be discovered during the integration phase which makes offshore development cost prohibitive.

2. *Planning, Policies, and Procedures*

Projects which are candidates for offshore development need to be planned as such from inception. All deliverables required by both onshore and offshore teams are well defined in our process documentation. Roles and responsibilities are also clearly defined and specified in the SLA (Service Level Agreement). In addition, suppliers are responsible for delivery of code as per our coding standards and guidelines.

All functional and technical artifacts are signed off and baselined prior to transitioning the project offshore. The offshore lead then moves back to the offshore site to conduct the knowledge transfer and to oversee the construction phase. The detailed resource planning and tracking during the construction phase and the overseeing of the offshore team is the responsibility of the offshore lead.

3. *Communication*

As has been noted by many others [9, 56], above all, both structured and unstructured communications are vital to the work effort in a global setting. We follow a set of key principles to keep communications effective:

- We maintain a direct line of communication between the onsite team and the offshore lead retaining the closeness developed in the initial co-located phases of the project.
- All issues are communicated via an issues tracking sheet keeping track of the date of initiation, originator, assignee, description, classification, status, and more. Most issues are resolved at the lead level. Critical issues are escalated to the management governance body as defined at the outset of the project.
- Weekly team meetings with the offshore team are held to monitor progress and discuss any open issues. These meetings also foster team spirit. The meetings are conducted mainly by conference call and recently by video conference.
- Depending on the project structure, there may be an onsite lead from offshore present throughout the project to answer any questions and address open issues.

4.7 A Micro Engineering Process
4.7.1 Infrastructure and Tools

Having the necessary infrastructure and tools to support a multicountry effort is critical as was pointed out by Carmel [8]. We have found as well that advanced planning and setup of supporting infrastructure and setting clear expectations of deliverables is critical to the delivery of a successful relationship. Items we concentrated on included the following:

- Physical connectivity
- Machine configuration and setup requirements
- Configuration management standards and guidelines
- Defect tracking standards and guidelines

Since our customers and product lines require a high level of security and protection, we established a dedicated LAN at the partner site for all developers working on WK projects. Over time, we also established WAN connectivity between NY and our primary partner in India. We developed detailed specifications on the machine configuration and software requirements for all developer machines. Templates and images were created which were used to replicate the onsite development environment offshore on a consistent and reproducible basis. This alleviated issues of a 'nonreproducible error' and ensured a stable environment for all developers.

At the outset we decided that all teams work off a common source control repository. The process and standards for configuration management were clearly documented. Similar processes were followed to extend and document the defect tracking repository to ensure its accessibility by all.

4.7.2 The Delivery Framework

Brand new infrastructure initiatives without an established foundation are generally not good candidates for offshore development. Since a large percentage of offshore engineers are new graduates, giving them a free reign could result in subquality code. Setting up a framework for offshore delivery to ensure consistent, repeatable results is crucial for successful offshore development.

This led us to the creation of our WDF (Web Delivery Framework) which encapsulated all the fundamental coding practices we wanted to insert in any project. The WDF comprises four pillars:

- Application building blocks (ABBs)
- Supporting coding standards and code review guidelines
- Value-added tools
- Best practices

4.7.2.1 Application Building Blocks.
Every software development company requires a core set of application building blocks which form the foundation layer for all development. A robust ABB is a necessity for the success of offshore initiatives.

We developed the application building blocks to be used by all offshore developers as a part of our first pilot initiative. This was done in parallel to the requirement gathering phases such that it did not impact the project timeline. While building the ABB, we had the option of adopting the Microsoft Application Building Blocks or other open source code. *However, our primary objective of the ABB was to provide a limited set of ways in which the offshore developers could implement.* Open source foundational components by their very nature provide multiple interfaces which allow developers the choice of approach to be adopted. In an offshore engagement, this can quickly lead to unmanageable code. Learning from the best practices of industry ABBs, we built the 'sandbox layer' of foundational components to be used by the offshore team at all times.

Establishing the application building blocks provided us the following benefits:

- *Code consistency:* The infrastructure code establishes a 'sand-box model' of application building blocks with supporting coding guidelines to provide consistency within the development architecture. Such a sand-boxed approach helps to improve consistency and reliability of the overall code base.

- *Approach consistency*: Developers used a consistent approach to writing code, always starting with the data access, layering the business objects on top of it, and finally plugging in the web pages.
- *Developer productivity*: Systematic ramp-up of all developers around the infrastructure layer plays a critical role in increasing developer productivity, thus reducing development timelines.
- *Safe code*: Critical components such as connection handling and caching were abstracted in the ABB and handled gracefully. Control over critical components ensured that all server and database resources were not abused.
- *Organizational standards*: We developed well-defined policies around the build, deployment, and management of runtime configurations across multiple development, test, and production environments. The infrastructure layer, serving as the sand-box, helps to enforce these policies and best practices.
- Clear guidelines and deliverables to team members.
- Consistent code quality in a cross-shore development model.
- *Estimation model*: The 'sandbox layer' and development approach provided a consistent framework which was used as the basis of our estimation model.

4.7.2.2 Supporting Standards and Code Review Guidelines.

In addition to the 'sandbox layer' of ABB, we created extensive coding standards, sample code, and ramp up documentation for the team. These encompassed the following primary areas:

- Concepts and Quick Start Guide for ABBs
- Sample Code and Unit Testing mechanism for Data Access and Business Layer
- Naming conventions and standard programming best practices
- Allowable data types and data model standards
- Query writing and optimization guidelines
- Component Inventory and Code Review corresponding to the coding standards

4.7.2.3 Value-Added Tools.

While the infrastructure code base and guidelines provide a foundation architecture, a set of supporting value-added tools and best practices ensure and validate application health on an ongoing basis. Key health indicators like performance monitoring and error tracking need to be baked into the overall delivery to ensure the quality and maintainability of the application.

We developed value-added tools to capture key metrics like page execution time starting from the development phase. The tools enabled developers to examine components comprising Page Execution time (Data Access, Business Object Load, and Save) to tune slow-performing pages early. Corresponding best practices were initiated to ensure all pages correspond to agreed-upon performance criteria prior to entering development shakeout.

Similar tools were developed to monitor other key criteria like Viewstate size, SQL execution time, and exception reports.

These tools enabled the onsite leads to keep an eye on key criteria and maintain code quality and performance while working with large offshore teams.

4.7.2.4 Best Practices. The ABBs, coding standards, and value-added tools together provided the necessary structure and tools in our arsenal which enabled us to define best practices required to identify issues well ahead of time during the development cycle and monitor the health of sites on an ongoing basis.

Below is a sample of reports created under our best practices (Fig. 10).

4.7.3 Interim Functional Delivery

To enable clear segregation of work between onsite and offshore, our model is based on functional separation of tasks between the two teams. While this provided clear separation of tasks and responsibilities, we faced numerous challenges when

#	Report name	Report description	Responsibility	Frequency - new projects	Frequency - production support
1	Error categorization report	Unique errors categorized by error description, no. of occurrences, defect number	Offshore lead	Daily starting from development shakeout through 2 weeks post production. Sign off required as entry criteria to QA	Weekly
2	Performance viewer report	Page execution time of 10 worst performing pages with exec time > 1 s	Offshore lead	Daily starting from development shakeout. Sign off from required as entry criteria to QA	Weekly
3	SQL execution time report	SQL execution time of all SQL's for identified customer accounts with large amount of data	Onsite lead	Deployment verification step while moving from one environment to another. Sign off required as entry criteria to each new environment	Weekly

FIG. 10. Sample management report.

attempting to integrate the code and then test end-to-end integration scenarios. Additionally, since the functionalities delivered back from offshore were delivered close to code freeze, it provided very little time to recover from integration issues. Teams attempted to address issues identified late in the project in panic mode, leading to patchy code.

To address these issues, a key milestone introduced during the construction phase is the Interim Functional Delivery (IFD). This represents a form of incremental development into the process. We learned to require that all main scenarios within the Use Cases are delivered in a functionally complete stage to the onsite team mid-way through the project. This enforces a practice of work allocation where all capabilities are functionally complete mid-way through the project as opposed to a subset of functionalities being fully complete while others have not even been started. This milestone is called the IFD and enforces incremental practice. Hence, if the project has an 8-week construction cycle, the IFD is received at the end of 4 weeks.

The IFD can be used in any environment irrespective of offshore or onsite. However, this is especially useful in an offshore delivery mode to ensure sufficient time for integration testing and review of code by the CLS onsite Technical Lead.

Figures 11 and 12 provide a pictorial comparison of a normal delivery cycle versus an IFD. As indicated in Table II, the IFD provides multiple benefits:

- The offshore team must ramp up faster since there is an onsite delivery mid-way into the project.
- The onsite lead starts reviewing code mid-way through the project providing sufficient time for corrections.
- Sufficient time is provided for end-to-end integration testing onsite.
- We have time to overcome the environmental and configuration challenges encountered while trying to deploy the new code onsite for the first time.
- Last but not the least, corrections can be done in a methodical manner as opposed to the panic-mode correction of last minutes issues.

4.8 Production Support

The production support or maintenance work which we have offshored follows a documented procedure as recommended by the literature [32]. The chief difference between new projects and production support arrangements is that knowledge transfer is done once upfront to transition the project offshore and thereafter no iteration takes place on moving the offshore lead to and from the onsite location. Apart from emergencies, PS is handled as bundled releases which follow a cyclic nature. Following is the procedure we follow for PS bundles:

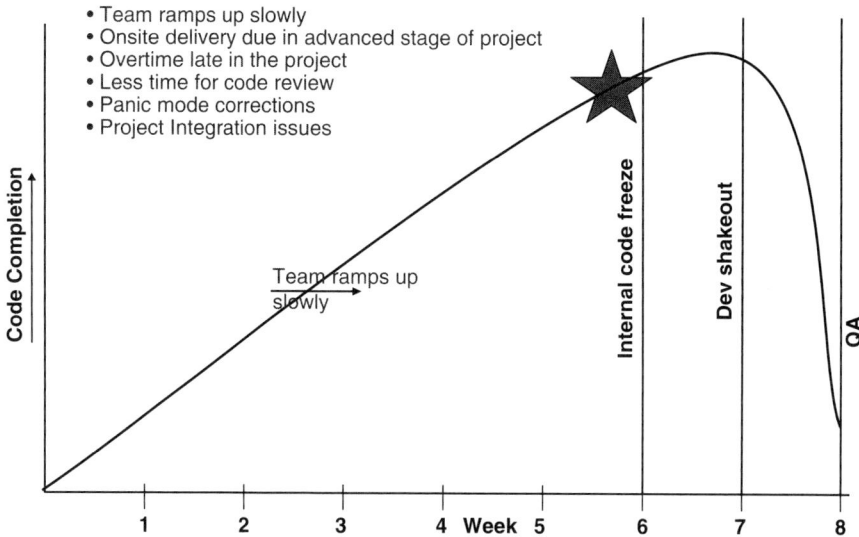

FIG. 11. A standard delivery model for code completion.

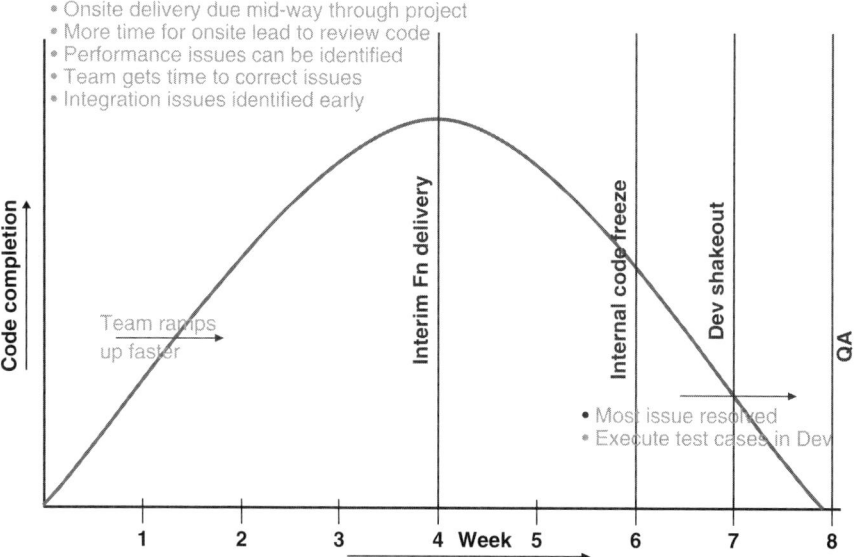

FIG. 12. An interim functional delivery model for code completion.

- The list of defects to be included in the bundle is reviewed and signed off by a business representative.
- Estimates for the bundle are completed jointly by onsite and offshore PS team.
- Analysis and design artifacts are updated by onsite and offshore PS team and signed off.
- Defects fixes are sent for construction offshore.
- Once offshore has finished construction and unit testing of the bundle, it is sent to the onsite PS lead for review and verification.

Key criteria for managing PS (Production Support) include the following:

- Since the PS team is typically smaller and less buffered than a project team, only experienced offshore members who have worked on a project are typically placed on the PS team. The team is continued through the year.
- An onsite PS lead is maintained at all times to review the code delivered from offshore which must adhere to coding standards.
- The list of defects/enhancements entering into bundled releases is closely monitored and reviewed by the onsite lead and technical stakeholders to ensure that big ticket items are not allowed into the PS track.

Once the new project execution process had been defined and stabilized, we initiated the process of transitioning the day-to-day maintenance of projects to the offshore team. The primary driver for transitioning Production Support to offshore was to free up valuable onsite development resources to focus on new project development.

We engaged our development partner to set up an Offshore Development Center (ODC) for ongoing Production Support activities. The profile of services included in the scope of the offshore development center included the following:

- Production Support
- Minor Application Enhancements
- Re-engineering and Maintenance Projects

Similar to new project development, the overall strategy was to maintain oversight of key Production Support tracks by employees managing large offshore teams.

4.8.1 Engagement Model

The engagement model for the ODC is composed of the following elements:

- A governance model that enables effective teaming of offshore resources with employees to provide high quality services.

- Key Result Areas: Cost Reduction; Capacity Ramp Up/Down; Free up employees for strategic initiatives.
- An organization structure of the ODC aligned with the internal organization structure of WK ensuring that every WK manager who gets services from the ODC has a partner counterpart from the ODC.
- Effective focused forums and mechanisms to provide the ODC with strategic direction, tactical management, and operational excellence.
- Clearly defined roles, responsibilities, and reporting and escalation procedures.

The processes and methodologies to be followed by the ODC were derived based on the Web Development Framework practices and best practices of partner. The processes covered the following:

- Knowledge and service transition processes to the ODC
- Service delivery processes
- Infrastructure requirements
- Software metrics for each service
- Business continuity and DR plans
- IT security policies
- IPR protection policies
- People development processes

4.8.2 Governance Model

The governance model for the ODC defines the key stakeholders and their roles and responsibilities. Key points in our governance model are as follows:

- Steering Committee (STC) comprising the CTO of the business unit and the partner Business Relationship Manager and WK Relationship Manager. Monthly scheduled meetings between the steering committee members to discuss key issues.
- Program Management Committee (PMC) responsible for monitoring the progress of all activities in the ODC.
- Application Leads comprising of the onsite employee and corresponding partner counterpart.

The organization structure also clearly segregates the responsibilities of the ODC across the strategic, tactical, and operational level ensuring a clear delineation of accountability within the ODC.

TABLE IV
STC AND PMC RESPONSIBILITIES

No.	Forum	Responsibilities
1	Steering Committee	• Provide strategic direction to the ODC • Approve scope, budget, resources, and schedule
2	Program Management Committee	• Identify and control necessary financial and personnel resources, information, etc. and to provide sufficient infrastructure and facilities to ensure unhindered progress • Identify and evaluate changes to the scope and content of work during the project which may impact the original budget and/or schedule • Approval of change management requests • Monitor the fulfillment of contractual terms • Compare project status against plan and identify corrective action to maintain progress toward meeting the objectives • Review significant pending and unresolved project-related issues and provide decisions or take suitable actions leading to resolutions • Ensure that all elements of the deliverables meet the specified quality objectives and goals • Approval of all the project deliverables

Each area (strategic, tactical, and operational) of the ODC organization has its own key result and key performance areas under which they are measured and these results and performance areas are aligned with the strategic vision of the ODC, ensuring an alignment of ODC organization with the vision for the ODC. The partner is responsible for measuring and tracking the performance of the ODC using a balanced scorecard approach.

Table IV summarizes the responsibilities of the STC and the PMC.

4.8.3 Infrastructure and Setup

Infrastructure and setup requirements for the ODC were defined and implemented with our offshore partner. These included the following:

- Sufficient VOIP (Voice over IP) lines for day-to-day communication.
- A dedicated link between our office and the offshore partner to provide secure access to intranet applications.

- Data provisioning strategy.
- Offshore staging setup for all production applications with remote connectivity to onsite leads. This was required for onsite leads to be able to test builds provided by offshore in their staging area prior to accepting onsite. In case of deployment issues such as application working in offshore staging while not working onsite, the onsite lead had the capability to remote into the offshore staging machine to validate configuration. This eliminated the necessity to wait for the offshore resource while it was night in India to troubleshoot deployment issues.

4.8.4 Knowledge Transition Process

The following diagram explains the knowledge transition process for maintenance (Fig. 13):

The various activities during the knowledge transition include the following:

- Identification of the provisioning of interfaces in the offshore development and staging environments.

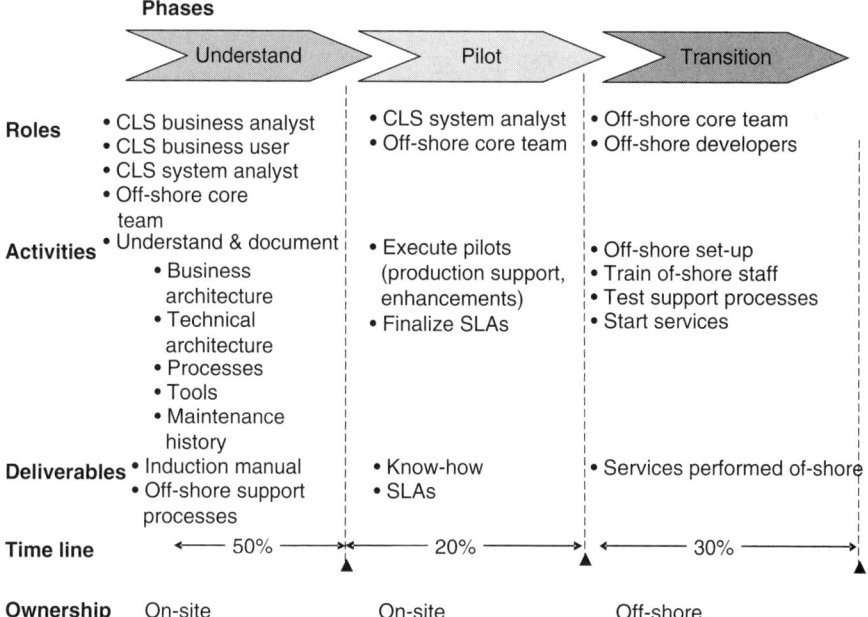

FIG. 13. Knowledge transition in practice.

- Identification of the scope of development and testing in view of the interfaces available at the offshore environments.
- Offshore data provisioning strategy.

4.8.5 Operational Process for Maintenance and Support

Maintenance and production support for mission-critical applications need perfect co-ordination between CLS and offshore teams, high levels of proactive monitoring, stringent service level adherence, quick reaction times, and well-defined processes.

The following diagram explains the overall context for the production support and maintenance of mission-critical applications (where CLS stands for Corporate Legal Services Division of Wolters Kluwer) (Fig. 14):

The Business Operations team works closely with the Call Center and provides Level 1 support for customer calls. It is composed of domain experts in the

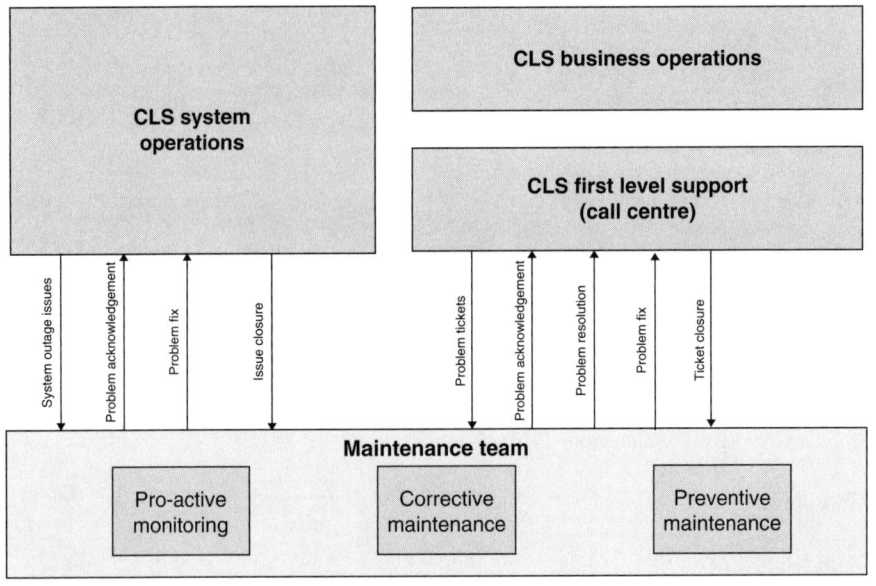

FIG. 14. Production support context.

individual applications. Approximately 80% of the issues can be addressed by this team without escalation to the next level. Issues which cannot be addressed by this team are prioritized and escalated to the System Operations team. This is composed of development resources from CLS and the partner organization. They work closely with the offshore development team to resolve the issues. The Systems Operations team is also responsible for proactive systems monitoring (performance, error logs, etc.) and performing corrective and preventive maintenance as required.

4.9 Knowledge Management

As more and more projects transition offshore, knowledge transition and management both onsite and offshore becomes a challenge. Domain expertise of onsite resources decreases over time as more and more work transitions to offshore. Higher churn of resources offshore creates challenges of domain knowledge continuity. To address these issues, we follow the following approaches:

- *Choose the critical projects where domain knowledge retention is mandatory:* With an increasing number of projects which need to be managed, IT needs to make a consumption choice between which projects can be completely outsourced versus which ones can only be offshored. In order to retain domain knowledge across key product lines, we consciously chose to focus our resources on the key projects. Other projects have been completely outsourced, typically on a fixed cost basis, and are monitored at a distance only.
- *Retain Offshore Engineering Manager across all projects:* An offshore Engineering Manager should be retained across all projects. The engineering manager is responsible for ensuring key best practices are followed across all projects. Over time, the engineering manager becomes an extension of the onsite development team and is a key participant in ensuring quality across all projects.
- *Retain at least one dedicated onsite lead for a project through project lifetime:* The onsite lead works closely with the partner lead while the project is in the concept/analysis/design phase. During the construction phase, the onsite lead retains a close eye on the project, at times also contributing to critical functionality. This ensures that the domain knowledge onsite is not compromised.
- *Keep a close eye on Production Support:* High volumes of enhancements/fixes moving through the Production Support pipeline creates the distinct possibility of degradation in code quality. Onsite leads should maintain oversight on Production Support at all times. Best practice guidelines should be followed to ensure errors and performance issues are proactively addressed. Automated code coverage reports and unit testing should be used where possible.

4.10 Critical Loose Ends

In addition to these fundamental process steps, there are several key points that must be kept in mind when deploying such an approach. These issues are discussed below:

- *Retain local domain experts:* With a larger number of offshore initiatives, dependence on the offshore team increases over time. However, the control and supervision of internal technical leads and domain knowledge should not be compromised.
- *Manage vendor experience levels:* Expect entry level talent to require some grooming. The principal offshore vendors are growing so fast that their experienced talent moves up or out quickly. Thus, they put a lot of junior people who need extra guidance to work effectively on projects.
- *Select the leads carefully:* The vendor leads for both onshore and offshore roles are critical to the success of the project. As the rank and file tends to be very junior, it is the leads who ensure delivery. They must have strong communication to interface with the onshore team. We look for a minimum of 5 years experience with at least 2 years in a lead capacity.
- *Forecast resource needs early:* Finally, it is more and more difficult to get good onsite staff from our vendors on short notice. Competing projects, resource shortages, and visa availability all need to be planned for in advance in order to get staff when required.

4.11 Things You Have to Live With

Through our work with offshore teams, we have also gained some key insights which we have learned need to be accepted and managed. These are things which are part and parcel of any offshore initiative which we do our best to accommodate. The list includes the following:

1. *Higher documentation overhead:* Projects for offshore implementation necessarily have higher documentation overhead. Detailed documentation needs to be produced to ensure clarity. All documents have to be signed-off prior to transitioning offshore.
2. *Locality:* You can no longer walk into a developer's cube to get status or make a request. You often need to wait for a full day to go by before you can get status. This means you have to plan ahead further and anticipate problems better. This leads to the issue tracking sheet which becomes an important tool in dealing with this reality.

3. *Higher management overhead:* Close management oversight is critical to the success of offshore initiatives. Lack of clear assignments or monitoring leads to gaps in delivery. Detailed planning, weekly checkpoints, team meetings, regular review of issues, and status reports are a necessity.
4. *Environments:* In our environment it has been difficult to set up a replica of some legacy or ERP infrastructure offshore. Projects which have only partial environments replicated offshore have had significant integration issues once the code was bought back onsite. Teams must evaluate their environments for portability and may need to execute tactical projects to convert code bases before offshoring.
5. *Quality:* We have learned through experience that quality levels are driven by the client – you get what you ask for. This has made us more careful in specifying our needs. This must be reiterated on every release to ensure compliance. We recommend specifying coding standards in detail and enforcing them. Also, technical measures like transaction throughput can be agreed to and monitored for achievement.
6. *Culture differences:* Cultural differences do play a role in running software projects between diverse locations. In our case, differences between American and Indian approaches often became apparent. We found that Indian engineers would infrequently push back or report problems. Further, if left to their own directions, they would rarely, if ever, take initiative and offer up creative solutions. These findings mirror those of Hofstede [28, 29] who characterized Americans as leading and Indians lagging in his individuality index. To manage these differences, we probed carefully on status and progress in status calls and maintained peer-to-peer dialogues with our on-staff engineers to work around hierarchies.
7. *Staff impacts:* In our model, there are new opportunities for people with excellent communication skills, good architecture ability, leadership, responsibility, and a keen sense of how to leverage partner teams onsite or offshore. Junior developers and senior software engineers need to actively develop the soft skills which are becoming increasingly crucial to their success. Managers need to communicate to staff regarding future opportunities to keep people engaged. In our experience, the job impacts have been neutral up to now as we always planned on using our offshore partners to expand capacity not decrease local staff.
8. *Customer data privacy:* Offshore development necessitates that we periodically transfer data offshore. Customers sometimes express concern over the security of data being transmitted offshore. All offshore vendors and employees working on the team are required to sign a confidentiality agreement with WK. Additionally, any private data is masked with a set of prepackaged scripts prior to sending offshore. This ensures the confidentiality of customer data.

4.12 Risks

Conducting offshore development brings with it some risks. The first risk is that of reduced productivity due to distributed team locations. Teasley [50] reported that in collocated teams, productivity is much higher as is job satisfaction. Further, Karolak [30] described common risks for Global Software Development projects to include decreased morale, loss of face-to-face interaction, and a lack of trust between teams. In our experience, we have been largely able to avoid these risks through collocating team leads with the onshore teams and circulating them to the offshore location periodically to provide a human bridge. From a sustained management perspective, it has proven much more difficult to communicate changes and get clear understanding of significantly modified approach and requirements with offshore teams as compared to co-located teams.

Over time, we also face the risk of domain knowledge and expertise diminishing onsite as most of the construction is done offshore. We mitigate this by ensuring that an onsite representative is involved in all key projects and work hands-on in defect fixing as necessary.

4.13 Collaborating with Vendors

Our offshore partners bring a variety of strengths to each project. First, they are eager for the business and the staff is willing to work long hours. In most cases, they bring adequate skills to the project. They routinely get the job done to match specifications. We count on them and they are very reliable.

On the flip side, we have found some drawbacks in working with our vendors. They normally work only on specific tasks as directed and can show limited creativity when faced with problems. This tendency forces more documentation on us than if the work were done onsite. Finally, infrastructure can pose problems and requires detailed planning for data sharing, intranet access, etc., which are all primarily logistical issues but can slow down projects on a tight schedule if not adequately planned for in advance.

4.14 Results

Over the last several years, we have made dozens of releases using our offshore model on three major platforms. While we do not have shareable metrics on the release performance, empirical evidence shows that these releases have all been delivered at or near their estimated release dates with schedule variance of less than 5%. Additionally, the defect count of the releases have been similar to what we encountered with onsite initiatives. Business and customer satisfaction for these releases have been high as we have been able to deliver more releases in a shorter time than we could have with onsite personnel only. These results offer proof that our offshore model has proven to be successful. As a testament to the success of the model, the model has been extended to other WK subsidiaries in the last two years.

4.15 Future Direction in the Program

Over time our offshore program has grown and matured. Starting from initial *ad hoc* project implementations, we are moving toward co-coordinated offshoring across all business units. Best practices and success stories within individual units are exchanged with others.

We have also initiated an audit with our partners to explore performance of sample projects across business units. This will be used to improve the processes being followed and continue growing up the value chain.

We began the Testing Center of Excellence (TCOE) and Offshore Development Center (ODC) in 2003. Over time these have evolved and now form the basic structure to enable co-ordination of processes and best practices across all business units. On the basis of successes in the past, we see our offshore program as continuing to grow and expand as we continue to explore ways to improve efficiency and provide better monitoring.

4.16 Practice Model Concluded

Our model allows for a truly globalized team. Today companies like ours find talent all over the world. The technical and intellectual infrastructure required to compete is relatively low cost and is transportable. *'Everything that can move down a wire is up for grabs'* [5].

In the onsite world, we are moving to a model that has a light developer core and a heavy project lead and architecture layer with development done offshore. The onsite staff needs to think at a system engineering level and offshoring allows engineers to focus on end-to-end problems if they are prepared to make that leap.

The offshore team will need strong management as well and improved architectural skills to design for completeness, modularity, and clarity. In this model standardization and communication of procedures is key, especially for consistency in requests and follow-up. Finally, awareness of cultural issues and clear expectations are paramount for all sides. In this way, offshore collaboration models such as ours can be even more successful in the future.

5. A Virtual Roundtable on Outsourcing

5.1 The Roundtable Mechanics

In order to highlight some of the issues discussed above, a survey was conducted on key aspects in outsourcing. The survey participants ranged from managers to executives in large US multinationals including Telecom Services, Telecom Equipment, an India-based services contractor, and an independent consultant. There were a total of 6 respondents out of 12 requested questionnaires. The surveys were responded to via e-mail. The survey was carried out in 2004 and first reported at a seminar conducted at Columbia University [11] but has never appeared in print until now.

The participants in this virtual roundtable were selected based on their broad industry experience and their direct work with offshore and outsourced environments. There answers to the survey questions are both revealing and to the point. It is hoped that their comments will put some additional context on this discussion of sourcing and GSD. The questions on the survey were as follows:

1. How has outsourcing affected your business?
2. What are the key criteria for a successful outsourcing deal?
3. What is the most successful blend of outsourcing (percentage or type of resources)?
4. What are the key business practices for managing an outsource deal?
5. What are the key engineering practices for running an outsource deal?
6. What types of process controls must be in place to manage outsourcing?
7. How is product quality affected by outsourcing?
8. How do you see the future of outsourcing, will it accelerate, change?
9. What is the net for US companies, overall beneficial?
10. What is the prospect for US IT professionals, how will careers be affected?

5.2 The Roundtable Responses

The verbatim responses are reproduced below.

1. *How has outsourcing affected your business?*
   - Allowed us to reduce costs from our internal operations; provided more process and measurement control of our systems initiatives.
   - Has also complicated getting new work done by having multiple suppliers to deal with when implementing new initiatives. Led to our needing to establish strong governance processes and controls over our suppliers.

- Very positively. While the cost structure continues to be a bonus, clients are also increasingly looking to outsource for reasons of specific skills and competence.
- Outsourcing has been challenging because the original structure of the contract had many areas that were not well defined in regard to roles, responsibilities, and deliverables. This led to much confusion between IT and the vendor and often created disputes on how work was to be performed and which organization was responsible for it. This situation has improved over a period of time but required much work by both parties to get it on the right path.

2. *What are the key criteria for a successful outsourcing deal?*
   - Picking a good partner to do business with. It isn't all just about cost. As the relationship continues over several years, the partnership aspect becomes key to resolving issues as well as to any changing business conditions.
   - It depends on from whose perspective you are asking the question. For client, cost, year-on-year productivity gains, flexibility in the contract structure, successful transition, risk minimization, trust, capability, infrastructure quality, etc. are some of the key factors. For the vendor, however, client lock-in and more business with a multiyear contract is a key consideration.
   - A good contract or statement of work that clearly defines responsibilities, deliverables, and pricing.
   - Must actively manage, and:
     - Must ask specifically what you want back.
     - Engineers will have to be better trained in writing and communicating.

3. *What is the most successful blend of outsourcing (percentage or type of resources)?*
   - If you mean onshore/offshore, it probably is good to have a 70/30 blend.
   - It depends on a company's business strategy, although a client would do well to keep a core IT organization for planning, architecting, coordination, and vendor management, as well as outsourcing in a way that risks are distributed.
   - Typically, application outsourcing, infrastructure outsourcing, and business process outsourcing are best candidates for outsourcing, although for BPO successful transition management is the key.
   - My perception is it should be 'all' or 'nothing.'. Operations should be outsourced or remain in-house but not split between IT and the vendor. A PMO organization should be established to oversee all vendor activities and staffed to provide sufficient quality control of vendor activities or in-house provided operations support.

4. *What are the key business practices for managing an outsource deal?*
   - Strong governance model that specifies how the contract will be governed. Strong SLAs that are defined and managed on-going and used to make corrections.
   - For a client, good RFP process, negotiation for flexibility in the contract, year-on-year productivity gains, tight SLAs, maintaining the control, a strong oversight process, asking for business innovation, competitive cost structures, IP protection, etc. are some of the key practices.
   - For a vendor, however, multiyear contract negotiation, good pricing, learning extraction, gaining domain expertise, productivity gain and optimal resource deployment, and cost management are some of the key considerations.
   - A PMO that lives with and understands the contract. This organization is responsible for quality and PM deliverables in addition to overseeing the vendor to ensure they deliver against the terms and conditions specified in the contract and/or statement of work.
   - Need to focus on legal requirements, SLAs, performance metrics more so than in the past. Pilot projects should be considered when ramping up. Communication is key. Indian firms, often CMM Level 5, can learn from them, may require more sophistication from US firm. Need good entry/exit criteria.
   - (For Global Development work) Crisp identification of responsibility for each subteam; constant communication.
5. *What are the key engineering practices for running an outsource deal?*
   - Strong control over architecture and where you are heading. You should direct suppliers. If you have multiple suppliers, you need to ensure they are all working together.
   - Good upfront analysis of new work and impacts that it has on the systems. We have a group call Solution Consultants that provide the first look at work to be done.
   - Good quality processes, reuse, and project management.
   - Discovery! IT needs to understand what makes up the IT environment, including 'shadow' functions that may need to be included in the deal, before considering outsourcing as an option.
   - Outsourcing companies are flexible, they will always say yes when asked to do something, not going to push back, especially in India social structure plays a role, less of a desire to report bad news.
   - Lack of face-to-face communications can limit effectiveness (onsite rep can help).

6. *What types of process controls must be in place to manage outsourcing?*
   - Strong governance with scheduled meetings to view issues. Stung review of SLAs and any corrective actions. Cross-vendor project management and release control is essential. We have a group that provides overall project management of cross-vendor activities and we also have a change control board that reviews any scheduled work going into production.
   - Service level agreement (SLA) based and quality of deliverables based, aligned to each delivery.
   - Financial, project, and quality controls must be in place to manage outsourcing. This includes mutually agreed upon metrics that are well defined and delivered at scheduled intervals to IT for review.
7. *How is product quality affected by outsourcing?*
   - I would stay initially it was worse' than over 2 years improved. However, we have still had some problems especially after the outsourcer has reduced staff. We have better mechanisms for tracking though than we did when it was internal.
   - The effect can be both positive and negative – depends on what kind of product, at what stage outsourced, the quality of the product before outsourcing, the outsourcing organizational model, the capability of the organization, complexity of the product, etc. Majority of instances I have seen have had positive impact on product quality.
   - Quality is possible if IT expectations are well defined up front. When expectations are not well defined, quality is jeopardized and often exploited.
8. *How do you see the future of outsourcing, will it accelerate, change?*
   - I believe that as companies seek to reduce costs and also devote more time to their core business, it will increase. I believe there will be increases in the business transformation outsourcing area, more than just in the IT area. Companies will look for the vendor to do more than just run things as they were; they are looking for improvements.
   - Of course, the trend is an irreversible one and will certainly accelerate, although the offshore geographies might change. The growth can be easily explained by transactions cost theory (in economics).
   - I think outsourcing will accelerate; many companies cannot afford to retain the required resources with current skills to do the work themselves.
   - Supply and demand will apply, trend will accelerate for now, will go too far, outsourcing wrong projects or against best practices, smaller projects may stay onshore. Will end up with the right mix of onshore and offshore projects.
   - Need to protect your core competencies to run your business.

9. *What is the net for US companies, overall beneficial?*
    - Yes – since it will reduce cost and provide more work to US companies (e.g., IBM, CSC, EDS). It also allows companies to focus more on their core competence and not worry about staffing areas that are not a core competence.
    - Yes, and there are studies confirming this. Notable studies are the McKinsey studies and a paper in recent issue of *Sloan Management Review*. But, certainly, it is a debatable topic.
    - It can be beneficial if expectations are set upfront, the contract is clearly defined, and both parties understand their roles within the specified parameters. The 'loyalty' factor is a concern with outsourcing as a salaried company employee stays with a problem after hours with no discussion about pay. A vendor will bring additional expense under this circumstance or perhaps have less interest in resolving the problem in a timely manner if service agreements are lacking. There are many pros/cons in this area and it really comes down to how well the 'deal' is structured and the cost.
    - Small, medium, large companies will adopt different strategies (onsite, low pay; outsource; global development, big problems).
10. *What is the prospect for US IT professionals, how will careers be affected?*
    - I believe the prospect for certain skill sets will diminish but there will be a need for other IT skill sets. For example, coding and testing will become more of a commodity skill set that can be done anywhere – offshore. I believe project management, business analysis, and IT vendor management will become more predominant skill sets in the US over time. There will a need to manage onshore or offshore outsourcing and project manage work across multiple suppliers.
    - How were US professionals' careers affected when significant US manufacturing was outsourced to China? Like manufacturing, IT is also increasingly becoming a commodity (see Nick Carr's 2003 controversial article in *Harvard Business Review*). For a professional, the key is to manage his professional value in the market and ensure that there will be demand for his skills and that those skills are not in abundance.
    - Some will move from the company they have been with and perhaps be absorbed or hired by the vendor. Some may have opportunities to join other vendors that provide services as this becomes more of a standard practice. Outsourcing to foreign-based entities may reduce the available domestic IT job base and potentially drive some to consider alternate career opportunities.

- Education is failing US industry, skills are below that of other countries, current outsourcing relies on architects to give specs to programmers, how will these architects be trained in the future?
- The search for talent is a global one. To succeed, individuals need breath (for flexibility) and depth (for unique value).

5.3 Roundtable Discussed

The expert roundtable discussion above underlines several of the points made throughout this chapter. Outsourcing brings lower cost but also greater complexity of operations. It can be challenging but rewarding for the company pursuing it. Running outsourced agreements requires strong governance and management by specific measures. From a technical point of view, one should retain strong architectural control and expect some initial drop in quality. Finally, this trend is seen as accelerating and it will have some impact on jobs in the countries that are offshoring work. Such observations are valuable in validating the research discussed as well as the GSD practice detailed above. To be successful with offshore outsourcing requires strong management and organization as well as appropriate technical capabilities.

6. Future Directions in Offshoring

Now that we have taken a broad look at sourcing and GSD as well as a deep dive into current offshore management practices, a look into the future is warranted. It should be emphasized that the practices described in this chapter are optimized for the current GSD environment. As the future unfolds, these practices will have to change to take into account the evolving GSD environment. There are some key topics related to GSD which bear discussion.

6.1 Political Factors Affecting Offshoring

The evolution of international political relationships will introduce risks and advantages that will influence the evolution of offshoring. The current international political climate is not all that stable, but all offshoring strategies demand stability.

6.1.1 Political Stability at the National Level

Change at an international level can occur with blinding speed. In January of 1978, the relationship between Iran and the US supported 4 billion dollars of imports by Iran. By January of 1979, only four months after a report by the CIA had asserted that the government of Iran was stable and would stand for another 10 years, the Shah had fled Iran and a new government had been formed. Sanctions in April 1980 left a number of companies sustaining major losses of revenue and investments [17].

Very few people predicted the fall of communism in Eastern Europe and the Soviet Union. Only a little more than two years passed from the date of Reagan's famous speech asking Gorbachev to tear down the Berlin Wall on June 12, 1987, to when the Berlin Wall was opened on November 9, 1989.

A number of countries that are currently sources in the GSD environment are considered moderately or significantly unstable.

6.1.2 Global Economic Stability

Like never before our world is tied closely together. Major upheavals in the world can now affect our technological infrastructure quickly and severely. An example is the monsoons in India. Years ago these were of local concern to the people directly affected. Now, if the streets of Chennai are flooded, our software and systems support teams cannot make it to work and their home computing infrastructure is not sufficient to allow them to work remotely. Imagine if political tensions or wars impacted key areas where US technical infrastructure is managed from. This could have a devastating effect. Current disaster planning does not generally take this into account.

6.1.3 Cross-Country Alliances

With a globally distributed IT infrastructure, the cooperation of multiple nations becomes required. Some code may reside in India and some in China or Russia. The economic relationships between these countries and the US become critical to sustain technical cooperation. If tariffs are established or trade barriers are erected, this could add cost to agreements and strain relations. This thought was spelled out recently as reported by Macintyre in discussing the 'triangular dance' between China, India, and the US. He reports that 'relations between the three big powers will outweigh all other ties' [37].

6.1.4 Local Effects

The economic multiplier effect will be active as new billions of dollars find their way into these diverse economies. With 10% of India's GNP tied to IT services, monetary benefits will spread out to the local economies. It is hard to predict how the economic growth in China and India will affect the populations there. Those people directly employed in foreign firms or local outsourcing firms will have relatively huge disposable income. This income will find its way into local goods and services and create jobs in construction, support, and new kinds of ventures. GSD is increasing the underlying economies of the major developing countries but may have a much more significant effect on the developing countries. For India and China, 40–60% of their labor forces are engaged in agriculture respectively as compared with 2.5% or less for the US and UK [18]. Changes in this proportion over the next 10 years due to advances in GSD are quite possible. In India, this expansion of the economy is already affecting small villages. In remote villages, major Indian companies are currently outsourcing their work to start-up firms with limited facilities but where English skills are adequate [22]. Naturally, there is a long way to go as over 300 million Indians live in undeveloped villages and literacy runs at only 33% but the effects may be profound [23]. It is not too far an exaggeration to believe that GSD can lead to a transformation in the economies of these countries.

6.2 Business Factors Affecting the Future of Offshoring

The business of software development is still relatively young compared to other manufacturing industries. This has introduced some business instabilities that require current practice to manage. It is a far different problem to outsource the manufacture of some widget to an overseas factory and to outsource the development of software to an overseas shop. With time, some of those management issues will be solved and implemented as standard practice.

6.2.1 Competition Among Suppliers

One can see a new landscape forming in global software. A bifurcate is developing between the US and many developing countries supplying software professionals. It appears that the US remains strong in new concepts, management, tools, languages, and domain engineering. The Far East is tending to develop language-specific applications, advanced manufacturing systems, and embedded systems. India and other developing countries are maturing in a wide range of systems development functions leveraging cost differentials [10]. Interestingly, costs are also adding layering into the supply chain. Russia is cheaper than India and Mexico is cheaper than Russia. This is shaping new relationships and partnerships in global development.

The specification and operation of many systems will occur in the US and the development will occur in multicountry development formats or completely offshore. Today some companies operate globally with R&D facilities throughout the world. The authors have witnessed technical interactions which have included engineers and managers from Ireland, France, Spain, England, India, China, and the US on a single project. Each of these sites was staffed by multidisciplinary teams who contribute when and where needed. At times the split in responsibilities is due to technical specialty, for example, database work versus network protocols. At other times, the split is on customer facing versus implementation grounds [10]. This changes the demand side of the equation for IT professionals around the world.

6.2.2 Talent Supply

In some countries, the supply of appropriately talented individuals cannot keep up with demand. In India, schools are struggling to provide enough qualified engineers. It has been stated that only 26% of graduates are employable. This could raise wages and also reduce the ability of offshore suppliers to meet demand in the developed countries. McKinsey reports that, among other factors, many professionals in developing countries may not live near major cities or be willing to relocate [18], which reduces the global talent pool. Specifically, among the 33 million university educated professionals in developing countries the majority were not suitable for the global workforce; in fact they found that only 6.4 million were suitable.

The numbers on the demand side are significant. Leading outsourcing firms such as Infosys and Tata Consultancy Services are hiring as many as 2,000 people a month. US-based companies are also hiring deeply [54]. Accenture plans to have more people in India than in the US (a total of 35 000 in India by year end 2007) and IBM has upwards of 53,000 staff and growing in India. A major fallout from this rush to increase staff is a decrease in qualified managers to lead these new armies of IT workers. Managers are grown over many years and require training and experience to manage well. In India, today some staffers can jump into management and even executive ranks rapidly rising with the tide [55]. It is not clear that these managers will have sufficient maturity to lead in difficult times or on challenging projects.

These wild increases in global IT staffing come after years of ups and downs in the US IT labor market. The authors have observed massive downsizings and outsourcings in both good and bad times.[2] Offshoring and other trends, including

---

[2] Based on [10]. © [2003] IEEE.

the commoditization of core infrastructure software such as in ERP and financial systems, have limited the need for new application creation skills. Vendors and integrators remain in demand of such skills but many mainstream IT shops have a much decreased need [10].

Throughout the 1990s, researchers like Howard Rubin reported on a glut of software jobs. At times this figure exceeded a million available positions in the US alone [47]. These kinds of statistics made the engineer confident that jobs were available and made employers shudder at the prospect of hiring replacements where none could be found. As hiring managers we experienced first hand the paucity of good talent at the height of the boom circa 1999. Some positions really did go unfilled at that time.

Immediately following these flush times, only one or two brief years later, there were tens of thousands if not hundreds of thousands of veteran developers out of work. Thus, the job surplus moved from a crisis for employers to a crisis for workers with a swing of at least a million jobs seemingly disappearing overnight. This may mean that many of these jobs never existed or were duplicate entries, or it may be that as R&D dried up, jobs truly disappeared. Further, the successful Y2K repairs that then surplused many maintenance programmers and the draw of the dot.com boom to thousands of new entrants into the profession also played a role. Either way, the mood among high-tech workers was bleak as reflective of the sparse opportunities available then. At that time, software professionals were spending six months or more out of work. It was not uncommon to hear tales of hundreds of job applications yielding only one interview.

Ed Yourdon predicted this scenario in his 1992 book *Decline and Fall of the American Programmer* [57]. Citing productivity gains, new tooling, quality, and the flight of jobs to low wage countries, especially India, Yourdon built a case for the declining future of software development in the US. The continued movement of both new application development and maintenance programming overseas provides ample testament to this prediction. Yourdon backtracked somewhat on this view a few years later citing the Internet boom, service systems, and embedded systems as new paradigms stalling this path of decline [58]. Interestingly, with the crash of the Internet stocks, much of the frenzied new development was halted and without unbounded capital low cost wages are again in favor.

At this writing, the market has changed again, rebounding in favor of IT staff. There are jobs available. For employers finding good people is difficult again. In our case, we are only looking for experienced staff with specific skills. We rely on our offshore partners for entry level programming talent which they have in abundance. This picture, while hard to project into the future, looks favorable for veteran engineers and those with in-demand skills. There appears to be continued demand for onsite project leads and architects.

6.2.3 Domain Knowledge Loss

As more and more work moves overseas, the detailed problem-related knowledge known as domain knowledge may move with it. Such knowledge is central to how a business operates, how its systems interoperate, how a service is provided. Without day-to-day deep work experience with this information, local staff will loose currency in this awareness. Dependence on foreign offshore developers who are not committed to the company can represent a serious risk to the ability to innovate on top of the existing infrastructure to meet competitive needs.

6.2.4 Reduction in Management and Technical Currency

Just as the domain knowledge drain hurts the understanding of the business, less hands on management work and technical tasks can reduce the capabilities of the in-house development team. The trend toward offshoring will also change the nature of careers for those in developed countries. As jobs move overseas '... workers in the [developed] countries [will] find other things to do' [6]. There will also be a shortage of qualified managers in offshore countries, which is already being observed [18].

6.3 Technology Factors Affecting Offshoring

The technologies upon which software is based evolve at a very fast rate. Innovations will start to emerge locally and propagate across political boundaries. The speed of adoption will make some localities more competitive than others. Thus, one can expect that the powerhouses of offshoring will change over time.

6.3.1 Innovation Emergence

Just as Japanese automakers at first were low budget entrants into the US market and now dominate in many ways and have become innovators along the way so too will the various undeveloped countries start to innovate and dominate. India's IT workforce will soon be about one third that of the US. GSD may lead to India and China producing start-ups that carve out all new markets in the US or Europe. Instead of being subcontractors, they may become technology leaders and move up the value chain in producing products and services far different from legacy maintenance projects. Naturally, this will take time as India is currently dependent on the US for 60% of its IT revenue [48].

Bangalore is known as India's Silicon Valley but is it destined to produce the same level of innovation as the original valley? Bangalore is home to over 1,200 technology firms and represents 35% of India's software exports [52]. There are other key software centers around the country such as Chennai, Pune, and others. The question is what is Silicon Valley really like, what makes it tick? Lee and his team [35] documented the essence of Silicon Valley and can it be mimicked overseas.

> Like Hollywood or Detroit, Silicon Valley is marked by a distinctive collection of people, firms, and institutions dedicated to the region's particular industrial activities. The Valley's focus on the intersection of innovation and entrepreneurship is evidenced by the many specialized institutions and individuals dedicated to helping start-ups bring new products to market.

Within this context, there were several key characteristics making Silicon Valley what it is; these include favorable rules, knowledge intensity, a high quality mobile workforce, results-oriented meritocracy, climate of risk taking and tolerance toward failures, open business environment, universities and research institutes, and a high quality of life. We might ask whether any of India's technical centers has produced much in the way of innovation thus far. By and large they have focused on picking up established sets of source code and building new releases on top of them. There have been few new products or breakthrough hits that created new markets. Up to now it has been primarily a game of subcontracting or working largely under the direction of US-based R&D in the case of the globalized development model. However, some India- and China-based researchers do appear in the literature and this trend is growing. Konana points out that the India firms are in a good position to begin such innovation [33]. He also points out that what got them to the dance, highly structured CMM-based development models, may not lead them far on this new path. That will require new models more open to risk taking, creativity, and experimentation.

In Fig. 15, the rise in patent applications from China and India is shown as a percentage of all patent applications [53]. Both China and India have grown in the last 10 years from as little as 142 patent applications per year to as many as 2,127 per year for China, reflecting a 14 times increase. This data should be understood in the backdrop of all the patent applications from abroad. Today, a full 46% of applications come from foreign countries. Japan alone accounts for 34% of all patent applications and has provided more than 25% of patent applications for over 20 years. In raw numbers, Japan has grown from 24,516 to 71,994 patents per year since 1987 and may set the model for China's growth in this area with a stronger manufacturing focus in the economy than India which often drives more patents. The rise in patents does not necessarily mean more innovation. It could

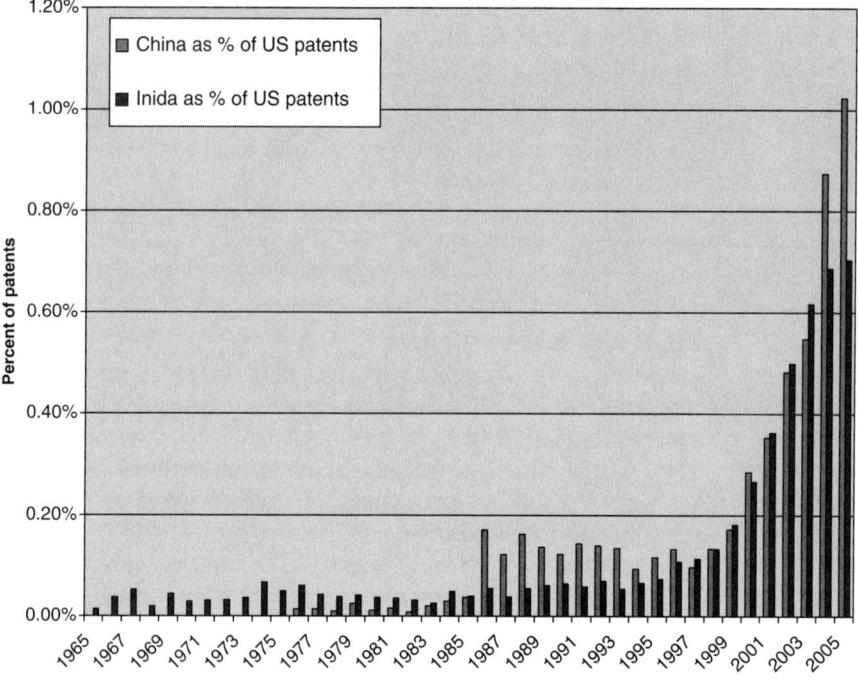

FIG. 15. China and India's share of US patent applications.

mean a move to protect commercial interests. While patents alone do not guarantee success in innovation, it is a reliable measure of where new ideas are coming from. The recent sharp increase coming from India and China is a useful barometer in tracking global innovation.

The return of Indians to India to start new businesses is today more talk than reality. The 'entrepreneurial culture is still embryonic in India' and failure is not generally accepted, 'in India, you tend to get one chance' [25]. Nevertheless, the innovation engine may kick on for India and other emerging technology countries. Such a switch from legacy support to forward-looking R&D could have a profound effect on global software development. New methods and tools could begin emerging from various locations around the world and not simply being developed by advanced countries and propagated outward. This could mark the switch from US and European countries being dominant technology leaders to being followers in some niche areas initially and then in broader technical sectors eventually.

6.3.2 Engineering Evolution

Just as business may change, Indian and Chinese engineers may begin contributing more broadly to the development of science and engineering. If the innovation engine does indeed start in these countries, it could greatly influence the future of software engineering. To date little fundamental research has been contributed from the developing countries to the advancement of the state of the art in software engineering. This has been largely a one-way flow of ideas from US and Europe to China, India, and other emerging countries. Nearly all major tools are produced in the developed countries as are advances in methodologies. No major languages have emerged from offshore. The core platforms of the major computing environments all originate in the developed world. Even the open source movement relies heavily on those in the developed world. Naturally, this could all change now that more than a million engineers are located in the emerging countries. But they must have the inclination to invent and the infrastructure or risk taking environment to do so.

6.4 A Future Target for GSD

Very few companies consider structural engineering and construction of their manufacturing facilities to be one of their core business capabilities. Most companies hire other companies to design, build, and maintain their factories. Architectural firms, structural engineering firms, contractors, subcontractors, and leasing companies are all involved in the effort to construct a new facility. In fact, many companies do not even own facilities, but rent them from holding companies. In essence, the development of the physical infrastructure of a company is completely outsourced.

The physical infrastructure is viewed as an engineering product that can be purchased. In many cases, a firm does not know the names of all of the companies that worked to build the facility they occupy. An American company can have a hotel constructed in Burma using local companies performing the construction under the guidance of German Structural Engineers who are working from an architecture specified by an Australian architect. It is truly a field that supports global development.

This situation does not hold for the software infrastructure of most large companies. Even in the GSD strategy presented within this chapter, the corporation retains a huge amount of control over the technologies, designs, and processes used to develop software in a global software development effort. It is not that corporations want to have total control, but at present they must have total control.

Despite the new corporate model presented in Fig. 2, companies have not restructured their IT management to reflect that model. There are several reasons for this and until these reasons are addressed it won't make sense for a company to fully restructure itself to that model. Once certain conditions are met, the whole nature of outsourcing will change.

Part of reason why companies have not fully embraced the new corporate model with regard to IT is because the software industry as a whole has not matured to the point where software engineering best practices have become standards. In the process presented here, this limitation is overcome by the specification of a delivery framework in which application building blocks are used, coding standards are specified, value-added tools are provided, and best practices are enumerated. It should be noted that these are defined for the corporation, not the industry. When the industry defines them, then corporations won't have to go through that step. This is the goal of the Software Engineering Institute [49].

The adoption of standardized business architectures, such as the Open Group Architecture Framework [43], Service Oriented Architectures [16], and Agent Oriented Architectures [3], will provide a basic framework against which software development shops can target their capabilities. These architectures help define the functional components that work together to achieve robust and uniform enterprise computing capabilities. It also gives a common language by which features can be specified and defined.

Adoption of the new corporate model is further hindered because corporations are hard pressed to identify and specify their existing software infrastructure in which future software must fit. At present, many corporations find that their software infrastructure is evolving at a rate that is faster than can be documented. This is most visible during attempts to migrate legacy systems to newer technologies.

Corporations will have to get a handle on their software infrastructure in order to remain competitive. Low-cost approaches for identifying the enterprise architecture of a corporation that can deal with the evolution of a software infrastructure have been developed [13]. Such a step will support the identification of application building blocks as well as dictate value-added tools.

Finally, the global adoption of a uniform means of documenting enterprise and system architectures will facilitate communications between clients and providers. While the Unified Modeling Language (UML) is widely used, it is not yet a universal standard particularly at the system and enterprise level. The Object Management Group is still developing UML profiles for Enterprise Application Integration (EAI), CORBA (Common Object Request Broker Architecture) Component Model (CCM), and Systems Modeling Language (SysML) [42].

Once appropriate engineering standards are adopted, standard architectures are defined, corporations have a handle on their enterprise architecture, and a global standard for documentation is adopted, then outsourcing will experience a fundamental change. New specializations will emerge that are intended to facilitate outsourcing. New types of sourcing companies will become commonplace and geography will be less of an issue.

A new and particularly important engineering position will emerge – the Information Engineer. These engineers will specialize in the identification of information and data processing needs of corporations. They will serve an equivalent role that industrial engineers have served with respect to the physical facilities of corporations. They will be critical in the successful acquisition of software capabilities from external sources. The Information Engineer will be the primary software specialist that remains within most corporations.

Information engineers will interact with Enterprise Architects. They will identify how best to incorporate elements into the existing enterprise architecture of a company that meet the information and data processing needs. Interestingly enough, corporations can follow the traditional path of contracting with Enterprise Architects on a competitive basis in which architects present their individual visions for the product and the architect whose vision best matches that of management is selected.

Enterprise Architects will oversee Systems Engineers who will modify existing systems or develop new systems consistent with the architecture. System engineers will deal with very concrete engineering concepts such as loading, throughput, accuracy, reliability, security, and cost. They will specify well-characterized software components in the same fashion that structural engineers specify well-known structural elements.

Systems Engineers will work with contractors who oversee subcontractors to implement new components or modify existing components. This will be the software construction activities that require the skills of detail-oriented coding specialists. This will be a truly global business with firms competing on a bid for work basis.

Testing firms will emerge to provide quality assurance and oversight in much the same way onsite inspectors operate in the world of physical structures. These firms will perform onsite construction inspections, code reviews, code audits, security testing, and integration testing.

It will take at least 20 years for GSD to evolve to this point. Although the pieces are slowly getting in place, professionals with the appropriate qualifications are scarce. Businesses will not change their structure overnight once an appropriate pool of professionals is in place. New businesses will not start up until the demand is sufficient to support them. All of this takes time. However, it is important to keep a final goal in mind when implementing a GSD strategy.

7. Conclusions

GSD is here to stay. Global trade has been with us for centuries and software development is just the latest industry to be pulled into the worldwide economy. Because of its intellectual nature instead of being physically oriented, GSD introduces new aspects to global trade. Work can be abstracted and transported across borders more readily than with localized and physicalized activities. While there will always be some local content to GSD, wide segments of the work can be transported to suitable locations. This will make the practices discussed here more critical. From vendor selection to software infrastructure development, successful global practices will be required to compete.

ACKNOWLEDGMENTS

The authors thank Dan Focazio of Wolters Kluwer Shared Services for his insights into the business process around outsourcing selection. Also, Subu Subramanian, Ravi Raghunathan, and Paddy Padmanabhan of TCS were generous in their help providing information from the vendor perspective. Support from the Wolters Kluwer management team is appreciated, especially from Venkatesh Ramakrishnan. Finally, we thank the reviewers and the editor for their many helpful comments.

Appendix 1. Interview with K. (Paddy) Padmanabhan of Tata Consultancy Services 3/9/07

This interview was conducted by telephone and was meant to provide perspective on the establishment, growth, and ongoing practice of offshore development. Mr Padmanabhan is an executive with Tata Consultancy Services based in India and has extensive experience with offshore projects and IT projects in general. The dialogue as captured follows:

1. When did you join TCS?
   I joined TCS in 1975 right after attaining my master's degree.
2. Is it true that TCS' first offshore project was in 1974?
   The company was very small at that time: about 100 professional staff, with about 100 support staff. We were doing what we called bureau service (which is now called BPO) for Indian companies, the electricity company, mutual funds company, and so on. Offshore came about because we wanted to get the latest computers. IBM was pushing the 1401 not the 360/370s. Getting the latest computers was very difficult and our CEO was keen on getting latest technology and disciplines to be able to build software for the best in the world. He used his IEEE connections to do a barter deal with Burroughs to develop software in order to get a computer. We acquired a Burroughs 1600 and then a 1700

and developed lots of software for the UK and other places in Europe and the US. I was involved in the 2nd or 3rd project which was for Builders and Plumbers Merchants. The approach was to leverage whatever computers we had to develop on what we had and deploy on the target machine even if it was different. We had to be good at migrations; we had to carry tape and load the applications (at that time we used telex to communicate). There might be a team of 10–15 in India in those days and 2–3 would go on site to do the deployment. Memory was also a constraint; Burroughs would allow 30K or 50K for a program that ran on a given machine that would require careful memory management.

3. Can you describe how those first projects were created and developed from a business perspective?

Early Business Development was done from India; Burroughs might suggest early projects for overseas work, they brought prospects or subcontracted back to us, we would develop and deliver. Eventually, the NY office was established in 1979; then we started doing our own sales development, and other joint venture partners were developed in Europe and Australia in the early 1980s. This is when we started to be independent of Burroughs from a business development perspective.

4. How was the offshore model conceived? Was there a vision for what you wanted to achieve with the business?

We had a vision – we wanted to 'transform the world using IT' – we thought that the time will come when IT will transform people's lives and it will become ubiquitous; we had a strong vision of IT transforming all of business, not a strong vision of what sectors or markets might be involved, but early it was in financial services and also in logistics as we had competencies there. The vision was that in 20 years we would transform the world; by 1978–1979 we recognized that specializing in Burroughs was too narrow, and so we set up joint a venture called Tata-Burroughs and then TCS started focusing on the IBM marketplace, then had to source new business without partner of Burroughs; this is when we, the business, started to grow independently.

5. In the early days, did you supply staff onsite more than conducting offshore projects or was it a mix?

We had numerous people who worked onsite with the customer, but significant work was done in India; for many reasons this was seen as more advantageous; we wanted to maintain roughly 80% of the work in India, sometimes less. It was also important to serve the India marketplace; we could not work only on overseas projects, and so a certain percentage of the business was from India; today less than 10% is India based but there is still a strong commitment on strategic projects such as the Indian stock exchange and some major bank projects. A key development was in 1982, establishment of the first R&D organization by a software company in India. We created numerous major sponsored projects to do significant efforts in waste management and process engineering and also tools for migrating, and so there was vision around R&D as well as the core business.

6. What were some of the barriers to developing the offshore model?

Credibility issues were always there. People would say 'Indian software? You must be joking!' People would say that we did not know what was needed. We used Burroughs as a big brother to get started, but success bred success, we started to get references and could go to other countries and multinationals to get more work, some things below the cutoff line for one customer is something we could do inexpensively and with high quality; for example, with AMEX we worked all over South America to develop systems for them and succeeded in building a strong relationship.

7. Was the government helpful?

Exchange regulations were problematic, even if you had the rupees you could not spend overseas without permissions, you needed guarantees on what would be imported on any foreign purchase, duties were also significant; the eventual export processing zones changed this to some extent in the early 1980s but these were very restricted to what you could bring out into India from these zones. However, there were tax benefits to the foreign exchange earned; major problems were also in the number of permits required.

8. What about the role of foreign governments?

The biggest problems were traditionally visas; initially this was difficult to do but not impossible; later job protectionists started and this became more complex and continues to be a challenge.

9. How did competition play a role in the development of the business?

In the 1970s and the 1980s, TCS created the software industry and set the path for many other companies, the competitors were not really considered until the 1990s, there was always a differentiator between TCS and others but India's capabilities were enhanced by the suite of companies doing business in IT services.

10. What have been some of the technical innovations required to run the business?

Ability to develop on one platform and use tools to migrate was a key. This migration challenge led to work which led to automation of program migration and automation; we also developed the poorly named 'addict' tool which was 'a data dictionary'; this was strong entry point to one client, also we developed a culture of building tools for everything, we thought productivity and quality could be enhanced through tool building and tool use, from small widgets to large project support, the ability to use standardized tools allowed for broad development capability and leverage, we also built performance modeling tools and quality prediction tools, we were not the first to build these things but the investment was critical to the business. We also started working with IBM Labs and raised our level of quality also through this relationship.

In all of this, Burroughs and other partnerships were key to development of business, relationships, and technology; some were key, for example, Citibank and AMEX; these relationships were key to our success.

11. What is the future of offshoring? What are the forecasts for growth? Are there new markets to be pursued?

We should be able to leverage a true global model in future, no longer a pure offshore-only model but use a global workforce to the benefit of the client. We will be able to put people where they need to be using collaboration tools, open source, web tools, wikis, you can get a totally different development environment so folding all this together for the client will be effective in new ways. It will no longer be solely offshore. We may see traditional models disappear.

Appendix 2. List of Acronyms

ABB	– Application Building Blocks
BPO	– Business Process Outsourcing
CAD	– Computer Aided Design
CCM	– CORBA Component Model
CEO	– Chief Executive Officer
CIA	– Central Intelligence Agency
CIO	– Chief Information Officer
CLS	– Corporate Legal Services
CMMI	– Capability Maturity Model Integrated
CORBA	– Common Object Request Broker Architecture
CTO	– Chief Technology Officer
EAI	– Enterprise Application Integration
ENIAC	– Electronic Numerical Integrator And Computer

ERP	–	Enterprise Resource Planning
GNP	–	Gross National Product
GPS	–	Global Positioning System
GSD	–	Global Software Development
ICGSE	–	International Conference on Global Software Engineering
IEEE	–	Institute of Electronic and Electrical Engineers
IFD	–	Interim Functional Delivery
ISV	–	Independent Software Vendors
IT	–	Information Technology
LAN	–	Local Area Network
MSA	–	Master Service Agreement
NY	–	New York
ODC	–	Offshore Development Center
PMC	–	Program Management Committee
PMO	–	Project Management Office
POS	–	Point of Sale
PS	–	Production Support
QA	–	Quality Assurance
R&D	–	Research & Development
RFID	–	Radio Frequency ID
RFP	–	Request for Proposals
ROI	–	Return on Investment
SEI	–	Software Engineering Institute
SLA	–	Service Level Agreements
SOA	–	Service Oriented Architectures
STC	–	Steering Committee
SysML	–	Systems Modeling Language
TCOE	–	Testing Center of Excellence
TCS	–	Tata Consultancy Services
UK	–	United Kingdom
UML	–	Unified Modeling Language
UNIVACI	–	Universal Automatic Computer
US	–	United States
VOIP	–	Voice over IP
WAN	–	Wide Area Network
WDF	–	Web Delivery Foundation
WDF	–	Web Delivery Framework
WK	–	Wolters Kluwer
Y2K	–	Year 2000

REFERENCES

[1] Battin R., et al., March/April 2001. Leveraging resources in global software development. *IEEE Software*, **18**(2).
[2] Belson K., Outsourcing, turned inside out. *New York Times*, April 11, 2004, section 3, Page 1.
[3] Bergenti F., Gleizes M.-P., and Zambonelli F., (Eds.), 2004. Methodologies and Software Engineering for Agent Systems: The Agent-Oriented Software Engineering Handbook. Springer.
[4] Berry J., 2006. Offshoring Opportunities: Strategies and Tactics for Global Competitiveness. John Wiley & Sons, Inc., Hoboken, NJ.
[5] Berryman K., et al., Software 2006 Industry Report, viewed 2/3/07, http://www.sandhill.com/conferences/sw2006_materials/SW2006_Industry_Report.pdf.
[6] Blinder A., Fear of Offshoring. Princeton University, December 16, 2005, viewed 1/31/07, www.princeton.edu/blinder/papers/05offshoringWP.pdf.
[7] *Bringing IT Back Home, CIO Magazine*, March 1, 2003, http://www.cio.com/archive/030103/home.html.
[8] Carmel E., and Agarwal R., March/April 2001. Tactical approaches for alleviating distance in global software development. *IEEE Software*.
[9] Carmel E., 1999. Global Software Teams: Collaborating Across Borders and Time Zones. Prentice-Hall.
[10] Cusick J., May/June, 2003. How the work of software professionals changes everything. *IEEE Software*, **20**(3): 92–97.
[11] Cusick J., August 5, 2004. Developing software (and careers) in a global IT market place: The realities of software work in today's offshore environment. In *Computer Technologies & Applications Seminar Series*. Columbia University.
[12] Cusick J., and Prasad A., Sept/Oct, 2006. A practical management and engineering approach to offshore collaboration. *IEEE Software*, **23**(5): 20–29.
[13] Cusick J., and Tepfenhart W., July, 2006. Creating an enterprise architecture on a shoestring: A light weight approach to enterprise architecture. In *IT Architecture Practitioners Conference*, The Open Group Architecture Forum, Miami, FL.
[14] Dossani R., Origins and Growth of the Software Industry in India. Asia-Pacific Research Center, Stanford University, http://iis-db.stanford.edu/pubs/20973/Dossani_India_IT_2005.pdf, viewed 2/10/2007.
[15] Drezner D., May/June 2004. The outsourcing bogeyman. *Foreign Affairs*. Council in Foreign Relations, http://www.foreignaffairs.org/20040501faessay83301/daniel-w-drezner/the-outsourcing-bogeyman.html.
[16] Erl T., 2005. Service-Oriented Architecture (SOA): Concepts, Technology, and Design. Prentice Hall.
[17] Estelami H., 1998. The evolution of Iran's reactive measures to US economic sanctions. *Journal of Business in Developing Nations*, **2**, ARTICLE 1, http://www.ewp.rpi.edu/jbdn/jbdnv201.htm.
[18] Farrell D., et al., June 2005. The Emerging Global Labor Market. McKinsey Global Institute, http://www.mckinsey.com/mgi/reports/pdfs/emergingglobalabormarket/MGI_executivesummaries_offshoring.pdf.
[19] *The First Commerical Computers*, http://physinfo.ulb.ac.be/divers_html/PowerPC_Programming_Info/intro_to_risc/irt2_history4.html, viewed 2/3/2007.
[20] Fowler M., April 2004. Using an Agile Software Process with Offshore Development, http://www.martinfowler.com/articles/agileOffshore.html., viewed 3/11/2006.
[21] Goldstine H., 1972. The Computer from Pascal to von Neumann. Princeton University Press, Princeton, New Jersey.

[22] Hamm S., Outsourcing heads to the outskirts. *Business Week*, January 22, 2007.
[23] Hempel J., The Indian paradox. *Business Week*, February 12, 2007.
[24] Herbsleb J., and Grinter R., Architectures, coordination, and distance: Conway's law and beyond. *IEEE Software*, September/October 1999.
[25] Hibbard J., A slow start for Indian startups? *Business Week Online*, March 8, 2005, http://www.businessweek.com/the_thread/dealflow/archives/2005/03/a_slow_start_fo.html.
[26] Hira R., Testimony to the US-China Economic Security Review Commission on Offshoring of Software & High Technology Jobs, January 13, 2005, http://www.ieeeusa.org/policy/POLICY/2005/021305.pdf.
[27] *History of computer hardware in Soviet Bloc countries*, http://en.wikipedia.org/wiki/History_of_computer_hardware_in_Soviet_Bloc_countries, viewed 2/3/07.
[28] Hofstede G., et al., June 1, 2004. Cultures and Organizations: Software of the Mind, 2nd edition. McGraw-Hill.
[29] Hofstede G., Cultural Dimensions, http://www.geert-hofstede.com/, viewed 3/19/2006.
[30] Karolak D., 1998. Global Software Development: Managing Virtual Teams and Environments. IEEE Computer Society, Los Alamitos, CA.
[31] Kelly T., A brief history of outsourcing. *Global Envision*, December 7, 2004, viewed on 1/28/07, http://www.globalenvision.org/library/3/702/.
[32] Kobitzsch W., et al., Outsourcing in India. *IEEE Software*, March/April, 2001.
[33] Konana P., July 2006. Can Indian software firms compete with the global giants? *Computer*, **39**(7): 43–47.
[34] Lackow H., Outsourcing Trends and Best Practices, http://www.cio.com/research/outsourcing/edit/trends/sld001.htm, November 6, 2005.
[35] Lee C.-M., et al., 2000. The Silicon Valley Edge: A Habitat for Innovation and Entrepreneurship. Stanford University Press, Stanford, California.
[36] Lewin A., and Mani M., Next generation offshoring: The globalization of innovation. *Offshore Research Network*, April 11, 2007.
[37] MaCintyre B., Midnight's grandchildren: A British correspondent reports on the emergence of modern India. *New York Times Book Review*, Sunday, February 04, 2007.
[38] McDougall P., Chase cancels IBM outsourcing deal, true to its President's Form. *Information Week*, September 15, 2004, viewed 2/3/07, http://www.informationweek.com/story/showArticle.jhtml?articleID=47208515.
[39] Mohagheghi P., Global Software Development: Issues, Solutions, Challenges, 21 September 2004, viewed 2/3/07, http://www.idi.ntnu.no/grupper/su/publ/parastoo/gsd-presentation-slides.pdf.
[40] Nissen H., Designing the inter-organizational software engineering cooperation: An experience report. In *Third International Workshop on Global Software Development*, ICSE Workshop: May 24, 2004, Edinburgh, Scotland.
[41] OAO Technology Solutions, Criteria for Making an Appropriate Outsourcing Selection: Insourcing, Nearshore, and Outsourcing, http://www.oaot.com/downloads/about/library/whitepapers/, November 6, 2005.
[42] Object Management Group, http://www.omg.org.
[43] The Open Group, TOGAF, http://www.opengroup.org/architecture/togaf, 2006.
[44] Paasivaara M., and Lassenius C., Using interactive & incremental processes in global software development. In *Third International Workshop on Global Software Development*, ICSE Workshop: May 24, 2004, Edinburgh, Scotland.
[45] Pacey A., Technology in World Civilization: A Thousand-Year History. The MIT Press, Edinburgh, Scotland, 1991.

[46] Regan K., Dell recalls tech support from India after complaints. *TechNewsWorld*, 11/25/03, viewed 2/3/07, http://www.technewsworld.com/story/32248.html.
[47] Rubin H., Global Software Engineering and Information Technology Competitiveness of the United States. Department of Commerce Presentation, March 17, 1997.
[48] Sahay S., Nicholson B., and Krishna S., 2003. Global IT Outsourcing: Software Development across Borders. Cambridge University Press.
[49] Software Engineering Institute, http://www.sei.cmu.edu.
[50] Teasley S. D., et al., July 2002. Rapid software development through team collocation. *IEEE Transactions of Software Engineering*, **28**(7): 671–683.
[51] Tepfenhart W., 2004. Discussions with the author.
[52] Tilak S., India's Silicon Valley hits Dirt Track, Aljazeera.net, Tuesday, November 29, 2005. http://english.aljazeera.net/news/archive/archive?ArchiveId=16357.
[53] U.S. PATENT AND TRADEMARK OFFICE, Number of Utility Patent Applications Filed in the United States, By Country of Origin, Calendar Years 1965 to Present, Electronic Information Products Division, Patent Technology Monitoring Branch (PTMB) http://www.uspto.gov/web/offices/ac/ido/oeip/taf/appl_yr.htm, viewed 2/18/2007.
[54] Weier M. H., As hiring soars in India, good managers are hard to find. *Information Week*, February 5, 2007, 33.
[55] Who Invented the Computer? Alan Turing's Claim. *The Alan Turing Internet Scrapbook*: http://www.turing.org.uk/turing/scrapbook/computer.html, viewed 2/3/2007.
[56] Yan Z., Efficient maintenance support in offshore software development: A case study on a global e-commerce project. In *Third International Workshop on Global Software Development, ICSE Workshop:* May 24, 2004, Edinburgh, Scotland.
[57] Yourdon E., 1992. Decline & Fall of the American Programmer. Yourdon Press: PTP Prentice Hall, Englewood Cliffs, NJ.
[58] Yourdon E., 1996. Rise & Resurrection of the American programmer. Yourdon Press, Upper Saddle River, N.J.
[59] Zhang J., SPECIAL REPORT: *Outsourcing to China, Part 1*, Sourceingmag.com, http://www.sourcingmag.com/content/c050802a.asp, viewed 1/29/07.

Author Index

Numbers in *italics* indicate the pages on which complete references are given.

A

Abadi, M., 54, *109*
Abran, A., 135, *171*
Abts, C., 126, 128, 134, 136, *170*
Agarwal, R., 230
Aguilar, J., 129, 131, *173*
Aguilar-Ruiz, J.S., 129–131, *169*
Agustin, J.M., 184, 198, *199*
Ahmed, F., 28, *37*
Ahmed, M.A., 135, 136, 140, *169*
Ahn, L., 54, *109*
Amirthajah, R., 31, *38*
Anderson, R., 84, *109*
Androutsopoulos, I., 68, 71, 72, 79, 85, 86, 89, 90, *109*, *110*
Angelis, L., 127, 128, 132, 135, 150, *170*
Arisholm, E., 197, *199*
Asada, G., 31, *38*
Asgari, S., 180, 185, *199*
Avramidis, A.N., 140, *170*

B

Basili, V.R., 150, *170*, 175–198, *199*
Battin, R., 216, *267*
Beck, K., 133, *170*
Beckman, R.J., 125, *172*
Belkin, N.J., 66, *110*
Bell, S., 98, *110*
Bellovin, S., 24, *38*
Belson, K., 208, *267*
Benini, L., 35, *38*
Bergenti, F., 260, *267*
Berghel, H., 1–15, *16*
Berry, J., 218, *267*
Berryman, K., 245, *267*
Bhardwaj, M., 36, *38*
Bickel, S., 85, *110*
Biggio, B., 108, *110*
Bleris, G.L., 127, 128, *173*
Blinder, A., 208, 211, 212, 256, *267*
Blom, R., 26, *38*
Boehm, B.W., 118, 126, 128, 133, 134, 136, *170*
Bojkovic, Z., 28, *38*
Borriello, G., 22, *38*
Brajkovska, N., 10, *16*
Bratko, A., 80–82, 86, 88, *110*
Breiman, L., 130, *170*
Briand, L.C., 117–119, 121, 122, 126, 131–136, *170*
Bringing IT Back Home, CIO Magazine, 209, *267*
Brochering, J.W., 133, *174*
Brodley, C.E., 81, 82, *114*
Brooks, E., 179, *199*
Buchenrieder, K., 22, 23, 35, *38*
Bult, K., 31, *38*
Burd, T., 36, *38*
Burgess, C.J., 130, 136, *170*
Byun, B., 109, *110*

C

Callaway, E., 32, *38*
Caloyannides, M.A., 10, *16*
Calzolari, F., 128, *170*
Campbell, D.T., 178, *199*
Canetti, R., 26, 33, *41*
Capretz, L.F., 136, *171*
Carelius, G.J., 197, *199*
Carlson, W., 179, *199*
Carmel, E., 216, 230, *267*
Caropreso, M.F., 72, *110*
Carreras, X., 72, 78, 85, 86, *110*
Carrier, B., 6, 16, *16*
Cartwright, M., 132, 133, *172*
Caruso, J., 98, *110*
Carver, J., 180, 184, 192, 196, 197, *199*
Carvey, H., 16, *17*
Casey, E., 16, *17*
Cates, P., 134, 135, *171*
Chan, P.W., 28, *38*
Chandrakasan, A.P., 31, 36, *38*
Chen, Z., 127, 128, *173*
Cheng, C.H., 133, *172*
Chotikakamthorn, N., 28, *38*
Choy, R., 179, *199*
Chris, T., 132, *172*
Christensen, D.S., 128, *171*
Chulani, S., 118, 126, 128, 134, 136, *170*
Chute, C.G., 77, *114*
Claburn, T., 105, *110*
Cleary, J.G., 82, *110*
Cohen, W.W., 74, *110*
Cole, E., 3, *17*
Conover, W.J., 125, *172*
Cormack, G.V., 56, 76, 80, 82, 86, 88, 92, *110*, *111*
Cosoi, C.A., 108, *111*
Cranor, L.F., 57, *111*
Croft, W.B., 66, 72, *110*, *112*, *114*
Cuadrado-Gallego, J.J., 129, 131, *173*
Cuelenaere, A.M.E., 126, *170*
Culler, D., 179, *199*
Cusick, J.J., 201–265, *267*
Czerwinski, S.E., 31, *38*

D

Dagum, L., 179, *199*
Dalvi, N., 58, 65, *111*
Dantin, U., 98, *111*
Davis, F.J., 140, *171*
De Lucia, A., 127, *170*
Deerwester, S., 72, *111*
Dittmann, J., 29, *38*
Dolado, J.J., 129, 130, *170*
Domingos, P., 74, *111*
Dongarra, J.J., 179, *199*
Dossani, R., 213, 214, *267*
Drezner, D., 208, *267*
Drucker, H., 72, 76, 84, *111*
Dugard, P., 129, *173*
Dumais, S., 65, 68, 71–73, 84, *111*
Dumke, R., 121, *172*
Dwork, C., 54, *111*

E

Eckelberry, A., 105, *111*
Edelman, A., 179, *199*
Elkjaer, M., 133, 135, *170*
Ellison, D., 129, *173*
Erl, T., 260, *267*
Essam, D.L., 129, 130, *173*
Estelami, H., 252, *267*

F

Farmer, D., 8, *17*
Farrell, D., 210, 211, 213, 214, 253, 254, 256, *267*
Fawcett, T., 65, 89–91, *111*
Feller, W., 145, *171*
Fenton, N., 134, 135, *171*
Ferens, D.V., 128, *171*

Findlater, L., 198, *200*
The First Commercial Computers, 213, *267*
Fisher, D., 130, *174*
Fleischmann, J., 22, 23, 35, *38, 39*
Focazio, D., *262*
Fowler, M., 133, *170*, 226, *267*
Friedman, J., 130, *170*
Fuhr, N., 74, *111*

G

Gan, O.P., 37, *43*
Garcia, F.D., 84, *111*
Garratt, P.W., 136, *171*
Gartner Inc., 117, *171*
Gee, K., 72, *111*
Ghose, A., 33, *39*
Gleizes, M.-P., 260, *267*
Goel, A.L., 129, *174*
Goldstine, H., 212, *267*
Gomez-Hidalgo, J.M., 68, 72, 76, 79, 80, 86, 87, 91, 92, *111, 112*
Gong, L., 26, *39*
Goodman, J., 36, *39* 51, *112*
Graham, P., 69, 73, 74, 84, 101, *112*
Graham-Cumming, J., 99, 102, 105, 106, 108, *112*
Grangetto, M., 25, *39*
Gray, A., 55, *112* 135, *171*
Green, S., 179, *199*
Grinter, R., 229, *268*
Guo-Tong, S., 140, *171*
Gutnik, V., 36, *39*

H

Haahr, M., 55, *112*
Haartsen, J., 32, *39*
Hac, A., 19–37, *39*
Hahn, J., 108, *112*
Hale, J.E., 128, *174*
Hall, R.J., 55, *112*
Hamm, S., 253, *268*
Hannay, J., 178, *200*
Hansen, O., 178, *200*
Hayashi, I., 145, *171*
He, M., 134, 135, *174*
Heckerman, D., 65, 68, 71–73, 84, *111, 113*
Heiat, A., 129, *171*
Heidrich, J., 134, 135, *174*
Heinzelman, W., 32, *39*
Heiser, J.G., 10, *17*
Helton, J.C., 140, *171*
Hempel, J., 253, *268*
Herbsleb, J., 229, *268*
Hesse, D., 29, *38, 39*
Hibbard, J., 258, *268*
Hihn, J., 127, 128, *173*
Hill, J., 32, *39*
Hira, R., 217, *268*
Hird, S., 98, *112*
History of Computer Hardware in Soviet Block Countries, 213, *268*
Hochstein, L., 175–198, *199*
Hodgkinson, A.C., 136, *171*
Hoelzer, D., 1–16, *16*
Hofstede, G., 243, *268*
Hollingsworth, J., 180, 184, 185, 192, *200*
Hörts, M., 135, *171*
Horvitz, E., 65, 68, 71–73, 84, *113*
Höst, M., 163, *171*
Hovemeyer, D., 181, 198, *200*
Hovold, J., 85, *112*
Hsu, C., 22, *39*
Huang, B.-B., 28, *39*
Huang, S.-J., 135, *171*
Huang, X., 136, *171*
Hudgins, W.R., 133, *174*

I

Idri, A., 135, *171*
InfoWorld Test Center, 84, *112*
Ishigai, Y., 115–174, *174*
Iwatsuki, N., 145, *171*

J

Jaccheri, L., 180, *199*
Jeffrey, E., 60, *112*
Jeffery, R., 126, 132, *171*
Jensen, C.D., 54, *114*
Jensen, R.W., 128, *171*
Joachims, T., 74, 76, *112*
Johnson, P.M., 134, *172* 184, *198, 199, 200*
Jones, T.C., 126, *172*
Jorgensen, M., 133–136, *171*

K

Kadoda, G., 132, 169, *171*
Kagawa, A., 184, 198, *199*
Kalavade, A., 22, 23, *39*
Kalos, M.H., 125, 164, *172*
Kampenes, V., 178, *200*
Karahasanovic, A., 178, *200*
Karolak, D., 244, *268*
Kelly, T., 212, 213, *268*
Keogh, E., 80, *112*
Kersten, M., 198, *200*
Khoshgoftaar, T.M., 136, *174*
Kiaei, M.S., 26, *39*
Kikuchi, N., 115–169, *174*
Kim, D.S., 37, *40*
Kirovski, D., 36, *40*
Kitchenham, B.A., 127, 128, 134–136, *172*
Kjiri, L., 135, *171*
Kläs, M., 115–169
Kobitzsch, W., 234, *268*
Kolcz, A., 55, *112*
Konana, P., 211, 257, *268*
Kong, X., 29, *40*
Kou, H., 184, 198, *199*
Koutsias, J., 68, 72, 86, 89, 90, *109, 110*
Kratzer, C., 29, *40*
Krishna, S., 210, 213, 256, *269*
Kruse, I.I., 10, *17*
Kundur, D., 26, 28, *40*

L

Lackow, H., 218, *268*
Langley, T., 131–133, *170*
Lanubile, F., 179, *199*
Larkey, L.S., 72, *112*
Larson, P.-A., 31, *40*
Lassenius, C., 228, *268*
L'Ecuyer, P., 140, *172*
Lee, A., 133, *172*
Lee, C.-M., 257, *268*
Lee, H., 37, *40*
Lefley, M., 130, 136, *170*
Lewin, A., 210, *268*
Lewis, D.D., 72–74, *112*
Li, H., 26, *40*
Li, J., 132, 135, 136, *172*
Li, M., 134, 135, *174*
Li, X., 28, *40*
Li, Y.H., 74, *113*
Li-Aiqun, S., 140, *171*
Liang, T., 135, *172*
Liborg, N., 178, *200*
Lie, D., 36, *40*
Lin Chieh-Yi, C., 135, *171*
Lin, D., 71, 72, 76, 89, *113*
Lindsjom, Y., 197, *199*
Linkman, S., 134–136, *172*
Liu, K.J.R., 30, *43*
Loève, M., 123, *172*
Lokan, C.J., 129, 130, *173*
Lother, M., 121, *172*
Lowd, D., 105, 106, *113*
Lu, D., 21, *39*
Lucas, M.W., 53, *113*
Luo, X., 34, *40*
Lynam, T.R., 56, 80, 86, *110*

M

MacDonell, S.G., 127, 128, 135, 136, *172*
MaCintyre, B., 252, *268*
Madden, S., 31, *40*
Mahadevan, S., 140, *174*

Mair, C., 130, 132, *172*
Maña-López, M., 68, 71, 72, 89, 90, *112*
Mani, M., 210, *268*
Marsh, W., 134, 135, *171*
Mason, J., 105, 106, *116*
Mastumoto, K., 198, *200*
McCallum, A., 73, *113*
McDougall, P., 209, *268*
McGarry, F., 177, *199*
McGill, M.J., 69, 77, *113*
McKay, M.D., 125, *172*
McKay, R.I., 129, 130, *173*
Meek, C., 105, 106, *113*
Meernik, P., 140, *174*
Memon, R., 179, *199*
Mendes, E., 132, *172*
Menezes, A.J., 24, *40*
Menzies, T., 127, 128, *173*
MessageLabs, 64, *113*
Meyer, T.A., 86, *113*
Miller, J., 180, *200*
Min, R., 36, *40*
Mitchell, T.M., 76, 77, *113*
Miyazaki, Y., 128, *173*
Mohagheghi, P., 210, *268*
Molokken-Ostvold, K.J., 133, 136, *173*
Monden, A., 132, *173*
Moore, C.A., 134, *172*
Morasca, S., 180, *199*
Morisio, M., 128, 150, *170*
Mourelatos, Z., 140, *174*
Mukhopadhyay, T., 118, 134, *173*
Münch, J., 115–169
Murphy, G.C., 198, *200*
Musilek, P., 135, *173*
My, S., 28, *37*
Myers, W., 128, 135, *173*

M

Nakakoji, K., 198, *200*
Nakamura, T., 175–198, *200*
Narang, S., 28, *40*
Nawab, S.H., 22, *40*

Needham, R., 34, *40*, *41*
Neil, M., 134, 135, *171*
Nelson, B., 10, *17*
Nicholson, B., 210, 213, 256, *269*
Nile, M., 132, *172*
Nissen, H., 226, *268*
NIST *see* U.S.National Institute of Standards and Technology
Noore, A., 135, *172*

O

OAO Technology Solutions..., 226, *268*
Object Management Group, 260, *268*
O'Brien, C., 84, *113*
O'Gorman, L., 3, *16*
Oh, G., 37, *41*
Ohlsson, M., 163, *174*
Ohsugi, N., 132, *173*
The Open Group, 260, *268*
Otta, S.W., 179, *199*
Owen, A.B., 138, 140, *173*
Ozaki, K., 128, *173*

P

Paasivaara, M., 228, *268*
Pacey, A., 212, *268*
Padmanabhan, Paddy, 262–265
Pajerski, R., 177, *199*
Paliouras, G, 68, 71, 72, 79, 86, 89, 90, *109*, *110*, *113*
Pampapathi, R., 83, *113*
Pantel, P., 71, 72, 76, 89, *113*
Parrish, A.S., 128, *174*
Pawlak, Z., 136, *173*
Pedersen, J.O., 71, 72, *114*
Pedrycz, W., 135, *173*
Pendharkar, P.C., 134, 135, *173*
Pérez, José Carlos Cortizo, 45–113
Pering, T., 36, *41*
Perrig, A., 26, 33, *41*
Perry, D.E., 178, *200*
Pfleeger, C.P., 24, *41*

Phillips, T.-Y., 178, *199*
Pickard, L., 134–136, *172*
Platt, J., 76, *113*
Pompella, E., 127, *170*
Port, D., 127, 128, *173*
Postini White Paper, 99, *113*
Pottie, G., 30, *41*
Prasad, A., 201–265, *267*
Provost, F., 89, 90, 91, *113*
Provost, J., 72, 76, 84, 89, *113*
Pugh, W., 181, 198, *200*
Putnam, L.H., 128, 135, *173*

Q

Quinlan, R., 75, *113*

R

Rabaey, J.M., 31, *41*
Raghunathan, R., 262
Raghunathan, V., 31, *41*
Ramos, I., 129–131, *169*
Regan, K., 209, *269*
Regnell, B., 163, *174*
Reifer, D.J., 126, *173*
Rekdal, A.C., 178, *200*
Ren, J., 136, *171*
Retsas, I., 28, *41*
Rigoutsos, I., 83, *113*
Riquelme, J.C., 129–131, *169*
@RISK, 140, *173*
Rivest, R.L., 25, 34, *41*
Robert, S., 135, *171*
Rodger, J.A., 134, 135, *173*
Rodriguez, D., 129, 131, *173*
Rohatgi, P., 26, *41*
Roussos, G., 37, *41*
Rubin, H., 255, *269*
Ruhe, G., 132, 135, 136, *171*
Runeson, P., 163, *174*

S

Saarinen, J., 98, *113*
Sahami, M., 65, 68, 71–73, 84, *113*

Sahay, S., 210, 213, 256, *269*
Sakkis, G., 71, 72, 79, 89, 90, *113*
Saliby, E., 138, *173*
Saliu, M.O., 135, 136, 140, *169*
Salton, G., 68, 69, 77, *113*
Samosseiko, D., 102, 108, *114*
Samson, B., 129, *173*
Sanz, E.P., 45–109, *111*, *112*
Schofield, C., 132, *174*
Sculley, D., 81, 82, *114*
Sebastiani, F., 63, 64, 68, 72, 88, *114*
Seigneur, J.-M., 54, *114*
Selby, W.R., 178, *199*
Sentas, P., 127, 128, *173*
Sergeant, M., 58, 60, *114*
Shan, Y., 129, 130, *173*
Shawbaki, W., 26, *41*
Shepherd, M., 105, 106, *114*
Shepperd, M., 130, 132, 133, 136, 169, *173*, *174*
Shih, F.Y., 28, *41*
Shima, K., 198, *200*
Shin, M., 129, *174*
Shukla, K.K., 129, *174*
Shull, F., 175–198, *200*
Singh, S., 3, *17*
Sinha, A., 31, *41*, *42*
Sjoberg, D., 178, 197, *200*
Slijepcevic, S., 34, *42*
Smith, R.K., 128, *174*
Snir, M., 179, *199*
Sobol, I.M., 123, *174*
Software Engineering Institute, 260, *269*
Song, Q., 132, *174*
Spacco, J., 181, 198, *200*
Spector, P., 137, *174*
Srinivasan, K., 130, *174*
Stamelos, I., 127, 128, 132, 135, 150, *174*
The Standish Group, 117, *174*
Stanley, J.C., 178, *199*
StatSoft Inc., 130, *174*
Staudenmayer, N.A., 178, *200*
Stern, H., 105, 106, *114*
Steve, C., 132, *172*

Sthultz, M., 1–17
Strecker, J., 181, 198, *200*
Stroud, A.H., 119, 145, 156, 166, 168, 169, *174*
Stukes, S., 127, 128, *173*
Su, K., 30, *42*
Subramanian, G.H., 134, 135, *173*
Subramanian, S., 262
Succi, G., 135, *173*
Sukhatme, G.S., 35, *42*
Swaminathan, A., 25, *42*

T

Taff, L.M., 133, *174*
Takada, Y., 198, *200*
Tang, S.-X., 28, *39*
Tang, Z., 134, 135, *174*
Taylor, B., 53, *114*
Teahan, W.J., 80, *114*
Teasley, S.D., 244, *269*
Tennenhouse, D., 35, *42*
Tepfenhart, W.M., 201–269, *269*
Terakado, M., 128, *173*
Theo, V.D., 60, *114*
Thomas, R., 102, 108, *114*
Tilak, S., 257, *269*
Tonella, P., 128, *170*
Torii, K., 198, *200*
Toro, M., 129–131, *169*
Trendowicz, A., 115–169, *174*
Tsunoda, M., 132, *173*
Turing, A., 54, *114* 212, *269*

U

Uppuluri, P., 35, *42*
U.S. National Institute of Standards and Technology (NIST), 24, *42*
U.S. Patent and Trademark Office, 257, *269*

V

van Genuchten, M.J.I.M., 126, *170*
Venema, W., 8, *17*

Vicinanza, S.S., 118, 134, *173*
Villan, R., 26, *42*
Voelp, M., 185, *200*
Vogel, C., 84, *113*
Vose, D., 141, 147, *174*
Votta, L.G., 178, *200*

W

Walker, D., 179, *199*
Walkerden, F., 126, 132, *174*
Wallace, D., 176, *200*
Wan, Y, 134, 135, *174*
Wang, A., 36, *42*
Wang, H., 3, *17*
Wang, S., 3, *17*
Wang, Z., 29, *40*
Warren, G., 10, *17*
Warren, K., 179, *199*
Watson, B., 52, *114*
Watson, I., 132, *172*
Wei, G., 31, *42*
Weier, M.H., 254, *269*
Weiser, M., 36, *42*
Wesslen, A., 163, *174*
Whateley, B., 86, *113*
Whitlock, P.A., 125, 164, *172*
Wickenkamp, A., 115–169
Wieczorek, I., 117, 126, 131–133, 136, *174*
Wiener, E.D., 72, *114*
Wilson, J.R., 140, *170*
Winter, B., 16, *16*
Wittel, G.L., 102, 105, 106, *114*
Witten, I.H., 79, *114*
Wohlin, C., 135, 163, *174*
Wu, F., 102, 105, 106, *114*
Wu, S., 134, 135, *174*
Wu, Y.-T., 28, *42*

X

Xiao, Y., 37, *42*
Xie, D., 26, *42*
Xu, Z., 136, *174*

Y

Yamashita, T., 184, 198, *199*
Yan, Z., 230, *269*
Yang, D., 134, 135, *174*
Yang, Y., 71, 72, 77, *114*
Yao, K., 30, *42*
Yellick, K., 179, *199*
Yerazunis, B., 69, 71, *114*
Yongliang, L., 26, *42*
You, X., 29, *40*
Younis, M.F., 35, *42*
Yourdon, E., 255, *269*
Yuan, C., 26, *43*
Yuval, G., 34, *43*

Z

Zadeh, L.A., 135, *174*
Zambonelli, F., 260, *267*
Zazworka, N., 175–198
Zdziarski, J., 69, *114*
Zelkowitz, M.V., 175–198, *200*
Zhang, J., 28, *43* 216, *269*
Zhang, Q., 184, 198, *199*
Zhao, H.V., 30, 37, *43*
Zhao, Y.Z., 37, *43*
Zhou, J., 26, *43*
Zimmermann, P.R., 53, *114*
Zou, T., 140, *174*

Subject Index

A

A/I efficiency measure, 153–69
A/ms efficiency measure, 153–69
ABBs *see* application building blocks
absolute error, 150–74
Accenture, 254
AccessData's Forensic Toolkit (FTK), 11–13
accuracy in collected data, software engineering experiments, 178, 181–2, 197–8
accuracy of simulation algorithms
  experimental results, 158–69
  sampling techniques for software cost estimation and risk analysis, 143, 149–74
acronyms' list, 265–6
active attacks, email spam filtering, 106
ADS *see* Alternative Data Streams
advance fee fraud, email spam, 49
advanced encryption standard (AES) 25, 35
advertising, IT uses, 204
AES *see* advanced encryption standard
aggregation, sensor networks, 31–2
AHP, 127, 133, 139
algorithms, learning-based spam filtering, 72–83
Alternative Data Streams (ADS), data hiding, 8–10, 13
ambiguity attacks, watermarking, 27–8
AMD AthlonXP3000+ 157–8
AMEX, 265
analogy-based methods *see* memory-based methods

analysis phase, GSD life cycle model, 226–8
analyst role
  experiment managers, 185
  GSD, 226–8
analytical approaches, simulation techniques, 143–5, 164–9
ANGEL, 127, 132–3, 139
ANN, 127–30, 133, 136, 139
anomaly detection capabilities, data hiding, 13
anonymity problems, email spam, 50
anonymization functions, sanitized data, 190
ANOVA, 127, 128, 139
antiforensic tools, data hiding, 16
application building blocks (ABBs), 231–4
Application Service Providers (ASPs), 218
applications
  ABBs, 231–4
  ASPs, 218
  multimedia security, 20–4
AQUA, 127, 132–3, 136, 139
ASCII, 107
ASPs *see* Application Service Providers
attacks on email spam filters, 98–109
audio
  see also multimedia
  concepts, 21–43
audits, GSD, 245
Australia, 210–11
authentication requirements
  multimedia, 24–6, 29–30
  sensor networks, 33–7
auto replies, email spam, 95–6, 99–101
AVN, 127, 133, 139

B

bad sectors, data hiding, 5, 7–8, 11–12
Bangalore, India, 257
basic evaluation metrics, email spam filtering, 88–90
Basili, Victor R. 175–200
batch evaluations, email spam filtering, 87–8
battery limitations, sensor networks, 32–4, 36, 37
Bayesian filters
  see also learning-based spam filtering
  concepts, 63, 71, 73–4, 96–8, 105–6
BBN, 127, 134, 135, 139
behavioral aspects, software engineering experiments, 176–7
Bell Labs, 212
Berghel, Hal, 1–17
best practices, GSD, 231, 233–4, 247–51
BIOS parameters, data hiding, 14
black lists, email spam, 51–3, 94–5, 98, 101
BLOC, Symantec Brightmail solution, 61–2
blocks, abstract addressability levels, 2, 4, 7–10
Bluetooth, 32–3
BMF, 25
BMP files, 9–10, 13–14, 28–9
boot sectors, partitions, 7–8
BorderWare MXtreme MX-200, 84
Boroughs, 213–14
bounce messages, 95–6, 99–101
BPO see Business Process Outsourcing
BRACE, 127, 132–3, 139
Brightmail, 58, 61–2
British National Corpus, 68
broadband, 23–4
Buffon, 197
Burroughs, 262–3
business change factors, GSD, 253–6
business drivers, GSD, 217–18, 220–2, 246–7, 253–6

Business Process Outsourcing (BPO), 207–8, 211, 215, 262–3

C

C#, 218
calculators
  see also computers
  historical background, 212–13
Camouflage, 2
Canada, 210
Carr, Nick, 250
CART, 127, 130–3, 139
CART+CBR, 127, 133, 139
CART+OLS, 127, 131–3, 139
case-based reasoning see memory-based methods
causal model, overhead costs, 120–2
CBM see control behavioral model
CBR, 127, 132–3, 139
CCM, 260
CD-ROMs, 4
CDMA, 27
CG&RT, 130–3, 139
chain letters, email spam, 48–9
chained tokens, learning-based spam filtering, 69–72, 83
Chase, 209
Chennai, India, 252, 257
Chief Information Officers (CIOs), 209
China, 204, 209, 210–12, 215, 216, 250, 252–6, 257–9
Chung-Kwei system, 83
CIA, 251–2
CIOs see Chief Information Officers
Citibank, 265
classifier committees, email spam filtering, 78–84
classroom resources
  IRB issues, 181–2, 192
  software engineering experiments, 175–7, 178–82
clean data, sanitized data, 188–90

client-side email spam filtering, 93–5
cluster analysis, 133
CMAC, 127, 129–30, 139
CMMI Level 2 framework, 218, 222, 248
CO *see* overhead...
Cobb-Douglas, 127–8
CoBRA
  see also hybrid methods; project risk analysis; software development cost estimation methods
  accuracy study, 149–74
  advantages, 122, 134, 138–9
  analytical approaches, 143–5, 164–9
  concepts, 116–17, 118–22, 134, 138–74
  critique, 122, 134, 138–9
  definition, 119–20, 134
  efficiency study, 149, 153–74
  empirical study, 119, 149–74
  experimental design, 155–7, 163–9
  experimental operation, 157–8
  experimental results, 158–69
  hypotheses, 154–69
  Latin Hypercube (LH) sampling, 116, 119, 138–40, 146–74
  Monte Carlo simulations, 119
  overhead costs, 119–74
  principles, 119–22
  research questions, 119, 140–74
  sampling techniques, 138–74
  stochastic approaches, 143, 145–9
  validity discussion, 162–9
  weaknesses, 139
CoBRA causal effort model, critique, 138, 153
COCOMO..., 127–8, 134–6, 139
COCONUT, 127–8, 139
code *see* software
coding, watermarking, 27–8
collaborative filtering, email spam, 55, 94–5
collecting data, software engineering experiments, 175–6, 177–8, 181–2, 186–90, 197–8
collusion attacks, forensics, 29–30

Colossus computer, 212
Columbia University, 246
commercial spam 49–50
  see also email spam
communications
  GSD, 216–17, 219–20, 226–30
  sensor networks security, 34–5
compact flash cards, 4
comparative studies, software development cost estimation methods, 136–40
competing suppliers, GSD, 253–6
compilers, Experiment Manager, 183–4, 194–5
complexity factors, software engineering experiments, 180–1
composite methods, software development cost estimation methods, 127, 133, 136–9
compression
  data hiding, 15
  email spam filtering, 56, 80–3, 109
  multimedia, 22–3, 25–6, 29, 37
Computer Sciences Corporation (CSC), 177–8, 250
Computer-Aided Design (CAD), 204
computers 204, 212–13
  see also IT
  historical background, 212–13
concept phase, GSD life cycle model, 226–8
conclusion validity, simulation techniques, 163–9
conference systems, multimedia, 21–2
confidentiality requirements
  data privacy GSD recommendations, 220, 243
  sensor networks, 33–7
configuration management, GSD, 219–20, 230–1, 259–61
confusion matrices, email spam filtering evaluations, 88–9
construct validity, simulation techniques, 163–9

construction phase, GSD life cycle model, 226–8
content-based spam filtering 51, 56–98, 109
see also email spam; heuristic...
concepts, 51, 56–98, 109
feature engineering, 67–72
implementation in practice, 92–8
learning-based spam filtering, 45, 56–7, 63–98
Machine Learning, 45, 56–7, 63–83, 109
processing structure, 65–7, 83
Text Categorization, 63–6, 72, 80–3
control behavioral model (CBM), 35
CORBA, 260
corporate models, outsourcing changes, 206–8, 213–14, 259–60
corpus of spam messages, SpamAssassin, 60, 84–5, 86
cost estimation methods
see also software development...
concepts, 115–74
cost overhead distributions 141–74
see also overheads
cost-sensitive learning, email spam filtering, 79–80
costs 47–8, 54, 98, 109, 115–74, 177–8, 205–6, 210–11, 213–14, 220–1, 222–6, 229–30, 253–4
see also overhead...; overheads
CoBRA, 116–17, 118–22, 134, 138–74
email spam, 47–8, 54, 98, 109
outsourcing, 205–6, 210–11, 213–14, 220–1, 222–6, 229–30, 253–4
software engineering experiments, 177–8
sourcing IT landscape, 205–6, 210–11, 213–14, 220–1, 222–6, 229–30, 253–4
country alliances, GSD, 252
crime
see also forensics
data hiding, 1–17, 29–30
email spam, 48–50
CRM-114, 105

cross-shore development model 226–8
see also global software development
cryptography
see also data hiding...
definition, 3
CSC see Computer Sciences Corporation
CSS tricks, 102
cultural issues, GSD, 203–4, 243
cumulative cost distributions, 121–2
current status, Experiment Manager, 190–7
Cusick, James, 201–69

D

Dalvi, 65
dark data
see also data hiding
concepts, 1–17
data
carving, 11–12, 15
collecting data, 175–6, 177–8, 181–2, 186–90, 197–8
mining paradigms, 80–3
self-reported data, 178–9
software engineering experiments, 176–200
data encryption standard (DES), 24–5
data hiding
abstract addressability levels, 2, 4
AccessData's Forensic Toolkit (FTK), 11–13
Alternative Data Streams, 8–10
anomaly detection capabilities, 13
antiforensic tools, 16
background, 1–2
bad sectors, 5, 7–8, 11–12
BIOS parameters, 14
compressed files, 15
concepts, 1–17, 29–30
conclusions, 16
definition, 3
digital storage, 4–10
disk slack, 5, 7–8

SUBJECT INDEX 283

DLL files, 14–15
encryption, 1–2, 10–11, 15–16
ExtX, 5, 8
FAT, 7–8, 10–12
file slack, 2, 3–4, 6–8, 16
file systems, 5, 7–10
forensics, 2, 10–16, 29–30
metadata manipulation, 5, 7–8, 15
Metasploit Project's Slacker tool, 16
methods, 14–16
Microsoft Office documents, 15
NTFS, 7–10, 12–13
partitions, 5, 6–8
philosophy, 3–4
physical aspects, 1, 3–4
redirected application executions, 14–15
registry entries, 14
relative volatility of hiding areas, 8–9
signature-based analysis of media, 12–13
steganography, 3, 15–16, 28–9, 37
superblock slack, 5, 8
swap files, 14
virtual files systems, 6–7
watermarking, 1–2
data management, software engineering experiments, 181–2
data privacy GSD recommendations, 220, 243
data-driven methods
 see also non-proprietary...
 software development cost estimation methods, 119–22, 126–34, 136–9
databases, 182, 183–4, 204
DCS, 32–3
DCT, 25, 27–8
debugging, Experiment Manager, 195
decision trees, 127, 130–3, 139
 email spam filtering, 74–5
 model-based data-driven methods, 127, 130–3, 139
decoy email boxes
 see also honeypotting (email traps)
 future spam filtering prospects, 98

Delivery Status Notifications (DSNs), 100–1
Dell, 209
denial of service (DoS), 23–4, 109
dependent variables, simulation techniques, 154–5
DES *see* data encryption standard
Descriptive Sampling (DS), 138–9
design phase, GSD life cycle model, 226–8
Device Configuration Overlay, 5–6
DFT 27
 see also Fourier transform
DHA *see* Directory Harvest Attacks
digital fingerprints, 30
digital rights management (DRM), 23–4
digital signatures
 concepts, 25–6, 53–4
 email spam, 53–4
digital storage
 concepts, 4–10
 data hiding, 4–6
 media types, 4
digital watermarking *see* watermarking
direct attacks, email spam filtering, 101–2
Directory Harvest Attacks (DHA), 100–1
directory slack, data hiding, 5–8
disasters, economic costs, 252
discretization functions, sanitized data, 189–90
disk slack, data hiding, 5, 7–8
disk structures, concepts, 4–6
disposable addresses, email spam, 54–5
distributed approach GSD details, 228–30
distributed systems, sensor networks, 31–2
distribution computation, 141–9
DLL files, data hiding, 14–15
DMC *see* Dynamic Markov Compression
DNA profiling, 29–30, 83
DNS Blacklists (DNSBL), 52, 59
DoCO, simulation algorithms, 150–74
Document Frequency, 72
documentation overhead, GSD, 242

domain knowledge loss, GSD, 256
DoP, simulation algorithms, 150–74
DoS *see* denial of service
DOS partitions, 6
dot.com boom, 255
Drezner, Daniel, 208
DRM *see* digital rights management
DS *see* Descriptive Sampling
DSNs *see* Delivery Status Notifications
Dspam, 70–1
DVDs, 4
DWT, 25
Dynamic Markov Compression (DMC), 82–3

E

EA, 127, 129–33
EA-MARS, 127, 131
EAI *see* Enterprise Application Integration
eavesdropping, 23–4
Eclipse, 198
ECML-PKDD Discovery Challenge, 84–5
economic costs
  disasters, 252
  email spam, 47–8, 54, 98, 109
economic multiplier effects, GSD, 252–3
EDS, 250
education systems
  India, 215, 221–2, 254
  US, 251
effectiveness issues, Experiment Manager, 191–2
efficiency of simulation algorithms
  experimental results, 158–69
  measures, 153–4
  sampling techniques for software cost estimation and risk analysis, 149, 153–74
electronic banking, 23
electronic blackboards 22
  see also multimedia

Emacs, 198
email spam
  advance fee fraud, 49
  anonymity problems, 50
  auto replies, 95–6, 99–101
  chain letters, 48–9
  commercial spam, 49–50
  concepts, 45, 47–113
  costs, 47–8, 54, 98, 109
  crime, 48–50
  definition, 47, 65
  economic costs, 47–8, 54, 98, 109
  evasive offenders, 50
  families of spam, 48–50
  features, 47
  Internet hoaxes, 48–9
  legal measures, 50, 98
  'phishing' attacks, 48, 49–50
  problems, 47–8
  pyramid schemes, 49
  spam bots, 52–3, 98
  third-party sources, 54–5
  Viagra, 49, 57–8, 65, 99, 101–4
  viruses, 48, 98
email spam filtering
  active attacks, 106
  attacks on filters, 98–109
  basic evaluation metrics, 88–90
  batch evaluations, 87–8
  Bayesian filters, 63, 71, 73–4, 96–8, 105–6
  black lists, 51–3, 94–5, 98, 101
  classifier committees, 78–84
  client-side filtering, 93–5
  collaborative filtering, 55, 94–5
  combined methods, 98
  compression filters, 56, 80–3, 109
  concepts, 45, 47–113
  confusion matrices, 88–9
  content-based filtering, 51, 56–98, 109
  cost-sensitive learning, 79–80
  critique, 83–92, 98
  decision trees, 74–5

SUBJECT INDEX

digital signatures, 53–4
direct attacks, 101–2
disposable addresses, 54–5
DMC, 82–3
evaluations, 83–92
feature engineering, 67–72
filtering concepts, 45–113
future prospects, 98, 109
graylistings, 52–3, 98
heuristic spam filtering, 51, 56–62
hidden text attacks, 102, 106–9
honeypotting (email traps), 55–6, 61–2, 98
HTML files, 69–71, 102–7
image-based spam, 107–9
implementation in practice, 92–8
indirect attacks, 99–101
invisible ink attacks, 106
k-Nearest Neighbors, 77–8
learning-based filtering, 45, 56–7, 63–98
legitimate emails, 47, 56, 83–96, 98
MDL, 81–3
metrics of evaluation, 84, 88–92
MIME encoding, 107
misclassification costs, 84–96, 98
Mozilla Thunderbird, 57, 71, 96–8
N-fold cross-validation, 87–8
obfuscation attacks, 102, 104–5, 108
passive attacks, 105–6
picospam, 101–2
postage techniques, 54
PPM, 82–3
primitive language analysis, 51
probabilistic approaches, 73–4
proxying methods, 96–8
quarantine folders, 58, 95–6
reputation controls, 53–4
ROCCH metrics, 88, 90–2
rule learners, 75–6
running test procedures, 84, 87–8
SCRIPT tags, 107
server-side filtering, 93–5

SpamAssassin filter, 58–60, 84–5, 86, 97–8
statistical attacks, 102, 105–6
summary, 83, 98, 109
support vector machines, 71–2, 76–7
Symantec Brightmail solution, 58, 61–2
tagging methods, 96–8
technical measures, 51–6
test collections, 84–7
Text Categorization, 63–6, 72, 80–3
tokenization attacks, 102–4, 109
tokens and weights, 68–72, 79–80, 83, 102–4, 109
TREC metrics, 45, 56, 80–3, 85–8, 92
Trend Micro InterScan Messaging Security Suite, 95–6
Turing Test, 54
white and black lists, 51–3, 59, 94–5, 98
embedded systems
    multimedia, 22–3, 27–8, 35
    sensor networks, 35
empirical studies
    GSD, 244–5
    software development cost estimation methods, 119, 149–74
employees *see* human resources
encryption
    see also steganography
    concepts, 24–6, 33–7, 53–4
    data hiding, 1–2, 10–11, 15–16
    email spam, 53–4
    keys, 24–5, 26, 33–7
    multimedia security, 24–6, 33–7
    sensor networks, 33–7
ENIAC, 212–13
Enterprise Application Integration (EAI), 260–1
Enterprise Architects, 261
Enterprise Resource Planning (ERP), 222, 255
Entropy, 76

epistemological uncertainty *see* possibilistic uncertainty
ERP *see* Enterprise Resource Planning
estimation of software efforts, related work, 126–40
Estimeeting, 127, 133, 139
ESTOR, 127, 134, 139
European Union (EU)
  email spam, 50
  GSD statistics, 210–11
evaluations
  email spam filtering, 83–92
  existing software effort estimation methods, 126–40
  Experiment Manager, 197
  software development cost estimation methods, 136–9
evolution, Experiment Manager, 192–3
existing software effort estimation methods, 126–40
expected error, 123–4
Experiment Manager
  see also software engineering experiments
  centralized database benefits, 182, 183–4
  collecting data, 186–90, 197–8
  compilers, 183–4, 194–5
  concepts, 176, 182–98
  consistency benefits, 183
  current status, 190–7
  effectiveness issues, 191–2
  evaluation, 197
  evolution, 192–3
  experiment managers, 182, 184–6
  future prospects, 192–3
  Hackystat, 182, 184, 185, 192, 198
  instrument package (UMDinst), 183–5, 192
  minimal disruption benefits, 183
  overview, 182–3, 190–1
  sanitized data benefits, 182, 183, 188–90, 193
  structure, 182–3
  subject views, 193–5, 196–7

supported analyses, 193–7
US studies, 190–1
validation of workflow heuristics, 195–6
web, 182, 183, 184–98
workflow cycles, 182, 185, 195–6
experiment managers
  analyst role, 185
  Experiment Manager, 182, 184–6
  professor role, 185
  roles, 184–5
  student role, 185
  technician role, 184–5
experiments
  replicated studies, 177, 180–1, 191–2
  software development cost estimation methods, 119, 149–74
  software engineering experiments, 175–200
expert systems, learning-based spam filtering, 63–4
expert-based methods, software development cost estimation methods, 119–22, 127, 133, 134, 136–9, 141–3
exports, 211–12
extended partitions, concepts, 5, 6–7
external validity, simulation techniques, 164–9
ExtX, data hiding, 5, 8

F

families of spam 48–50
  see also email spam
far shore outsourcing, IT resources, 205–9
Fast Fourier Transformation, 145
FAT *see* File Allocation Table
FDMA, 27
feature engineering, email spam, 67–72
feature selection and extraction, learning-based spam filtering, 71–2
Feistel Cipher Structure, 24–5
FGS, 26
File Allocation Table (FAT), 7–8, 10–12

file carvers, 11–12, 15
file slack, data hiding, 2, 3–4, 6–8, 16
file systems, concepts, 5, 7–10
filtering
    see email spam...
fingerprints, forensics, 30
finite state machine (FSM), 82–3
fitness function, 129–30
floppy disks, data hiding, 1, 3–4
Focazio, Dan, 262
focus groups, 204
forensics see also crime
    collusion attacks, 29–30
    concepts, 29–30
    data hiding, 2, 10–16, 29–30
    definition, 29–30
    multimedia security, 20–1, 29–30
Fourier transform, 25–6, 27
France, 254
Fraunhofer Institute for Experimental
    Software Engineering 116, 118
    see also CoBRA
free-form diaries, 178
freshness requirements, sensor networks, 33–7
FSM see finite state machine
FTK see AccessData's Forensic Toolkit
FTTH, 26
full text indexing, 11–12
future prospects
    email spam filtering, 98, 109
    Experiment Manager, 192–3
    GSD, 201–2, 204, 245, 249–62
fuzzy decision trees, 127, 130–3, 135–6, 139

G

Gandhi, Rajiv, 213–14
GANN, 127, 129–30, 139
GATE, 68
Germany, 259
GGGP, 130–3

GIF files, 11, 24–5, 29
Ginger2, 198
global software development (GSD)
    see also *outsourcing; software...*
    ABBs, 231–4
    audits, 245
    best practices, 231, 233–4, 247–51
    blend statistics, 247
    business change factors, 253–6
    business drivers, 217–18, 220–2, 246–7, 253–6
    China, 204, 209, 210–12, 215, 216, 250, 252–6, 257–9
    communication problems, 216–17
    communications/management recommendations, 219–21, 226–30
    competing suppliers, 253–6
    concepts, 201–69
    configuration management, 219–20, 230–1, 259–61
    controls, 249
    costs, 205–6, 210–11, 213–14, 220–1, 222–6, 229–30, 253–4
    country alliances, 252
    critical loose ends, 242
    critique, 244–5, 246–62
    cross-shore development model, 226–8
    cultural issues, 203–4, 243
    current GSD practice, 216–45
    definition, 210, 211–12
    distributed approach details, 228–30
    documentation overheads, 242
    domain knowledge loss, 256
    economic disasters, 252
    economic multiplier effects, 252–3
    empirical study, 244–5
    future prospects, 201–2, 204, 245, 249–62
    governance model, 237–9
    human resources, 208–9, 214–15, 219–26, 243, 250–1, 254–6
    IFD, 233–5

global software development (GSD) (Continued)
- India, 201, 204, 208–9, 210–12, 213–16, 221–2, 243, 246, 252–65
- industry concepts, 210–11
- infrastructural recommendations, 219–21, 229–31, 238–9, 259–61
- innovation leaders, 256–9
- job losses, 208–9, 220, 250–1, 254–5
- knowledge management, 241
- knowledge transition process, 239–40
- life cycle model, 226–8
- locality issues, 242
- McKinsey study, 210–11, 250, 254
- management reports, 233–4
- management shortages, 254–6
- managing-development recommendations, 219–21, 229–34, 248–51, 254–5, 259–61
- micro engineering process, 230–4, 248–9
- models, 218–20, 226–8
- ODC, 236–8, 245
- operational process for maintenance and support, 240–1
- organizational recommendations, 219, 229–30
- origins, 201–2, 203–4, 211–14
- overheads, 242–3
- planning guidelines, 219–20, 229–30
- PMO, 207–8, 247–8
- politics, 203–4, 251–3
- practical tutorial, 203–4
- practices, 201–2, 203–4, 216–45
- privacy recommendations, 220, 243
- problems, 216–17
- production (maintenance) support, 234–41
- quality recommendations, 220, 222–8, 232–4, 243, 249
- recommendations, 217–22, 226–30
- results, 244–5
- retention issues, 241–2
- RFPs, 221–5, 248
- risks, 244
- statistics, 210–11, 247
- suitability issues, 229–30
- suppliers, 217–18, 222–6, 244, 253–6
- supporting standards and code review guidelines, 231, 232–4
- surveys, 246–51
- talent supply, 254–6
- targets for the future, 259–62
- TCS, 213–14, 254, 262–5
- teams, 216–18, 226–30
- technology change factors, 256–9
- testing, 226–8, 245, 261
- things you have to live with, 242–3
- trends, 210–11
- US, 201, 206–9, 210–11, 243, 250–6
- value-added tools, 231, 232–3
- virtual roundtable, 246–51
- WDF, 226–8, 231–4

Gmail, 53, 54, 101
GNU Privacy Guard (GPG), 53–4
Gorbachev, Mikhail, 252
governance model, production (maintenance) support, 237–9
GP, 127, 131–3, 139
GPG *see* GNU Privacy Guard
GPS, 204
GRACE, 127, 132–3, 139
graylistings, email spam, 52–3, 98
GSD *see* global software development
GSFC *see* NASA Goddard Space Flight Center
Guesstimation, 127, 133, 139

H

Hac, Anna, 19–43
hackers, data hiding, 3
Hackystat, 182, 184, 185, 192, 198
hard drive architecture, concepts, 4–6
hardware
  see also IT
  historical background, 212–13
  sensor networks, 32–4, 35, 36–7

TCPA, 32, 36–7
trusted hardware, 36–7
Harvard's Mark I computer, 212
Helius Project, 48
heuristic spam filtering
   see also content-based...; email spam
   concepts, 51, 56–62, 83
   problems, 62, 83
   SpamAssassin filter, 58–60
   Symantec Brightmail solution,
      58, 61–2
Hidalgo, José María Gómez, 45–113
hidden data *see* data hiding
hidden text attacks, email spam filtering,
   102, 106–9
HIDER, 127, 131–3
hijacking attacks, Internet security
   challenges, 23–4
histograms, 123–4
Hochstein, Lorin, 175–200
Hoelzer, David, 1–17
honeypotting (email traps), email spam,
   55–6, 61–2, 98
Host Protected Area (HPA), 5–6
Hotmail, 52, 53, 54, 101
HPA *see* Host Protected Area
HPC classroom environments, 182–91,
   193, 196–8
HTML files, 15, 28–9, 69–71, 102–5
   data hiding, 15, 28–9
   email spam filtering, 69–71, 102–7
Huffman coder, 81–2
human resources
   see also teams
   GSD, 208–9, 214–15, 219–26, 243,
      250–1, 254–6
   India, 214–15, 221–2, 243, 254–6
   outsourcing job losses, 208–9, 220,
      250–1, 254–5
   programmers, 176–98, 213–14, 226–8,
      231–4, 243, 250–1, 254–6
   talent supply, 254–6
hybrid methods

see also CoBRA; data-driven...;
   expert-based...
software development cost estimation
   methods, 116, 119–22, 127, 134,
   136–9
hybrid sourcing, IT resources, 205–9
hypertextus interruptus tokenization attacks,
   103
hypotheses
   sampling techniques for software cost
      estimation and risk analysis, 154–69
   software engineering experiments, 176–7

I

IBM, 83, 209, 212–13, 216, 250, 254, 262–5
ICGSE *see* International Conference on
   Global Software Engineering
ICMP options field, 3
identification requirements,
   multimedia, 29–30
IEEE, 32, 217, 262–3
IFD *see* interim functional delivery
image-based spam attacks, 107–9
images
   see also multimedia
   concepts, 21–43
Imap, 97
implementation in practice, email spam
   filtering, 92–8
imports, 211–12
in-house development, IT resources, 205–9
in-house offshoring, IT resources, 205–9
in-sourcing, 208
Independent Software Vendors (ISVs), 214
independent variables, simulation
   techniques, 154–5
India
   background, 213–16, 221–2, 246,
      252–65
   Bangalore, 257
   Chennai, 252, 257
   economic multiplier effects, 252–3

India
(Continued)
  education system, 215, 221–2, 254
  GSD, 201, 204, 208–9, 210–12, 213–16, 221–2, 243, 246, 252–65
  historical background, 213–16
  human resources, 214–15, 221–2, 243, 254–6
  innovation leaders, 256–9
  monsoons, 252
  New Computer Policy, 213–14
  patent applications, 257–8
  software engineering experiments, 259
  strengths of IT industry, 214–16, 221–2
indirect attacks, email spam filtering, 99–101
Information Engineers, 261
information gain (IG), 71–2, 76
Infosys, 213, 254
infrastructural GSD recommendations, 219–20, 229–31, 238–9, 259–61
innovation leaders, GSD, 256–9
Instant Messaging, 65
instrument package (UMDinst), Experiment Manager, 183–5, 192
intellectual property, multimedia security, 23–4
interim functional delivery (IFD), GSD, 233–5
internal validity, simulation techniques, 163–9
International Conference on Global Software Engineering (ICGSE), 217
Internet
  see also email...; web
  hoaxes, 48–9
  multimedia security, 20–4, 29
  security challenges, 23–4
  VoIP, 29, 238–9
invisible ink attacks, email spam filtering, 106
Iran, 211–12, 251–2
IRB issues, software engineering experiments, 181–2, 192
Ireland, 210–11, 254

Ishigai, Yasushi, 115–74
ISM band, 32
ISPs, email spam, 49, 52
Israel, 210–11
ISVs see Independent Software Vendors
IT
  see also hardware; software...
  corporate models, 206–8, 213–14, 259–60
  costs, 205–7
  GSD change factors, 256–9
  historical background, 201–4, 211–14
  importance, 201–2, 204
  India, 201, 204, 208–9, 210–12, 213–16, 221–2, 243, 246, 252–65
  LANs/WANs, 230–1
  PMO, 207–8
  RFID, 21, 37, 204
  sensor networks, 30–7
  sourcing landscape, 203–9

J

Jacquard's paper tape-driven loom, 212
Japan, 204, 210–11, 256
JavaScript, 102
Jaynes, Jeremy, 50, 54
Jensen's regression models, 128
JND, 27
job losses
  see also human resources
  outsourcing, 208–9, 220, 250–1, 254–5
JPEG files, 24, 28–9
jUnit, 198
junk email see email spam

K

k-Nearest Neighbors (kNN), email spam filtering, 77–8
K9, 97
keys, encryption, 24–5, 26, 33–7
Kiev Institute of Electrotechnology in the Ukraine, 213
Kikuchi, Nahomi, 115–74

Kläs, Michael, 115–74
kNN *see* k-Nearest Neighbors
knowledge management, GSD, 241
knowledge transition process, production (maintenance) support, 239–40
Korea, 210–11

L

LANs, 230–1
Latent Semantic Indexing, 72
Latin Hypercube (LH) sampling
  accuracy study, 154–74
  concepts, 116, 119, 124–6, 138–40, 146–74
  DS comparison, 138–9
  efficiency study, 154–74
  empirical study, 154–74
  experimental results, 158–69
  hypotheses, 154–69
  Monte Carlo simulations, 149, 164–5
  research questions, 119, 140–74
  stochastic approaches, 146–9
  validity discussion, 162–9
learning-based spam filtering
  algorithms, 72–83
  classifier committees, 78–84
  concepts, 45, 56–7, 63–98
  cost-sensitive learning, 79–80
  decision trees, 74–5
  evaluations, 83–92
  feature engineering, 67–72
  feature selection and extraction, 71–2
  k-Nearest Neighbors, 77–8
  multi-word features, 69–71
  probabilistic approaches, 73–4
  processing structure, 65–7, 83
  rule learners, 75–6
  support vector machines, 71–2, 76–7
  Text Categorization, 63–6, 72
  tokens and weights, 68–72, 79–80, 83, 102–4
leetspeak, 104–5
legal measures
  see also crime
  email spam, 50, 98
legitimate emails, email spam filtering, 47, 56, 83–96, 98
Lempel-Ziv algorithm (LZ), 82–3
Lewis, 72
LH sampling *see* Latin Hypercube sampling
LHARC, 15
LHRO, 148–9, 157–69
life cycle model, GSD, 226–8
Likert scale, 137–9
Lingspam, 84, 86, 87
Linux, 8
locality issues, GSD, 242
log-linear form, 127
Loki, 3
lossless compression, 25, 29
lossy compression, 25
lost in space tokenization attacks, 104
LR-WPAN, 32–3
LSB insertion, 29

M

Machine Learning 45, 56–7, 63–83, 109
  see also content-based spam filtering
McKinsey study, GSD, 210–11, 250, 254
macrosensor nodes, sensor networks, 30–7
Mail Abuse Prevention System, 52
Mail Delivery Agent, email spam filtering, 93–5
maintenance *see* production (maintenance) support
management reports, GSD, 233–4
management shortages, GSD, 254–6
managing-development GSD recommendations, 219–21, 229–34, 248–51, 254–5, 259–61
Markov models, 82–3
Marmoset, 198
Master Boot Record (MBR)
  concepts, 5, 6–7
  data hiding, 6–7
Master File Table (MFT), 10

Matlab*P, 179
Matvec, 197
MBR *see* Master Boot Record
MC *see* Monte Carlo simulations
MD5, 25
MDL *see* Minimum Description length Principle
memory-based methods, software development cost estimation methods, 127, 131–3, 136–9
MEMS, 31
MessageLabs, 64–5
metadata manipulation, data hiding, 5, 7–8, 15
Metasploit Project's Slacker tool, 16
metrics of evaluation, email spam filtering, 84, 88–92
Mexico, 211, 253
MFBP, 129
MFT *see* Master File Table
MHT, 26
micro engineering process, GSD, 230–4, 248–9
microsensor nodes, sensor networks, 30–7
Microsoft
  C#, 218
  Office documents, 15
  Outlook, 71, 96–7
  Windows, 1–17
  Windows XP (SP2), 157–8
  Word, 198
MIKEY, 26
MIME encoding, email spam filtering, 107
Minimum Description length Principle (MDL), 81–3
misclassification costs, email spam filtering, 84–96, 98
mobile phones, 23
model-based data-driven methods
  see also nonparametric...; parametric...; semiparametric...
  software development cost estimation methods, 127–33, 136–9

modified data, sanitized data, 188–90
modulation, watermarking, 27–8
monsoons, India, 252
Monte Carlo simulations 116, 119, 121–4, 138–74
  see also simulation techniques
  accuracy study, 154–74
  basic principles, 123–4
  CoBRA, 119
  concepts, 123–4, 138–40, 145–6, 149, 154–69
  efficiency study, 154–74
  empirical study, 154–74
  experimental results, 158–69
  hypotheses, 154–69
  Latin Hypercube (LH) sampling, 149, 164–5
  stochastic approaches, 145–6, 149
Monty Python, 47
mortgages, email spam, 50, 58
Mozilla Thunderbird, 57, 71, 96–8
MPEG standards, 23, 26
MPI, 179
multi-word features, learning-based spam filtering, 69–71
multifunction networked embedded systems, concepts, 22–3
multilevel 2D bar codes, 26
multimedia
  see also sensor security
  applications, 20–4
  authentication requirements, 24–6, 29–30
  compression, 22–3, 25–6, 29, 37
  concepts, 19–43
  conference systems, 21–2
  definition, 20–1
  digital signatures, 25–6
  DRM, 23–4
  embedded systems, 22–3, 27–8, 35
  encryption, 24–6, 33–7
  forensics, 20–1, 29–30
  MPEG standards, 23, 26
  networks, 21–4, 30–7
  SPEF, 36

SUBJECT INDEX 293

steganography, 20–1, 28–9
trusted hardware, 36–7
trusted software, 32, 35–6
types, 20–1
watermarking, 20–1, 24, 26–30, 37
multimode networked embedded systems, concepts, 22–3
multivariate regression, 127–30, 139
Münch, Jürgen, 115–74
Mutual Information, 72
Mylyn, 198

N

N-fold cross-validation, email spam filtering, 87–8
Naïve Bayes, 65, 73–4
Nakamura, Taiga, 175–200
NASA Goddard Space Flight Center (GSFC), 177
National Library of Medicine, 63
NDRs *see* Non-Delivery Reports
near shore outsourcing, IT resources, 205–9
Nearest Neighbors, 71–2
networks
    multimedia, 21–4, 30–7
    neural networks, 128–40
    sensor networks, 30–7
neural networks, 128–40
New Computer Policy, India, 213–14
NFS, 127, 134, 136, 139
Nigerian fraud, 49
noise, image-based spam attacks, 108–9
nominal project costs
    CoBRA, 119–74
    definition, 120, 121
Non-Delivery Reports (NDRs), 100–1
non-proprietary data-driven methods
    see also composite...; memory-based...; model-based...
    software development cost estimation methods, 126–40

nonparametric model-based data-driven methods
    see also model-based...
    software development cost estimation methods, 127, 130–3, 136–9
normalization functions, sanitized data, 189–90
NTFS, data hiding, 7–10, 12–13
Nucleus Research, 48

O

obfuscation attacks, email spam filtering, 102, 104–5, 108
Object Management Group, 260–1
OCR, 107–8
Offshore Development Center (ODC), 236–8, 245
on-site contractors, IT resources, 205–9
onshore outsourcing, IT resources, 205–9
open source, 231–2
OpenMP, 179, 196–7
operational process for maintenance and support, GSD, 240–1
organizational GSD recommendations, 219, 229–30
OSR, 127, 132–3, 139
OSR+OLS, 127, 133, 139
outsourcing
    see also global software development
    best practices, 231, 233–4, 247–51
    blend statistics, 247
    business drivers, 217–18, 220–2, 246–7, 253–6
    concepts, 203–9, 212–16
    controls, 249
    corporate models, 206–8, 213–14, 259–60
    costs, 205–7, 210–11, 213–14, 220–1, 222–6, 229–30, 253–4
    critique, 244–5, 246–51
    failures, 209
    functional areas, 206–8
    future prospects, 201–2, 204, 245, 249–62

outsourcing
(Continued)
  India, 201, 204, 208–9, 210–12, 213–16, 221–2, 243, 246, 252–65
  job losses, 208–9, 220, 250–1, 254–5
  options, 205
  origins, 212–16
  PMO, 207–8, 247–8
  principles, 220–2
  RFPs, 221–5, 248
  supplier selection, 217–18, 222–6, 244, 253–6
  surveys, 246–51
  testing, 208, 221, 226–8, 245, 261
  virtual roundtable, 246–51
overheads
  see also costs
  causal model, 120–2
  CoBRA, 119–74
  global software development (GSD), 242–3
  GSD, 242–3
  software engineering experiments, 180–8

P

P2MP, 26
P2P, 26
packets, Internet security challenges, 23–4
Padmanabhan, K. (Paddy), 262–5
paired t-tests, 157, 160–1
parallel coding, Experiment Manager, 195–7
parallel programming languages, 176, 179–98
parametric model-based data-driven methods
  see also model-based. . .
  software development cost estimation methods, 127–33, 136–9
partition slack, data hiding, 5, 7–8
partitions
  boot sectors, 7–8
  concepts, 5, 6–8
  data hiding, 5, 6–8

Pascal's calculating machine, 212
passive attacks, email spam filtering, 105–6
patent applications, India and China, 257–8
PDF files, 106–7, 109
percentiles, definition, 151
Pérez, José Carlos Corizo, 45–113
Perl, 60
PGP *see* Pretty Good Privacy
pharmaceutical industry, outsourcing, 209
Philippines, 210–11
philosophy, data hiding, 3–4
'phishing' attacks, 48, 49–50
physical aspects, data hiding, 1, 3–4
picospam, 101–2
pilots, knowledge transition process, 239–40
Planning Game, 127, 133, 139
planning GSD guidelines, 219–20, 229–30
PLUM, 198
PMO *see* Project Management Office
PNG files, 25, 28–9
point estimates, 116, 141–4
politics, GSD, 203–4, 251–3
POP, 97
pornography, email spam, 50, 58–9, 96
POS systems, 204
possibilistic uncertainty, software development cost estimation methods, 134–6
postage techniques, email spam, 54
power constraints, sensor networks, 32–4, 36, 37
PPM *see* Prediction by Partial Matching
practices, current GSD practice, 201–2, 203–4, 216–45
Prasad, Alpana, 201–69
predator-prey models, 127–9, 139
Prediction by Partial Matching (PPM), 82–3
Pretty Good Privacy (PGP), 53–4
PRICE-S, 126, 127
Price-to-Win efforts, 126

primitive language analysis
    see also content-based...; email spam
    concepts, 51
privacy GSD recommendations, 220, 243
probabilistic approaches, email spam
    filtering, 73–4
probabilistic uncertainty, software
    development cost estimation methods,
    134–6
probability density function, 123–4, 141–9
processing structure, learning-based spam
    filtering, 65–7, 83
product designs, IT uses, 204
production (maintenance) support
    concepts, 234–41, 261
    governance model, 237–9
    GSD, 234–41
    knowledge transition process, 239–40
    management issues, 236–7
    new projects, 234–5
productivity issues, programmers, 176,
    179–98, 231–2
professor participation overheads, software
    engineering experiments, 180–3
professor role, experiment managers, 185
programmers 176–98, 208–9, 213–15, 220,
    226–8, 231–4, 243, 250–1, 254–5
    see also global software development;
        human resources
    historical background, 213
    job losses, 208–9, 220, 250–1, 254–5
    productivity issues, 176, 179–98, 231–2
    software engineering experiments, 176–98
prohibited data, sanitized data, 188–90
project management committee (PMC),
    237–8
Project Management Office (PMO), 207–8,
    247–8
project risk analysis
    CoBRA, 116–17, 118–22, 134, 138–74
    concepts, 116–74, 244
    empirical study, 119, 149–74
    experimental results, 158–69

GSD, 244
simulation techniques, 115–74
Proofpoint P800 Message Protection
    Appliance, 84
proprietary/non-proprietary data-driven
    methods, software development cost
    estimation methods, 126–34, 136–9
proxying methods, email spam filtering,
    96–8
PS see production (maintenance) support
psychological aspects, software engineering
    experiments, 176–7
PU1, 84–5, 87
Public Key Cryptography, 53–4
purchased applications, IT resources, 205–9
pX-effort approach, 135
pyramid schemes, email spam, 49

Q

QKI, 169
quality GSD recommendations, 220, 222–8,
    232–4, 243, 249
quarantine folders, email spam, 58, 95–6
Quasi-MC, 138–9

R

R-DCS, 32–3
RAM slack, 7
random sampling
    concepts, 116, 118–19, 138–9
    overview, 138–9, 190–1
randomized arithmetic coding, 25–6
RBFN, 129
RBL see Realtime Blackhole List
RC4, 34
RC5, 34
RC6, 34–5
Reagan, Ronald, 252
Realtime Blackhole List (RBL), 52
redirected application executions, data
    hiding, 14–15
reference data, definition, 150

Register of Known Spam Offenders (ROKSO), 50
registry entries, data hiding, 14
regression analysis, 127–31
related work
  simulation techniques, 126–40
  software engineering experiments, 197–8
relative volatility of data hiding areas, 8–9
Relevancy Score, 72
replicated studies, experiments, 177, 180–1, 191–2
reputation controls, email spam, 53–4
Request for Proposals (RFPs), 209, 221–5, 248
research questions (RQs), CoBRA, 119, 140–74
research/pedagogy balance, software engineering experiments, 180–1
retention issues, GSD, 241–2
Reverse NDR Attacks, 100–1
RF, 31–4
RFC2821, 100
RFC3834, 100
RFID, 21, 37, 204
RFPs *see* Request for Proposals
Ripper, 76
risks
  see also project risk analysis
  GSD, 244
ROCCH metrics, 88, 90–2
ROKSO *see* Register of Known Spam Offenders
Rolex, 95
roundtable *see* virtual roundtable
RQs *see* research questions
RSA, 25
RTF files, 106–7, 109
Rubin, Howard, 255
rule induction, 127, 130–1, 139
rule learners, email spam filtering, 75–6
rules, heuristic spam filtering, 51, 56–62
running test procedures, email spam filtering evaluations, 84, 87–8
Russia, 210–11, 213, 252, 253

S

sampling techniques for software cost estimation and risk analysis *see also* Latin Hypercube (LH) sampling; simulation techniques
  accuracy study, 143, 149–74
  concepts, 116, 118–19, 124–6, 138–74
  dependent variables, 154–5
  efficiency study, 149, 153–74
  empirical study, 149–74
  experimental design, 155–7, 163–9
  experimental operation, 157–8
  experimental results, 158–69
  hypotheses, 154–69
  independent variables, 154–5
  Monte Carlo simulations, 116, 119, 121–4, 138–74
  validity discussion, 162–9
'sand-box model' of ABBs, 231–4
sanitized data
  classifications, 188–9
  Experiment Manager, 182, 183, 188–90, 193
  functions, 189–90
Sanz, Enrique Puertas, 45–113
SARI, 25
SAwin32, 97
SBS, 127, 133, 139
Schickard's calculating machine, 212
SCRIPT tags, email spam filtering, 107
SEAL, 34
Sebastiani, 64, 72
sectors, data hiding, 7–10
security
  see also data hiding; sensor...
  concepts, 19–43
security protocols, sensor networks, 33–7

SEER-SEM, 127–9, 139
SEL *see* Software Engineering Laboratory
self-reported data, 178–9
semantic-based multimedia retrieval, 23, 25–6
semiparametric model-based data-driven methods *see also* model-based...
   software development cost estimation methods, 127, 131–3, 136–9
Sender Policy Framework (SPF), 53–4
sensor networks
   aggregation, 31–2
   battery limitations, 32–4, 36, 37
   communications security, 34–5
   concepts, 30–7
   DCS benefits, 32–3
   encryption, 33–7
   hardware, 32–4, 35, 36–7
   RFID, 21, 37, 204
   security protocols, 33–7
   software design, 35–6
   SPEF, 36
   threats, 32–3, 34–5
   trusted software, 32, 35–6
sensor security *see also* multimedia
   concepts, 19–20, 22–4, 30–7
   encryption, 33–7
   needs, 21, 22–4, 33–7
   SEAL, 34
   TEA, 34
   XXTEA, 34
serial coding, Experiment Manager, 195
server-side email spam filtering, 93–5
Service Level Agreements (SLAs), 209, 229, 239–40, 248–9
SESE system, 197–8
Shull, Forrest, 175–200
signature-based analysis of media, data hiding, 12–13
Silicon Valley, 257
simulation techniques
   see also Latin Hypercube...; sampling...
   accuracy study, 143, 149–74

   analytical approaches, 143–5, 164–9
   concepts, 119, 121–6, 138–74
   definition, 122–3
   dependent variables, 154–5
   efficiency study, 149, 153–74
   empirical study, 119, 149–74
   experimental design, 155–7, 163–9
   experimental operation, 155–7
   experimental results, 158–69
   hypotheses, 154–69
   independent variables, 154–5
   Monte Carlo simulations, 116, 119, 121–4, 138–74
   software development cost estimation, 115–74
   stochastic approaches, 143, 145–9
   validity discussion, 162–9
single-layer perception 129
   see also ANN
slack space *see* file slack
SLAs *see* Service Level Agreements
slice and dice tokenization attacks, 103
SLIM, 127–9, 139
Sloan Management Review, 250
Smith, Adam, 212
SMTP, 93, 97, 100–1
SNEP, sensor security, 33–4
socio-economic aspects, software engineering experiments, 176–7
SoftCost-R, 126, 127
software 115–74, 175–200, 201–69
   see also global software development; IT
   ABBs, 231–4
   costs, 205–7
   development resources needed, 204–5
   email spam, 50, 96
   engineering, 175–200, 259–61
   historical background, 201–4, 211–14
   importance, 201–2, 204
   life cycle model, 226–8
   open source, 231–2
   production (maintenance) support, 234–41

software
(Continued)
  programmers, 176–98, 213–15, 226–8, 231–4, 243, 250–1
  sourcing landscape, 203–9
  WDF, 226–8, 231–4
software development cost estimation methods
  accuracy study, 143, 149–74
  analytical approaches, 143–5, 164–9
  CoBRA, 116–17, 118–22, 134, 138–74
  comparative studies, 136–40
  composite methods, 127, 133, 136–9
  concepts, 115–74
  data-driven methods, 119–22, 126–34, 136–9
  difficulties, 117–18
  efficiency study, 149, 153–74
  effort estimation methods, 126–40
  evaluation criteria, 136–9
  existing effort estimation methods, 126–40
  experimental design, 155–7, 163–9
  experimental operation, 157–8
  experimental results, 158–69
  experimental study, 119, 149–74
  expert-based methods, 119–22, 127, 133, 134, 136–9
  hybrid methods, 116, 119–22, 127, 134, 136–9
  memory-based methods, 127, 131–3, 136–9
  model-based data-driven methods, 127–33, 136–9
  nonparametric model-based data-driven methods, 127, 130–3, 136–9
  parametric model-based data-driven methods, 127–33, 136–9
  possibilistic uncertainty, 134–6
  probabilistic uncertainty, 134–6
  proprietary/non-proprietary data-driven methods, 126–34, 136–9
  related work, 126–40
  research questions, 140–74
  semiparametric model-based data-driven methods, 127, 131–3, 136–9
  simulation techniques, 116, 119, 121–6, 138–74
  stochastic approaches, 143, 145–9
  uncertainty concepts, 134–6, 137–9
  validity discussion, 162–9
software engineering experiments
  see also Experiment Manager
  accuracy in collected data, 178, 181–2, 197–8
  behavioral aspects, 176–7
  classroom resources, 175–7, 178–82
  collecting data, 175–6, 177–8, 181–2, 186–90, 197–8
  complexity factors, 180–1
  concepts, 175–200
  consistent replication across classes, 180–1, 183, 191–2
  costs, 177–8
  data management, 181–2
  experiment managers, 182, 184–6
  Hackystat, 182, 184, 185, 192, 198
  hypotheses, 176–7
  India, 259
  IRB issues, 181–2, 192
  PLUM, 198
  professor participation overheads, 180–3
  related work, 197–8
  research/pedagogy balance, 180–1
  SEL, 177–82
  SESE system, 197–8
  socio-economic aspects, 176–7
  student participation overheads, 181–8
Software Engineering Institute, 260
Software Engineering Laboratory (SEL), 177–82
Sophos, 49, 108
sourcing IT landscape
  see also outsourcing 203–9, 212–16
  concepts, 203–9, 212–16
  conclusions, 209

corporate models, 206–8, 213–14, 259–60
costs, 205–6, 210–11, 213–14, 220–1, 222–6, 229–30, 253–4
failures, 209
job losses, 208–9, 220, 250–1, 254–5
options, 205–6
origins, 212–16
Soviet Union 213, 252
see also Russia
Spain, 254
spam *see* email spam
spam bots, 52–3, 98
SpamAssassin, 58–60, 84–5, 86, 97–8
SpamBase, 76–7, 84–5, 87
SpamBayes, Microsoft Outlook, 96–7, 105–6
The Spamhaus Project, 50
Sparse Binary Polynomial Hash (SBPH), 71
SPEF, 36
SPF *see* Sender Policy Framework
SPINS, sensor security, 33–4
SPQR/100, 126, 127
spread spectrum, watermarking, 27–8
SQL, 31, 182, 233–4
SRTP, 26
Stacked Generalization, 79
staff *see* human resources
statistical attacks, email spam filtering, 102, 104–6
steering committees (STC), 237–8
steganalysis, concepts, 29
steganography
   see also data hiding; encryption
   concepts, 3, 15–16, 19, 20–1, 28–9, 37
   definition, 3, 28
   multimedia security, 20–1, 28–9, 37
stepwise analysis of variance 127, 128, 139
   see also ANOVA
Sthultz, Michael, 1–17
stochastic approaches
   Latin Hypercube (LH) sampling, 146–9
   Monte Carlo simulations, 145–6, 149
   simulation techniques, 143, 145–9

stock scams, email spam, 50, 58
strata concepts, 124–6
stratification, concepts, 124–6
streaming video, 23–4, 30
string matching, data hiding, 11–12, 15
strong law of large numbers, 123–4
structural engineers, 259–60
student participation overheads, software engineering experiments, 181–8
student role, experiment managers, 185
subject views, Experiment Manager, 193–5, 196–7
suitability issues, GSD, 229–30
summarization functions, sanitized data, 189–90
superblock slack, data hiding, 5, 8
suppliers, GSD, 217–18, 222–6, 244, 253–6
support vector machines (SVMs), email spam filtering, 71–2, 76–7
supported analyses, Experiment Manager, 193–7
supporting standards and code review guidelines, GSD, 231, 232–4
SVMs *see* support vector machines
swap files, data hiding, 14
Symantec Brightmail solution, 58, 61–2
Symantec Corporation, 58
syntax fixes, Experiment Manager, 195
SysML, 260
systems analysts/designers, GSD, 226–8
Systems Engineers, 261

T

tagging methods, email spam filtering, 96–8
talent supplies
   see also human resources
   GSD, 254–6
targets for the GSD future, 259–62
Tata Consultancy Services (TCS), 213–14, 254, 262–5

TCOE *see* Testing Center of Excellence
TCP/IP, 3
TCPA 32, 36–7
 see also hardware
TCS *see* Tata Consultancy Services
TDMA, 27
TEA, 34
teams
 see also human resources
 GSD, 216–18, 226–30
technician role, experiment
  managers, 184–5
technology change factors, GSD, 256–9
Tepfenhart, William M., 201–69
Term Clustering, 72
term frequency (TF), learning-based spam
  filtering, 68–9
TESLA, sensor security, 33–4
test collections, email spam filtering
  evaluations, 84–7
testing
 Experiment Manager, 195
 GSD life cycle model, 226–8,
  245, 261
 outsourcing, 208, 221, 226–8, 245, 261
Testing Center of Excellence
  (TCOE), 245
Text Categorization, email spam, 63–6,
  72, 80–3
TF *see* term frequency
things you have to live with, GSD, 242–3
third-party sources, email spam, 54–5
Thunderbird, 57, 71, 96–8
TIFF, 25
tokens and weights
 learning-based spam filtering, 68–72,
  79–80, 83, 102–4, 109
 tokenization attacks, 102–4, 109
Translog form, 127
TREC metrics, 45, 56, 80–3, 85–8, 92
Trend Micro InterScan Messaging Security
  Suite, 95–6
Trendowicz, Adam, 115–74

trust, sensor networks, 32, 35–6
trusted hardware, multimedia, 36–7
trusted software, multimedia, 32, 35–6
Turing Test, email spam, 54

U

UK
 Colossus computer, 212
 computing history, 212–13
 GSD, 201, 204
UMDinst, Experiment Manager,
  183–5, 192
UML *see* Unified Modeling Language
uncertainty concepts, software development
  cost estimation methods, 134–6, 137–9
Unified Modeling Language (UML), 260
United Nations Conference on Trade and
  Development, 47
UNIVAC I, 213
univariate regression, 127–30, 139
University of Maryland, 177–8, 184–5,
  192–3
University of Pennsylvania, 213
Unix
 data hiding, 1–17
 UMDinst, 183
unsolicited emails *see* email spam
UPC, 179
URL, 102
US
 computing history, 212–13
 education system, 251
 email spam, 47–8, 50
 Experiment Manager studies, 190–1
 GSD, 201, 206–9, 210–11, 243, 250–7
 patent applications, 257–8
 Silicon Valley, 257
USB flash drives, 4
USSR
 see also Russia
 computing history, 213
 political changes, 252

SUBJECT INDEX

V

validity discussion, simulation techniques, 162–9
value-added tools, GSD, 231, 232–3
Viagra, email spam, 49, 57–8, 65, 99, 101–4
video
    see also multimedia
    concepts, 21–43
video teleconferencing
    see also multimedia
    concepts, 21–2
virtual files systems
    concepts, 5, 6–7
    data hiding, 6–7
virtual roundtable, outsourcing, 246–51
viruses, email spam, 48, 98
VLSI, 36
VoIP, 29, 238–9
volume slack, data hiding, 5, 6–7
VQ, 25

W

WANs, 230–1
watermarking
    ambiguity attacks, 27–8
    coding, 27–8
    collusion attacks, 29–30
    concepts, 1–2, 3, 19, 20–1, 24, 26–30, 37
    definition, 3, 26–7
    fragile watermarks, 28
    modulation, 27–8
    multimedia security, 20–1, 24, 26–30, 37
    types, 26–7
WAV files, 28–9
WDF *see* Web Delivery Foundation
weather disasters, economic effects, 252

web 182, 183, 184–98, 226–8, 231–4
    see also Internet
    Experiment Manager, 182, 183, 184–98
Web Delivery Foundation (WDF), 226–8, 231–4
weights, learning-based spam filtering, 68–9, 79–80
white and black lists, email spam, 51–3, 59, 94–5, 98
Wickenkamp, Axel, 115–74
Wideband Delphi, 127, 133, 139
Windows, data hiding, 1–17
Windows XP (SP2), 157–8
Wipro, 213
wireless sensor networks
    see also sensor networks
    concepts, 30–7
Wolters Kluwer (WK), 218, 220–1, 237, 243–5, 262
words, abstract addressability levels, 2, 4
workflow cycles, Experiment Manager, 182, 185, 195–6

X

XTEA, 34
XXTEA, 34

Y

Yourdon, Ed, 255

Z

Zazworka, Nico, 175–200
Zelkowitz, Marvin V., 175–200
zip disks, 4

Contents of Volumes in This Series

Volume 42

Nonfunctional Requirements of Real-Time Systems
    TEREZA G. KIRNER AND ALAN M. DAVIS
A Review of Software Inspections
    ADAM PORTER, HARVEY SIY, AND LAWRENCE VOTTA
Advances in Software Reliability Engineering
    JOHN D. MUSA AND WILLA EHRLICH
Network Interconnection and Protocol Conversion
    MING T. LIU
A Universal Model of Legged Locomotion Gaits
    S. T. VENKATARAMAN

Volume 43

Program Slicing
    DAVID W. BINKLEY AND KEITH BRIAN GALLAGHER
Language Features for the Interconnection of Software Components
    RENATE MOTSCHNIG-PITRIK AND ROLAND T. MITTERMEIR
Using Model Checking to Analyze Requirements and Designs
    JOANNE ATLEE, MARSHA CHECHIK, AND JOHN GANNON
Information Technology and Productivity: A Review of the Literature
    ERIK BRYNJOLFSSON AND SHINKYU YANG
The Complexity of Problems
    WILLIAM GASARCH
3-D Computer Vision Using Structured Light: Design, Calibration, and Implementation Issues
    FRED W. DEPIERO AND MOHAN M. TRIVEDI

Volume 44

Managing the Risks in Information Systems and Technology (IT)
    ROBERT N. CHARETTE
Software Cost Estimation: A Review of Models, Process and Practice
    FIONA WALKERDEN AND ROSS JEFFERY
Experimentation in Software Engineering
    SHARI LAWRENCE PFLEEGER
Parallel Computer Construction Outside the United States
    RALPH DUNCAN
Control of Information Distribution and Access
    RALF HAUSER

Asynchronous Transfer Mode: An Engineering Network Standard for High Speed Communications
　　Ronald J. Vetter
Communication Complexity
　　Eyal Kushilevitz

Volume 45

Control in Multi-threaded Information Systems
　　Pablo A. Straub and Carlos A. Hurtado
Parallelization of DOALL and DOACROSS Loops—a Survey
　　A. R. Hurson, Joford T. Lim, Krishna M. Kavi, and Ben Lee
Programming Irregular Applications: Runtime Support, Compilation and Tools
　　Joel Saltz, Gagan Agrawal, Chialin Chang, Raja Das, Guy Edjlali, Paul Havlak, Yuan-Shin Hwang, Bongki Moon, Ravi Ponnusamy, Shamik Sharma, Alan Sussman, and Mustafa Uysal
Optimization Via Evolutionary Processes
　　Srilata Raman and L. M. Patnaik
Software Reliability and Readiness Assessment Based on the Non-homogeneous Poisson Process
　　Amrit L. Goel and Kune-Zang Yang
Computer-Supported Cooperative Work and Groupware
　　Jonathan Grudin and Steven E. Poltrock
Technology and Schools
　　Glen L. Bull

Volume 46

Software Process Appraisal and Improvement: Models and Standards
　　Mark C. Paulk
A Software Process Engineering Framework
　　Jyrki Kontio
Gaining Business Value from IT Investments
　　Pamela Simmons
Reliability Measurement, Analysis, and Improvement for Large Software Systems
　　Jeff Tian
Role-Based Access Control
　　Ravi Sandhu
Multithreaded Systems
　　Krishna M. Kavi, Ben Lee, and Alli R. Hurson
Coordination Models and Language
　　George A. Papadopoulos and Farhad Arbab
Multidisciplinary Problem Solving Environments for Computational Science
　　Elias N. Houstis, John R. Rice, and Naren Ramakrishnan

Volume 47

Natural Language Processing: A Human–Computer Interaction Perspective
　　Bill Manaris

Cognitive Adaptive Computer Help (COACH): A Case Study
    EDWIN J. SELKER
Cellular Automata Models of Self-replicating Systems
    JAMES A. REGGIA, HUI-HSIEN CHOU, AND JASON D. LOHN
Ultrasound Visualization
    THOMAS R. NELSON
Patterns and System Development
    BRANDON GOLDFEDDER
High Performance Digital Video Servers: Storage and Retrieval of Compressed Scalable Video
    SEUNGYUP PAEK AND SHIH-FU CHANG
Software Acquisition: The Custom/Package and Insource/Outsource Dimensions
    PAUL NELSON, ABRAHAM SEIDMANN, AND WILLIAM RICHMOND

Volume 48

Architectures and Patterns for Developing High-Performance, Real-Time ORB Endsystems
    DOUGLAS C. SCHMIDT, DAVID L. LEVINE, AND CHRIS CLEELAND
Heterogeneous Data Access in a Mobile Environment – Issues and Solutions
    J. B. LIM AND A. R. HURSON
The World Wide Web
    HAL BERGHEL AND DOUGLAS BLANK
Progress in Internet Security
    RANDALL J. ATKINSON AND J. ERIC KLINKER
Digital Libraries: Social Issues and Technological Advances
    HSINCHUN CHEN AND ANDREA L. HOUSTON
Architectures for Mobile Robot Control
    JULIO K. ROSENBLATT AND JAMES A. HENDLER

Volume 49

A Survey of Current Paradigms in Machine Translation
    BONNIE J. DORR, PAMELA W. JORDAN, AND JOHN W. BENOIT
Formality in Specification and Modeling: Developments in Software Engineering Practice
    J. S. FITZGERALD
3-D Visualization of Software Structure
    MATHEW L. STAPLES AND JAMES M. BIEMAN
Using Domain Models for System Testing
    A. VON MAYRHAUSER AND R. MRAZ
Exception-Handling Design Patterns
    WILLIAM G. BAIL
Managing Control Asynchrony on SIMD Machines—a Survey
    NAEL B. ABU-GHAZALEH AND PHILIP A. WILSEY
A Taxonomy of Distributed Real-time Control Systems
    J. R. ACRE, L. P. CLARE, AND S. SASTRY

Volume 50

Index Part I
Subject Index, Volumes 1–49

Volume 51

Index Part II
Author Index
Cumulative list of Titles
Table of Contents, Volumes 1–49

Volume 52

Eras of Business Computing
    ALAN R. HEVNER AND DONALD J. BERNDT
Numerical Weather Prediction
    FERDINAND BAER
Machine Translation
    SERGEI NIRENBURG AND YORICK WILKS
The Games Computers (and People) Play
    JONATHAN SCHAEFFER
From Single Word to Natural Dialogue
    NEILS OLE BENSON AND LAILA DYBKJAER
Embedded Microprocessors: Evolution, Trends and Challenges
    MANFRED SCHLETT

Volume 53

Shared-Memory Multiprocessing: Current State and Future Directions
    PER STEUSTRÖM, ERIK HAGERSTEU, DAVID I. LITA, MARGARET MARTONOSI, AND MADAN VERNGOPAL
Shared Memory and Distributed Shared Memory Systems: A Survey
    KRISHNA KAUI, HYONG-SHIK KIM, BEU LEE, AND A. R. HURSON
Resource-Aware Meta Computing
    JEFFREY K. HOLLINGSWORTH, PETER J. KELCHER, AND KYUNG D. RYU
Knowledge Management
    WILLIAM W. AGRESTI
A Methodology for Evaluating Predictive Metrics
    JASRETT ROSENBERG
An Empirical Review of Software Process Assessments
    KHALED EL EMAM AND DENNIS R. GOLDENSON
State of the Art in Electronic Payment Systems
    N. ASOKAN, P. JANSON, M. STEIVES, AND M. WAIDNES
Defective Software: An Overview of Legal Remedies and Technical Measures Available to Consumers
    COLLEEN KOTYK VOSSLER AND JEFFREY VOAS

CONTENTS OF VOLUMES IN THIS SERIES

Volume 54

An Overview of Components and Component-Based Development
  ALAN W. BROWN
Working with UML: A Software Design Process Based on Inspections for the Unified Modeling Language
  GUILHERME H. TRAVASSOS, FORREST SHULL, AND JEFFREY CARVER
Enterprise JavaBeans and Microsoft Transaction Server: Frameworks for Distributed Enterprise Components
  AVRAHAM LEFF, JOHN PROKOPEK, JAMES T. RAYFIELD, AND IGNACIO SILVA-LEPE
Maintenance Process and Product Evaluation Using Reliability, Risk, and Test Metrics
  NORMAN F. SCHNEIDEWIND
Computer Technology Changes and Purchasing Strategies
  GERALD V. POST
Secure Outsourcing of Scientific Computations
  MIKHAIL J. ATALLAH, K. N. PANTAZOPOULOS, JOHN R. RICE, AND EUGENE SPAFFORD

Volume 55

The Virtual University: A State of the Art
  LINDA HARASIM
The Net, the Web and the Children
  W. NEVILLE HOLMES
Source Selection and Ranking in the WebSemantics Architecture Using Quality of Data Metadata
  GEORGE A. MIHAILA, LOUIQA RASCHID, AND MARIA-ESTER VIDAL
Mining Scientific Data
  NAREN RAMAKRISHNAN AND ANANTH Y. GRAMA
History and Contributions of Theoretical Computer Science
  JOHN E. SAVAGE, ALAN L. SALEM, AND CARL SMITH
Security Policies
  ROSS ANDERSON, FRANK STAJANO, AND JONG-HYEON LEE
Transistors and 1C Design
  YUAN TAUR

Volume 56

Software Evolution and the Staged Model of the Software Lifecycle
  KEITH H. BENNETT, VACLAV T. RAJLICH, AND NORMAN WILDE
Embedded Software
  EDWARD A. LEE
Empirical Studies of Quality Models in Object-Oriented Systems
  LIONEL C. BRIAND AND JÜRGEN WÜST
Software Fault Prevention by Language Choice: Why C Is Not My Favorite Language
  RICHARD J. FATEMAN
Quantum Computing and Communication
  PAUL E. BLACK, D. RICHARD KUHN, AND CARL J. WILLIAMS
Exception Handling
  PETER A. BUHR, ASHIF HARJI, AND W. Y. RUSSELL MOK

Breaking the Robustness Barrier: Recent Progress on the Design of the Robust Multimodal System
    SHARON OVIATT
Using Data Mining to Discover the Preferences of Computer Criminals
    DONALD E. BROWN AND LOUISE F. GUNDERSON

Volume 57

On the Nature and Importance of Archiving in the Digital Age
    HELEN R. TIBBO
Preserving Digital Records and the Life Cycle of Information
    SU-SHING CHEN
Managing Historical XML Data
    SUDARSHAN S. CHAWATHE
Adding Compression to Next-Generation Text Retrieval Systems
    NIVIO ZIVIANI AND EDLENO SILVA DE MOURA
Are Scripting Languages Any Good? A Validation of Perl, Python, Rexx, and Tcl against C, C++, and Java
    LUTZ PRECHELT
Issues and Approaches for Developing Learner-Centered Technology
    CHRIS QUINTANA, JOSEPH KRAJCIK, AND ELLIOT SOLOWAY
Personalizing Interactions with Information Systems
    SAVERIO PERUGINI AND NAREN RAMAKRISHNAN

Volume 58

Software Development Productivity
    KATRINA D. MAXWELL
Transformation-Oriented Programming: A Development Methodology for High Assurance Software
    VICTOR L. WINTER, STEVE ROACH, AND GREG WICKSTROM
Bounded Model Checking
    ARMIN BIERE, ALESSANDRO CIMATTI, EDMUND M. CLARKE, OFER STRICHMAN, AND YUNSHAN ZHU
Advances in GUI Testing
    ATIF M. MEMON
Software Inspections
    MARC ROPER, ALASTAIR DUNSMORE, AND MURRAY WOOD
Software Fault Tolerance Forestalls Crashes: To Err Is Human; To Forgive Is Fault Tolerant
    LAWRENCE BERNSTEIN
Advances in the Provisions of System and Software Security—Thirty Years of Progress
    RAYFORD B. VAUGHN

Volume 59

Collaborative Development Environments
    GRADY BOOCH AND ALAN W. BROWN
Tool Support for Experience-Based Software Development Methodologies
    SCOTT HENNINGER
Why New Software Processes Are Not Adopted
    STAN RIFKIN

Impact Analysis in Software Evolution
  MIKAEL LINDVALL
Coherence Protocols for Bus-Based and Scalable Multiprocessors, Internet, and Wireless Distributed Computing Environments: A Survey
  JOHN SUSTERSIC AND ALI HURSON

Volume 60

Licensing and Certification of Software Professionals
  DONALD J. BAGERT
Cognitive Hacking
  GEORGE CYBENKO, ANNARITA GIANI, AND PAUL THOMPSON
The Digital Detective: An Introduction to Digital Forensics
  WARREN HARRISON
Survivability: Synergizing Security and Reliability
  CRISPIN COWAN
Smart Cards
  KATHERINE M. SHELFER, CHRIS CORUM, J. DREW PROCACCINO, AND JOSEPH DIDIER
Shotgun Sequence Assembly
  MIHAI POP
Advances in Large Vocabulary Continuous Speech Recognition
  GEOFFREY ZWEIG AND MICHAEL PICHENY

Volume 61

Evaluating Software Architectures
  ROSEANNE TESORIERO TVEDT, PATRICIA COSTA, AND MIKAEL LINDVALL
Efficient Architectural Design of High Performance Microprocessors
  LIEVEN EECKHOUT AND KOEN DE BOSSCHERE
Security Issues and Solutions in Distributed Heterogeneous Mobile Database Systems
  A. R. HURSON, J. PLOSKONKA, Y. JIAO, AND H. HARIDAS
Disruptive Technologies and Their Affect on Global Telecommunications
  STAN MCCLELLAN, STEPHEN LOW, AND WAI-TIAN TAN
Ions, Atoms, and Bits: An Architectural Approach to Quantum Computing
  DEAN COPSEY, MARK OSKIN, AND FREDERIC T. CHONG

Volume 62

An Introduction to Agile Methods
  DAVID COHEN, MIKAEL LINDVALL, AND PATRICIA COSTA
The Timeboxing Process Model for Iterative Software Development
  PANKAJ JALOTE, AVEEJEET PALIT, AND PRIYA KURIEN
A Survey of Empirical Results on Program Slicing
  DAVID BINKLEY AND MARK HARMAN
Challenges in Design and Software Infrastructure for Ubiquitous Computing Applications
  GURUDUTH BANAVAR AND ABRAHAM BERNSTEIN

Introduction to MBASE (Model-Based (System) Architecting and Software Engineering)
  DAVID KLAPPHOLZ AND DANIEL PORT
Software Quality Estimation with Case-Based Reasoning
  TAGHI M. KHOSHGOFTAAR AND NAEEM SELIYA
Data Management Technology for Decision Support Systems
  SURAJIT CHAUDHURI, UMESHWAR DAYAL, AND VENKATESH GANTI

Volume 63

Techniques to Improve Performance Beyond Pipelining: Superpipelining, Superscalar, and VLIW
  JEAN-LUC GAUDIOT, JUNG-YUP KANG, AND WON WOO RO
Networks on Chip (NoC): Interconnects of Next Generation Systems on Chip
  THEOCHARIS THEOCHARIDES, GREGORY M. LINK, NARAYANAN VIJAYKRISHNAN, AND MARY JANE IRWIN
Characterizing Resource Allocation Heuristics for Heterogeneous Computing Systems
  SHOUKAT ALI, TRACY D. BRAUN, HOWARD JAY SIEGEL, ANTHONY A. MACIEJEWSKI, NOAH BECK, LADISLAU BÖLÖNI, MUTHUCUMARU MAHESWARAN, ALBERT I. REUTHER, JAMES P. ROBERTSON, MITCHELL D. THEYS, AND BIN YAO
Power Analysis and Optimization Techniques for Energy Efficient Computer Systems
  WISSAM CHEDID, CHANSU YU, AND BEN LEE
Flexible and Adaptive Services in Pervasive Computing
  BYUNG Y. SUNG, MOHAN KUMAR, AND BEHROOZ SHIRAZI
Search and Retrieval of Compressed Text
  AMAR MUKHERJEE, NAN ZHANG, TAO TAO, RAVI VIJAYA SATYA, AND WEIFENG SUN

Volume 64

Automatic Evaluation of Web Search Services
  ABDUR CHOWDHURY
Web Services
  SANG SHIN
A Protocol Layer Survey of Network Security
  JOHN V. HARRISON AND HAL BERGHEL
E-Service: The Revenue Expansion Path to E-Commerce Profitability
  ROLAND T. RUST, P. K. KANNAN, AND ANUPAMA D. RAMACHANDRAN
Pervasive Computing: A Vision to Realize
  DEBASHIS SAHA
Open Source Software Development: *Structural Tension in the American Experiment*
  COSKUN BAYRAK AND CHAD DAVIS
Disability and Technology: Building Barriers or Creating Opportunities?
  PETER GREGOR, DAVID SLOAN, AND ALAN F. NEWELL

Volume 65

The State of Artificial Intelligence
  ADRIAN A. HOPGOOD
Software Model Checking with SPIN
  GERARD J. HOLZMANN

Early Cognitive Computer Vision
  Jan-Mark Geusebroek
Verification and Validation and Artificial Intelligence
  Tim Menzies and Charles Pecheur
Indexing, Learning and Content-Based Retrieval for Special Purpose Image Databases
  Mark J. Huiskes and Eric J. Pauwels
Defect Analysis: Basic Techniques for Management and Learning
  David N. Card
Function Points
  Christopher J. Lokan
The Role of Mathematics in Computer Science and Software Engineering Education
  Peter B. Henderson

Volume 66

Calculating Software Process Improvement's Return on Investment
  Rini Van Solingen and David F. Rico
Quality Problem in Software Measurement Data
  Pierre Rebours and Taghi M. Khoshgoftaar
Requirements Management for Dependable Software Systems
  William G. Bail
Mechanics of Managing Software Risk
  William G. Bail
The PERFECT Approach to Experience-Based Process Evolution
  Brian A. Nejmeh and William E. Riddle
The Opportunities, Challenges, and Risks of High Performance Computing in Computational Science and Engineering
  Douglass E. Post, Richard P. Kendall, and Robert F. Lucas

Volume 67

Broadcasting a Means to Disseminate Public Data in a Wireless Environment—Issues and Solutions
  A. R. Hurson, Y. Jiao, and B. A. Shirazi
Programming Models and Synchronization Techniques for Disconnected Business Applications
  Avraham Leff and James T. Rayfield
Academic Electronic Journals: Past, Present, and Future
  Anat Hovav and Paul Gray
Web Testing for Reliability Improvement
  Jeff Tian and Li Ma
Wireless Insecurities
  Michael Sthultz, Jacob Uecker, and Hal Berghel
The State of the Art in Digital Forensics
  Dario Forte

Volume 68

Exposing Phylogenetic Relationships by Genome Rearrangement
    YING CHIH LIN AND CHUAN YI TANG
Models and Methods in Comparative Genomics
    GUILLAUME BOURQUE AND LOUXIN ZHANG
Translocation Distance: Algorithms and Complexity
    LUSHENG WANG
Computational Grand Challenges in Assembling the Tree of Life: Problems and Solutions
    DAVID A. BADER, USMAN ROSHAN, AND ALEXANDROS STAMATAKIS
Local Structure Comparison of Proteins
    JUN HUAN, JAN PRINS, AND WEI WANG
Peptide Identification via Tandem Mass Spectrometry
    XUE WU, NATHAN EDWARDS, AND CHAU-WEN TSENG

Volume 69

The Architecture of Efficient Multi-Core Processors: A Holistic Approach
    RAKESH KUMAR AND DEAN M. TULLSEN
Designing Computational Clusters for Performance and Power
    KIRK W. CAMERON, RONG GE, AND XIZHOU FENG
Compiler-Assisted Leakage Energy Reduction for Cache Memories
    WEI ZHANG
Mobile Games: Challenges and Opportunities
    PAUL COULTON, WILL BAMFORD, FADI CHEHIMI, REUBEN EDWARDS, PAUL GILBERTSON, AND
        OMER RASHID
Free/Open Source Software Development: Recent Research Results and Methods
    WALT SCACCHI

Volume 70

Designing Networked Handheld Devices to Enhance School Learning
    JEREMY ROSCHELLE, CHARLES PATTON, AND DEBORAH TATAR
Interactive Explanatory and Descriptive Natural-Language Based Dialogue for Intelligent Information Filtering
    JOHN ATKINSON AND ANITA FERREIRA
A Tour of Language Customization Concepts
    COLIN ATKINSON AND THOMAS KÜHNE
Advances in Business Transformation Technologies
    JUHNYOUNG LEE
Phish Phactors: Offensive and Defensive Strategies
    HAL BERGHEL, JAMES CARPINTER, AND JU-YEON JO
Reflections on System Trustworthiness
    PETER G. NEUMANN

CONTENTS OF VOLUMES IN THIS SERIES

Volume 71

Programming Nanotechnology: Learning from Nature
    BOONSERM KAEWKAMNERDPONG, PETER J. BENTLEY, AND NAVNEET BHALLA
Nanobiotechnology: An Engineer's Foray into Biology
    YI ZHAO AND XIN ZHANG
Toward Nanometer-Scale Sensing Systems: Natural and Artificial Noses as Models for Ultra-Small, Ultra-Dense Sensing Systems
    BRIGITTE M. ROLFE
Simulation of Nanoscale Electronic Systems
    UMBERTO RAVAIOLI
Identifying Nanotechnology in Society
    CHARLES TAHAN
The Convergence of Nanotechnology, Policy, and Ethics
    ERIK FISHER

Volume 72

DARPA's HPCS Program: History, Models, Tools, Languages
    JACK DONGARRA, ROBERT GRAYBILL, WILLIAM HARROD, ROBERT LUCAS, EWING LUSK, PIOTR LUSZCZEK, JANICE MCMAHON, ALLAN SNAVELY, JEFFERY VETTER, KATHERINE YELICK, SADAF ALAM, ROY CAMPBELL, LAURA CARRINGTON, TZU-YI CHEN, OMID KHALILI, JEREMY MEREDITH, AND MUSTAFA TIKIR
Productivity in High-Performance Computing
    THOMAS STERLING AND CHIRAG DEKATE
Performance Prediction and Ranking of Supercomputers
    TZU-YI CHEN, OMID KHALILI, ROY L. CAMPBELL, JR., LAURA CARRINGTON, MUSTAFA M. TIKIR, AND ALLAN SNAVELY
Sampled Processor Simulation: A Survey
    LIEVEN EECKHOUT
Distributed Sparse Matrices for Very High Level Languages
    JOHN R. GILBERT, STEVE REINHARDT, AND VIRAL B. SHAH
Bibliographic Snapshots of High-Performance/High-Productivity Computing
    MYRON GINSBERG

Volume 73

History of Computers, Electronic Commerce, and Agile Methods
    DAVID F. RICO, HASAN H. SAYANI, AND RALPH F. FIELD
Testing with Software Designs
    ALIREZA MAHDIAN AND ANNELIESE A. ANDREWS
Balancing Transparency, Efficiency, AND Security in Pervasive Systems
    MARK WENSTROM, ELOISA BENTIVEGNA, AND ALI R. HURSON
Computing with RFID: Drivers, Technology and Implications
    GEORGE ROUSSOS
Medical Robotics and Computer-Integrated Interventional Medicine
    RUSSELL H. TAYLOR AND PETER KAZANZIDES